ACRYLIC
FIBER
TECHNOLOGY
AND
APPLICATIONS

ACRYLIC FIBER TECHNOLOGY AND APPLICATIONS

edited by

JAMES C. MASSON
JCM Consulting
Mooresville, North Carolina

CRC Press
Taylor & Francis Group
Boca Raton London New York

CRC Press is an imprint of the
Taylor & Francis Group, an **informa** business

CRC Press
Taylor & Francis Group
6000 Broken Sound Parkway NW, Suite 300
Boca Raton, FL 33487-2742

First issued in paperback 2019

© 1995 by Taylor & Francis Group, LLC
CRC Press is an imprint of Taylor & Francis Group, an Informa business

No claim to original U.S. Government works

ISBN-13: 978-0-8247-8977-0 (hbk)
ISBN-13: 978-0-367-40185-6 (pbk)

Visit the Taylor & Francis Web site at
http://www.taylorandfrancis.com

and the CRC Press Web site at
http://www.crcpress.com

This volume is dedicated to three friends,
all of whom passed away in 1993,
long before their time

Dr. Hartwig Bach, Monsanto Company
Mr. Roger Cordle, BASF Corporation
Dr. Giorgio Doria, Montefibre

Preface

The synthetic fiber industry is now over 50 years old. In this period it has progressed from being viewed as a minor miracle (nylons in World War II) to contempt (polyester leisure suits) to acceptance as a small but useful component of everyday life. Four major product types—nylon, polyester, acrylic and polypropylene—have emerged and stood the test of time. At this stage the basic technology and products have been developed and most ongoing technical effort is centered on refinement: productivity improvement, energy reduction, automation, quality enhancement, and the development of product variants. This is an excellent time to examine what has been generated technically in the synthetic fiber industry.

In contrast to the development of nylon 66 by duPont and polyester by ICI, acrylic fibers were developed independently by many companies. Although important patents were issued covering aspects of the technology, no patent or patents preempted the field. This has resulted in a wealth of processes, for both producing the base polymer and converting the polymer into fiber. For example, three different modifying comonomers are used in "commodity" polymer compositions, and at least six distinct solvent systems are currently used to spin acrylic fibers.

Spinning methods include both dry and wet spinning. Acrylic fibers are produced in many variants, including staple, tow and continuous filament, natural and precolored (dyed and pigmented), mono- and bicomponent, in a range of deniers from less than 1 (microfiber) to 50 (wig fiber); they may be mildly (carpet fiber) or strongly (modacrylic) flame retardant and may serve as a precursor for another fiber (carbon).

Because of the proprietary nature of the work, most information on actual commercial practices in acrylic fiber technology is buried in internal company reports or appears in the patent literature where one has the task of discerning the wheat from the chaff. Several lengthy encyclopedia articles have attempted to survey the field of acrylic technology, but their scope was limited. In this volume the authors present a broad yet detailed view of acrylic fiber technology and products.

The information in this monograph is the most comprehensive yet assembled on acrylic fiber technology. It should prove useful to a wide audience, including textile and macromolecular engineering students, industrial researchers in acrylic and competitive fibers, persons in the dyes and pigments industry, downstream users such as sales yarn manufacturers, knitters, weavers, nonwovens manufacturers, and reinforcement fibers users. The book will be of value for its detailed treatments and also for the critical assessment of the technology and the supporting references to the original literature.

As detailed in Chapter 2, the acrylic fiber industry is in a mature phase in the developed world, but an expanding phase in the developing countries. This expansion is occurring almost completely through the purchase of technology from the major companies in Europe, Japan, and the United States. DuPont and Cytec (formerly American Cyanamid) have been particularly active in technology sales. The major companies are no longer attempting to develop wholly new technology for the polymerization or spinning processes, because, for the most part, they no longer have sufficient staff to mount such an effort and the profitability of the business will not support it. The last such "new beginning" was Bayer's development of a one-step dry spinning process in the late 1980s; this is reported on in Chapter 5. It is too early to say whether the new acrylic enterprises in developing countries will mount substantial R & D efforts that will lead to step changes in processes. I doubt it, as most are relatively small businesses producing 30,000 to 60,000 metric tons. It is more likely that any new technology will be of rather limited scope and, for the foreseeable future, will come from companies in the developed world.

Production of this volume has been a long process. I believe, though, that the results justify the effort.

James C. Masson

Contents

Contributors

G. J. Capone Monsanto Company, Decatur, Alabama

Hubert Emsermann Miles Inc., Rock Hill, South Carolina

Reimund Foppe Bayer AG, Leverkusen, Germany

Bruce G. Frushour Monsanto Company, St. Louis, Missouri

Filon A. Gadecki* E. I. duPont de Nemours & Co., Camden, South Carolina

Raymond Knorr Monsanto Company, Pensacola, Florida

Arthur Lulay Consultant, Sun City, Arizona

James C. Masson JCM Consulting, Mooresville, North Carolina

Robert R. Matzke, Jr. Monsanto Company, St. Louis, Missouri

B. von Falkai Bayer AG, Dormagen, Germany

Bruce Wade Monsanto Company, Decatur, Alabama

Gary Wentworth The C. P. Hall Company, Chicago, Illinois

---

* *Current affiliation*: Monsanto Company, Decatur, Alabama

1

History

Filon A. Gadecki*

E. I. duPont de Nemours & Co.
Camden, South Carolina

I. INTRODUCTION

Synthetic fibers based on synthetic polymeric materials have been known only a little more than 50 years. When one considers that over 32 billion pounds of synthetic fibers were consumed in 1990 [1], one begins to appreciate the amazingly rapid growth of fiber technology in that very short time.

Prehistoric man used animal furs to make clothes and furnishings to provide him shelter from the elements. The transition to synthetic fabrics was a gradual one involving developing the spinning and weaving crafts. The spinning of natural fibers to yarn and weaving to fabric were noted in the early civilizations in China, Egypt, and South America as far back as 3000 B.C. Fibers used for spinning came from cotton, other cellulosic fibers from plants, silk, and wool. From those early times until the nineteenth century most development came in improving yarn-spinning systems and means of weaving or knitting fabrics. With the onset of the industrial revolution, yarn spinning and fabric production was mechanized. In the twentieth century many new sophisticated spinning systems were developed in concert with new methods for making fabric.

While the evolution of spinning and fabric making was taking place, man began to develop new methods for producing fiber. In the latter part of the nineteenth century, regenerated fibers were being introduced on a commercial scale [2]. In 1846 Freidrich Schonbein produced nitrocellulose by treating cellulose,

---

Current affiliation: Monsanto Company, Decatur, Alabama

derived from wood, with nitric acid. Nitrocellulose was soluble in common solvents, whereas cellulose was not. In 1855, George Audermars found that if he dipped a needle into a solution of nitrocellulose and then drew it away a fiber could be formed. However, not until 1884 when Hilaire Chardonnet squirted his nitrocellulose solution through tiny holes and then treated the fibers with chemicals to regenerate the cellulose was a practical material produced. This led to the first commercial process for making synthetic fibers. Various forms of regenerated cellulose followed, including the cuprammonium fiber and viscose rayon. A chemically modified version, cellulose acetate, became popular as a silk substitute. In 1892 the English group of C. Beadle, E. J. Bevan, and C. F. Cross developed the viscose process for regenerating cellulose without first making nitrocellulose. In 1904, the rights to the process were purchased by Courtaulds Ltd., who developed it into a world-class method of manufacturing rayon. The commercial introduction of truly synthetic fibers based on synthetic polymers did not occur until a half-century later. The pioneering work of W. H. Carothers with the discovery of nylon is credited with the beginning of the era of synthetic fibers. The commercialization of nylon by E. I. du Pont in 1938 opened the door to a worldwide industry which now produces over 32 billion pounds of fiber, employs several million people, and represents an industry of very high economic value.

The major growth in these fibers took place with four generic types based on nylon, polyester, polyhydrocarbon, and polyacrylonitrile. Fibers were made from nylon, polyester, and polypropylene using melt-spinning technology. These polymeric materials had adequate stability in the molten stage to permit extrusion through the spinnerette holes to form fibers. Avoiding the use of solvents greatly simplified the fiber manufacturing process. The conversion of polyacrylonitrile to fiber was more troublesome as the polymer would not melt without decomposition and solvents were not known which would permit wet or dry spinning. Acrylonitrile was first prepared by the French chemist C. Moureau before the turn of the twentieth century [3]. It was not until the late 1920s and early 1930s that German chemists first used polymeric acrylonitrile in oil and gasoline-resistant rubber [4]. Because of the ready availability of the polymer at low cost as well as attractive physical properties as exhibited in plastics and rubber, intensive research to find a solvent for polyacrylonitrile was initiated by a number of American and German companies in the 1940s. In the United States, Du Pont was active in evaluating a number of organic solvents and eventually selected dimethylformamide for development and scale-up. In approximately the same time period several European companies including I. G. Farben found that aqueous ammonium thiocyanate or zinc chloride would dissolve polyacrylonitrile [5]. The race to the marketplace entailed developing the appropriate spinning technology to convert the solution to fiber having the most desirable physical properties. The use of organic solvents permitted spinning the fiber using either a wet-spinning process where the solution was extruded into an aqueous bath or a dry-spinning process where the solution was extruded into a hot-gas environment that removed solvent. Du Pont elected to pursue the dry-spinning route, and in early 1950 they introduced Orlon® to the market. Since the polymer was well known, patent protection similar to that afforded nylon was not possible and numerous other companies entered the acrylic fiber business. Interestingly, most other companies elected a wet-spinning process to produce fiber. Even today the debate exists among those in the acrylic fiber business regarding the relative merits of product value and economics of dry-spinning versus wet-spinning acrylics.

II. DEVELOPMENT OF ACRYLIC FIBERS IN THE UNITED STATES

The first acrylic fiber to be made on a commercial scale was Orlon [6]. It was introduced into the market by Du Pont in 1950. It was the second major fiber type commercialized by Du Pont, following nylon. After the discovery of nylon by W. H. Carothers and its subsequent commercialization, the Du Pont Company launched a systematic exploration of a range of polymeric materials. As part of this overall effort, the Rayon Department Pioneering Research section in Buffalo was studying the possible development of a fiber from polyacrylonitrile. The polymer itself was well known in the patent literature. Similarly, fibers produced from drastic solvents which seriously degraded the polymer were disclosed in the patent art. The major obstacle to commercializing a fiber made from polyacrylonitrile was to find a suitable solvent for the polymeric material. R. C. Houtz, a research chemist in Pioneering Research, embarked on a research program to identify such a solvent [7]. Stimulated by the work of C. S. Marvel, a company consultant, Houtz identified a number of organic solvents that were of potential commercial value. The attractive properties of dimethylformamide were rediscovered by Houtz in October 1941, some eight months after G. H. Latham of Du Pont's chemical department had made the initial discovery. Dimethylformamide had characteristics which made it attractive for commercial development. It could be produced at reasonable cost and had solubility characteristics and a boiling point which permitted several process options for its use in making fibers. The finding of a suitable solvent for polyacrylonitrile made possible the development of Orlon, but not until after many years of intensive work.

Early in the development of Orlon, it was recognized that the new fiber was difficult to dye [8]. Because of the difficulty of dyeing the fiber, it was decided to build a continuous yarn plant where there appeared to be sufficient industrial uses to support commercialization. These end uses would capitalize on the high acid and sunlight degradative resistance of the fiber. Test results on fiber were so promising that in 1944 Du Pont announced that a pilot plant was being built at Waynesboro, Virginia, for experimental production of the fiber. The research and development effort was also shifted to Waynesboro where a newly formed group was organized in 1946. The development of Orlon required the identification of different process steps which had to be integrated to permit a smooth working manufacturing system. Individual process steps could be conducted in alternative ways. At the onset the decision had to be made between wet spinning and dry spinning. Process and economic evaluations frequently dictated one approach only to discover that it could not be integrated with the rest of the process. In 1948 a process for manufacturing Orlon was developed to the point where a project could be submitted for a plant to be built in Camden, South Carolina. After project approval, intense effort was directed at resolving identified process issues. Some of the critical technology was not agreed upon until late in the design stages. Solvent removal was an issue that was not readily resolved. Design concepts vacillated between drying the fiber versus extracting with water. The decision to use water was made when it was found that drying aggravated an already bothersome problem of fiber discoloration. Drawing and subsequent drying the fiber were also areas where the technology swung back and forth between alternative schemes.

In July 1950 the plant at Camden was in commercial production of Orlon acrylic fiber in continuous filament form. Although the fiber was well suited for use in awnings, tarpaulins and filters because of its resistance to degradation from sunlight and its mildew resistance, insufficient sales materialized to sustain this initial venture. The fact that Orlon was difficult to dye initially prevented it from entering the major apparel area. The dyeing problem led to the investigation of copolymers and with the discovery and development of new cationic dyes, a product which could be dyed was developed. The development leading to the selection of the appropriate copolymer was difficult, as some of the attractive physical properties of polyacrylonitrile had to be compromised to achieve commercial dyeability. The selection of a suitable comonomer evolved to a choice between methyl acrylate and vinyl acetate. Du Pont selected methyl acrylate, and this ended up being the composition most generally used. Orlon acrylic fiber in staple form suitable for use in apparels was commercialized in a new unit at Camden in March 1952. By the end of the year, Orlon was appearing in various wearing apparel for men, women, and children. By the fifth year of production more than 70 million women's sweaters alone were made from the fiber [9]. In 1957, a second plant was constructed at Waynesboro, Virginia, to make staple products. In that same year, production of Orlon acrylic filament yarn was discontinued for lack of sufficient sales. The first start-up of Orlon acrylic staple and tow production outside the United States took place in August 1957, at DuPont of Canada's Maitland Plant. Research effort aimed at duplicating the desirable features of wool while correcting some of its performance deficiencies resulted in the introduction in 1960 at Camden of Type 21 Orlon bicomponent with a water-reversible crimp [10]. Sayelle was adopted as the trademark for the fiber that became quite popular in the sweater and craft yarn markets. Other major fiber variants of Orlon included Type 43 cashmere-like fiber in 1968, Type 24 bicomponent with Shetland-like aesthetics in 1973, an acid-dyeable family of deniers, and a variety of surface-modified fibers aimed at emulating the various types of wool. Trademarks or Certification Marks included Sayelle, Wintuk, Nomelle, Civona, and Ardina.

As the use of synthetic fibers was growing rapidly on a worldwide basis, Du Pont decided to build plants in Europe to manufacture Orlon. In February of 1962, an Orlon plant went into operation at Dordrecht, The Netherlands, and a second one started up in December 1968, at Maydown, Northern Ireland. A joint venture with an Iranian group, Polyacryl, led to a plant start-up in the production of acrylic fibers in August 1978.

With the commercial introduction of Orlon in 1950, it was not long before other companies in the United States entered the market. By 1951, six other companies were known to be working on the development of acrylic or modacrylic fibers, and rather promptly there appeared Union Carbide's Dynel®, Monsanto's Acrilan®, American Cyanamid's Creslan®, Dow Chemical's Zefran®, Tennessee Eastman's Verel®, and Goodrich's Darvan®.

Work on Acrilan started in 1942 at Monsanto's Central Research department in Dayton, Ohio [11]. In 1949, Monsanto and American Viscose announced that the two companies had agreed to enter a joint venture to create a new company called Chemstrand which would produce Acrilan acrylic fiber. The fiber invented at Dayton moved to Marcus Hook, Pennsylvania, and then to Decatur, Alabama. The solvent that was selected was dimethylacetamide, based on the work of A. B. Craig [12]. When Acrilan was introduced into the market, a serious problem of fibrillation was encountered. When fabric was abraded, the fibrils of each damaged fiber would, because of greater reflectance, result in a whitening of the fabric. Additionally, the fiber was not as soft and wool-like as Orlon. So while Du Pont was struggling with improving the dyeability of its fiber, Chemstrand was working to solve the fibrillation problem. The latter was serious enough to shut down the plant for two years until a solution was in hand. It was found that an annealing step involving treating the fiber with above-atmospheric steam was required to make a fiber that would not fibrillate. "By 1955 Chemstrand's faith in the soundness of its acrylic property was restored as evidenced by the fact that Chemstrand took its first big step abroad—to England, where it formed a wholly-owned subsidiary, Chemstrand, Ltd. to make and market Acrilan [13]." By the end of 1959 Chemstrand employed some 9000 people to produce and sell 130 million annual pounds of Acrilan fiber. In 1961, Monsanto acquired American Viscose's equity in the Chemstrand Company. Chemstrand built acrylic fiber plants at Coleraine, Northern Ireland, Lingen,

West Germany, and Ashdod, Israel. It also entered joint ventures in Italy and Japan, spawning the products Leacryl and Vonnel. Monsanto brought the Chemstrand operation under the parent name in 1964.

Acrilan's early success in the market was largely in carpets, where it displaced a high portion of the wool market. The fiber also enjoyed success in the pile, circular knit, and blanket markets. It later built a large business in sweaters and upholstery based on its producer-dyed fiber.

Two other companies that developed and commercialized acrylic fibers in the United States included American Cyanamid's Creslan and Dow Chemical's Zefran. While American Cyanamid had earlier developed a strong business based on materials derived from acrylonitrile chemistry, it was not until 1958 that Creslan was introduced [14]. American Cyanamid employed the aqueous sodium thiocyanate wet-spinning system similar to that used by Courtaulds [15]. Its composition was also different from Orlon and Acrilan, and it found general acceptance in the circular knit market.

In the 1980s American Cyanamid, as well as DuPont, developed modified melt-spinning processes for making acrylic fibers [16]. Although the processes were of considerable technical interest, they were not developed adequately for general commercial exploitation by either company. The hydrated melt-spinning process required significant capital investment but provided potential for considerable cost reduction. The fibers that were produced had characteristics more reminiscent of melt-spun fibers rather than those obtained in dry or wet spinning. The use of "melt-spun" acrylics as precursors to carbon fiber will be discussed in Chapter 6.

III. WORLDWIDE PRODUCTION OF ACRYLIC FIBERS

At the same time as acrylic fibers were being introduced in the United States, considerable effort was expended throughout the world, especially in Western Europe and Japan. Three dominant technologies were the basis for the worldwide expansion of acrylic production: the dry-spinning process commercialized initially by DuPont using dimethyl formamide as solvent and methyl acrylate as dye modifier, the wet spinning process commercialized by Chemstrand involving the use of dimethyl acetamide as solvent and vinyl acetate as dye modifier, and the wet-spinning process as used by Courtaulds, where an aqueous sodium thiocyanate system is used and methyl methacrylate is the dye site modifier. Most new installations are based on one of these systems.

In Europe the early work within the I. G. Farben facilities to find solvents for polyacrylonitrile eventually led to the develonment of wet-spinning or dry-spinning processes for making fibers [17]. Prominent among the German Companies was the commercialization of Bayer's Dralon, Hoechst's Dolan, and the East German Chemie Faserwerk Freidrich Engels' Wolpryla. The French company

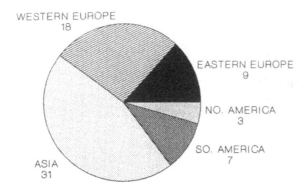

Figure 1 Acrylic plants worldwide (1990).

Rhone Poulenc became a significant participant in the acrylic fiber business. Other prominant acrylic fiber manufacturing companies included Courtaulds from Great Britain, and Anic, Montefibre, and Snia Viscosa from Italy. By 1984 European acrylic fiber capacity had peaked out at slightly more than 2 billion annual pounds as compared to the United States capacity of 800 million annual pounds [18].

The acrylic fiber capacity in Japan was also growing at a rapid rate. Companies such as Asahi, Nippon Exlan, Kanegafuchi, Mitsubishi Rayon, Toho Beslon, Toyo Rayon, and Kanebo had installed over 800 million pounds of capacity by 1980. Including the facilities located in China, India, and South Korea, the area had slightly in excess of 1.4 billion pounds of capacity. In the late 1980s, major new facilities for making acrylics were being built in India and China. As these countries were building their economies they found that the textile industry was an effective way of utilizing their abundant labor force with reasonably modest capital investment.

This was also true in the developing countries such as in South and Central America. About 200 million pounds of acrylic capacity was in place by 1976 in Argentina, Brazil, and Mexico.

By 1990 there were 68 plants worldwide producing 5.1 billion pounds of acrylic or modacrylic fibers (see Figure 1). Nameplate capacity of these facilities was in excess of 6 billion pounds. A breakdown by area is presented in Figure 2.

IV. ACRYLIC FIBER LIFE CYCLE

As the acrylic fiber base has grown over the past 40 years it has now become a mature business in the developed countries. In the life cycle of most mature businesses, retraction in market size occurs as new preferences and needs along

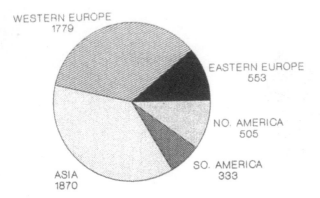

Figure 2 Acrylic fiber production—1990 (in million lbs).

with new products are introduced. The first major decline of the acrylic fiber market occurred in the carpet area in the 1970s. Overcapacity was experienced in Europe and the United States. Producers which had significant carpet fiber production shifted to apparel fibers. Acrylic fiber usage in the United States reached its peak in 1980 and has declined since then (Figure 3). This decline is associated with the rebirth in popularity of cotton in sweaters and socks. Some of the early companies to enter the acrylic/modacrylic fiber business have

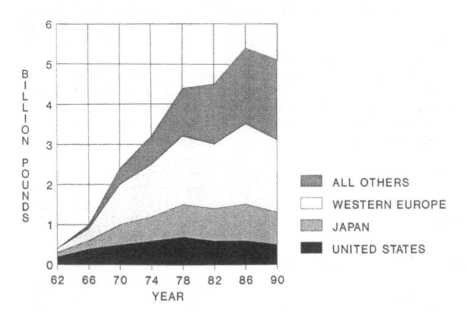

Figure 3 Acrylic fiber production.

terminated participation in the market. This includes Du Pont, which announced in 1990 that it was exiting the Orlon business after having commercialized the first acrylic fiber 40 years earlier. As world capacity exceeded market demand, Du Pont began closing its Orlon manufacturing facilities. The initial plants that were closed were located outside the United States. The plant in Dordrecht, The Netherlands, closed in 1978, in Maydown, Northern Ireland, in 1980 and in Maitland, Canada, in 1981. In the United States the plant in Waynesboro was shut down in 1990, and in 1991 the last plant in Camden (the initial site of commercialization) was closed. New capacity to produce acrylic fibers for the apparel and industrial markets is now primarily focused in developing countries. Because of its unique range of useful properties it still offers potential in specialized applications that have high value. The acrylic fiber's resistance to degradation from sunlight and the environment has not yet been fully exploited. The fiber's ability to transport moisture without excessively absorbing it is now being promoted for use in athletic socks. With its low density and attractive insulating properties, it provides a fiber which should have more usage in garments where thermal control is desired. The fiber has some antimicrobial characteristics which have only recently been recognized [19] and which may prove useful for some applications. It is known as one of the best precursor fibers used in the production of carbon fiber—a highly desirable reinforcement fiber in the growing composites market.

Acrylics are still the fiber of choice as a wool replacement. As long as the wool type of performance and aesthetics are desired, it is expected acrylic fibers will be used commercially for a long time to come.

REFERENCES

1. *Fiber Organon*, Vol. 62, No. 7, Fibers Economics Bureau, Inc., Roseland, NJ, 1991.
2. Cook, J. G., *Handbook of Textile Fibers*, Merrow, Durham, England, 1984.
3. Pearce, E. M., *Acrylonitrile in Macromolecules, Applied Polymer Symposium*, Wiley, New York, 1974; Moureau, C., *Ann. Chem. Phys.*, [7], 2, 186 (1893).
4. Raheel, M., Abstracts of Papers, 201st ACS National Meeting, Atlanta, Ga., 1991.
5. Rein, H., Survey of synthetic fiber developments, *Angew. Chem.*, 60, 1948, pp. 159–61.
6. Baker, B. A., Chronicle-Independent, Camden, South Carolina, 1991.
7. Houtz, R. C., *Textile Res. J.*, 786–801, (Nov. 1950).
8. Cook, J. G., *Handbook of Textile Fibers*, Merrow, Durham, England, 1984, p. 398.
9. *Guide to Fibers*, DuPont, P. R. Dept., New York.
10. DuPont Public Relations Publications.
11. Forrestal, D., *Faith, Hope and $5000, The Story of Monsanto*, Simon and Schuster, New York, 1977, pp. 121–134.
12. Craig, A. B., to Chemstrand, U.S. Patent 2,613,194, October 7, 1952.
13. Forrestal, D., *Faith, Hope and $5000, The Story of Monsanto*, Simon and Schuster, New York, 1977, p. 128.
14. Bendigo, C. W., Creslan, *Textile Res. J.*, 27, 701–702 (1957).

15. Creswell, A., to American Cyanamid Company, U.S. 2,558,730, U.S. 2,558,731, U.S. 2,558,732, July 3, 1951.
16. Blickenstaff, R. A., to DuPont, U.S. Patent 4,094,948, October 2, 1972. Logullo, F. M. to DuPont, U.S. Patent 4,238,440, December 29, 1978. Cline, E. T. and Cramer, F. B., to DuPont, U.S. Patent 4,238,442, December 29, 1978. Porosoff, H., to American Cyanamid, U.S. Patent 4,163,770, August 7, 1979. Pfeiffer, R. E., Peacher, S. E., and Roberts, R. W., to American Cyanamid, U.S. Patent 4,264,076, March 3, 1980. Pfeiffer, R. E. and Roberts, R. W., to American Cyanamid, U.S. Patent 4,220,616, September 2, 1980. Pfeiffer, R. E. and Peacher, S. E., to American Cyanamid, U.S. Patent 4,220,617, September 2, 1980.
17. Rein, H., Survey of synthetic fiber developments, *Angew. Chem.*, *60*, 159–161 (1948).
18. Textile Organon, Textile Economics Bureau, Inc., New York, June 1963, June 1970, June 1976, June 1985.
19. Pardini, S. P. to DuPont, U.S. Patent 4,708,870, November 24, 1987.

2

The Acrylic Fiber Industry Today

Robert R. Matzke, Jr.

Monsanto Company
St. Louis, Missouri

I. OVERVIEW

As detailed in Chapter 1, the acrylic fiber industry has been in existence for over 40 years. In the economically developed nations—primarily the United States, West Europe and Japan—acrylic fiber consumption is in a mature phase where growth has slowed, stopped or reversed. In addition to the maturing of demand, local acrylic producers in the developed nations have seen a growing share of that demand supplied by imports from developing nations of final products made with acrylic fiber produced in the exporting region.

In many developing nations, the acrylic fiber industry is growing. Growth has been stimulated by the need for acrylic fiber to produce products like sweaters for export. In addition, there is growing home market demand in the developing world. Per capita fiber consumption remains relatively low in these countries today, but empirical studies show that demand for textile products rises sharply with gains in income. This income elasticity of demand for textile fibers promises significant potential for future growth of demand for acrylic fiber in the developing nations of the world if their governments succeed with economic development plans.

Over the past 20 years, acrylic fiber production has grown at a trend rate of 2.5%. In the 1970s and 1980s through 1987 the trend rate for acrylic was about 4%, but forces since then, especially economic recession in West Europe, Japan and the United States, lowered acrylic fiber demand.

Acrylic fiber usage has averaged 5.2 billion pounds per year from 1989 to 1993, below the peak of 5.5 billion pounds in 1987. Acrylic accounts for about 6% of total world fiber consumption, which will approach 95 billion pounds in 1994. Manufactured fibers—mainly acrylic, polyester, nylon, polypropylene, rayon and acetate—account for half of world fiber usage, while cotton accounts for 46% and wool 4%. World total fiber demand is growing at a trend rate of about 3%. Approximately 60% of the growth rate is a function of population growth, and the balance is due to increased per capita consumption.

Acrylic fiber demand is a function of its price and properties relative to substitutes. There are a number of major properties that differentiate acrylic fiber: (1) low specific gravity of 1.18 (cotton's is 1.58), which gives bulk, apparent value and warmth at reduced weight; (2) soft and wool-like aesthetics; (3) dyes atmospherically to bright colors; (4) resists ultraviolet fading; (5) wicks moisture for quick drying; (6) retains shape after machine washing. The main markets for acrylic fiber are those where its properties match well with the end-use requirements. Among such uses are sweaters, socks, fleecewear, hand knitting yarns, blankets, upholstery and others.

The following key words communicate the essential characteristics of the acrylic fiber industry today:

1. *Global*: In today's global economy, *demand* for the acrylic fiber final products consumed in one nation is often *supplied* by acrylic fiber and/or products manufactured therefrom in one or more different nations. A significant percentage of world production of acrylic fiber is exported to another country to be advanced into yarn and final products. A significant percentage of world production of acrylic fiber end products is exported from producing nations for final sale to consumers in an importing nation.

2. *Restructuring*: Acrylic fiber producing capacity in the economically developed regions is being downsized, while capacity expansion investments are being made in the developing regions. Some acrylic producers are merging to reduce cost, and some fiber capacity is being relocated.

3. *Interfiber competition*: Acrylic fiber faces competition in most key end products from other generic fibers that are acceptable substitutes.

4. *Mature (developed world)*; *growing (developing world)*: From the late 1950s to the mid-1970s, acrylic fiber usage grew rapidly in the developed world. Demand was spurred by pent up consumer demand, economic growth and the substitution of acrylic for wool in carpets, sweaters, hand knitting yarns, and so on. Acrylic fiber is now in a mature phase in the developed world, but the developing regions of the world are in a growth phase.

5. *Cyclical demand*: Acrylic fiber demand is cyclical for several reasons. First, in the developed world, which accounts for nearly half of acrylic fiber consumption, most textile purchases are postponable and suffer in economic

recessions. Second, inventory corrections made in response to lower consumer demand multiply the effect. Third, acrylic is widely used in apparel products like sweaters that are subject to fashion cycles.

6. *Concentration of markets and fabrics*: Compared to other fibers such as cotton or polyester, acrylic fiber usage is concentrated in relatively few end use markets. In addition, acrylic fiber is not often used in woven fabrics. Its use, therefore, is concentrated in knitted fabrics, which account for less than half the world production of textile fabrics.

This chapter will develop a perspective on the acrylic industry in the mid-1990s by detailing the six characteristics summarized above. The industry will be examined first from the demand side and then the supply side. Detailed analysis of the present situation will build a foundation for the industry outlook that will conclude the chapter.

It is logical to begin the discussion with *demand*, because in the long run supply is generally a response to demand, not vice versa. This is true notwithstanding the fact that fiber producers, individually or collectively in industry associations, routinely engage in activities to influence demand. Such activities include new product development, pricing, promotion, advertising and so forth. Typically these activities impact the producers' share of final demand but do not increase total demand.

In the short term, however, the styling of the end product in which fiber is supplied can influence demand. In the economically developed nations, many consumers have the discretionary income to make unplanned impulse purchases of items that appeal to them but which they do not "need." The distinction here can be thought of as the difference between a planned purchase to replace a pair of worn-out socks and the impulse purchase of your 25th sweater because the look is appealing. We have no measure of the impact of fashion influences and impulse buying on acrylic fiber demand. The effects would be greatest in the developed world, but certainly not limited to the economically advanced nations. Ultimately demand is demand, whether due to fashion impulse or true need to satisfy one of the basic necessities of life. The issue is that the former is more volatile.

Unless otherwise specified, the word *acrylic* should be taken to embrace modacrylic as well as acrylic fiber. There are relatively few producers of modacrylic fiber; separate data for modacrylics by country are not published to avoid disclosing capacity or production of individual companies.

II. DEMAND SIDE

Acrylic fiber is a raw material for yarns and fabrics that become apparel, home furnishings or industrial end products. Final demand typically means the purchase by a citizen consumer of an end product at a store, but governments, institutions and businesses also are substantial purchasers of acrylic fiber products. Such organizations include the military, schools, sports teams, hospitals, office buildings, airlines, etc.

A. Acrylic Fiber Markets in the United States

Total acrylic fiber consumption in the United States peaked in 1973 at 662 million pounds. The previous year was the high water mark for acrylic carpet fiber shipments, which totalled 169 million pounds in 1972 [1]. As shown in Fig. 1, from 1973 to 1980, U.S. domestic acrylic carpet fiber demand declined steadily. Textile acrylic fiber demand shows a sharp cyclical decline bottoming at 380 million pounds in 1974–1975, then recovery to the 515-550-million-pound range in 1977–1980.

Figure 2 shows that domestic acrylic fiber shipments have trended downward since 1980 due to erosion of textile denier acrylic markets in apparel and home textiles. Domestic acrylic fiber shipments grew 1% in 1992 and 3% in 1993, reversing five consecutive years of declines.

In 1991, domestic shipments of U.S. acrylic fiber producers were 42% below 1980, but U.S. apparent consumption of acrylic fiber (domestic fiber shipments plus imports of fiber, yarn and garments) in 1991 was only 28% below the 1980 level (Fig. 3). This differential reflects the growth of imports. In 1980, domestic acrylic fiber accounted for 82% of apparent consumption, but this share had fallen to 66% in 1991 (Fig. 4) [2]. The overall decline in domestic acrylic fiber shipments is the summary result of different forces in each of the end-use markets. The decline can only be understood properly, therefore, by understanding the individual end market situations. Table 1 displays the changes that have occurred.

Sweater market shipments declined due to several factors:

* Imports of acrylic sweaters grew steadily and reached a penetration of about 60% of the market.
* Imports of yarn and fiber grew and took some spinner level share from U.S. producers.
* Imports of ramie blend sweaters took share.
* Cotton sweaters, largely made in the United States, took market share.
* The sweater category reached a women's wear fashion cycle peak in 1986; unit sales have declined about 30% since. Females buy two-thirds of all sweaters.

In 1990, the U.S. Department of Commerce (International Trade Commission) determined that sweater knitters in Korea, Taiwan and Hong Kong had been dumping acrylic sweaters in the U.S. market. Dumping duties were levied, and as a result imports' share of acrylic sweaters sold at retail dropped seven

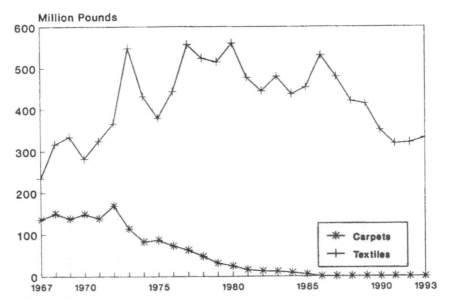

Figure 1 U.S. domestic acrylic fiber shipments; carpets vs. textiles.

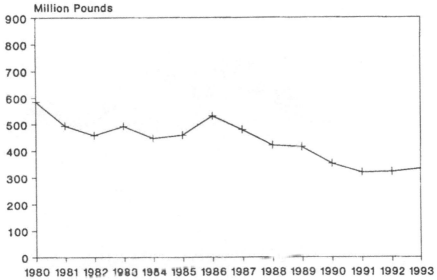

Figure 2 U.S. total domestic acrylic fiber shipments.

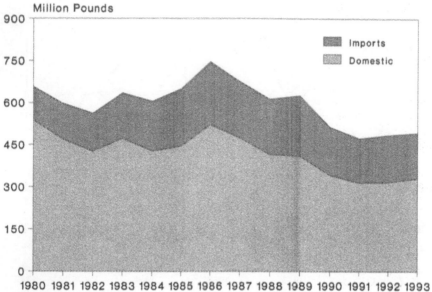

Figure 3 U.S. apparent consumption of textile denier acrylic fiber.

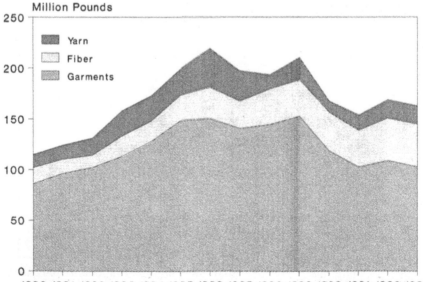

Figure 4 U.S. acrylic fiber imports by type.

TABLE 1 U.S. Acrylic Fiber Producers' Domestic Shipments
(million pounds)

	1980	1991
Apparel	466	242
Sweaters	88	63
Socks	93	36
Craft yarns	95	36
Circular knits	125	95
High pile	60	11
Woven fabrics	5	1
Home furnishings	113	55
Carpet	32	n.a.
Blankets	45	23
Upholstery	9	20
Other	27	12
Industrial	7	24
Total domestic	585	319

Source: Fiber Economics Bureau, Monsanto estimates.

share points in 1991 and one additional point (to 49%) in 1992 [3]. This was a boost for U.S. acrylic sweater knitters, but only time will tell if the share gain was permanent. In July 1992, the U.S. Court of Trade ruled that the ITC based its 1990 decision on faulty information and ordered the agency to rehear the case. In November 1992, the ITC voted unanimously to set aside its 1990 ruling. The National Knitwear and Sportswear Association appealed the reversal, but on June 15, 1994, the United States Court of Appeals (Federal Circuit) ruled against the NKSA and affirmed the ITC set aside. Acrylic sweaters from Korea, Hong Kong and Taiwan can now enter the U.S. market without payment of dumping duties [4]. The impact of this change on the U.S. acrylic sweater business remains to be seen.

Sock market shipments have declined due to share losses to cotton socks [5]. The U.S. Acrylic Council, which was formed in 1989 to promote industry interests, targeted socks for a concentrated promotion effort in 1992–1993. The sock market was targeted, because this is a product where acrylic fiber offers superior performance, especially moisture transport for blister free comfort in athletic socks. In addition, following consumer research that identified a need for socks that last longer, Monsanto, the largest U.S. acrylic fiber producer, introduced a more durable sock fiber with the trade name Duraspun®. These efforts are expected to increase the use of acrylic fiber in the sock market.

Craft yarn shipments have declined because lifestyles changed in the 1980s. Acrylic has retained a high share of the market, but the market itself declined. Reasons for that decline include the increasing number of women entering the work force and their growing interest in fitness activities such as aerobic sports. These trends reduced time available for knitting or crocheting. In 1990–1992, retail sales data from the MRCA Information Services consumer panel showed a plateauing or halting the decline of the previous decade.

Circular knit fabric accounted for 125 million pounds of acrylic fiber in 1980 and 95 million pounds in 1991. There was actually greater change in this market segment than is apparent in the 1980–1991 comparison. By 1986, acrylic usage in circular knits had grown to 195 million pounds spurred by the rapid growth of sweatshirts for casual wear and active sports. It fell sharply thereafter as polyester replaced a substantial share of the acrylic used in blends with cotton in sweatshirts, which in 1986 was the largest end use for acrylic circular knit fabrics. A main reason why polyester supplanted most of acrylic's position in sweatshirts is that polyester's better whiteness and thermal stability made it a superior background material for heat tranfer printing, which was growing rapidly. A second factor was market share growth by sweatshirt makers like Russell and Champion, which have a polyester/cotton orientation. Their success with polyester/cotton products forced acrylic/cotton producers to match their offerings.

High pile fabrics are used in apparel products like robes, fake furs and raincoat liners. Domestic shipments of acrylic fiber to the high pile market have declined for several reasons:

1. The main U.S. supplier of modacrylic to the fake fur market was Eastman Chemical Co., which exited the Verel business in 1983. Eastman's business was picked up by imports, primarily the Japanese fiber Kanecaron, which is available in the numerous deniers and cross sections required by the fake fur business.
2. Raincoat liners was a major market for acrylic high pile in 1980. The pile segment of this market was eroded by other liner fabrics, and acrylic's share of the remaining pile business was reduced by gains by cheaper polyester staple fiber.
3. In 1985, BASF announced it was withdrawing from the high pile market segment.

Carpet shipments for acrylic fiber had declined to 32 million pounds in 1980, sharply below the peak of 169 million pounds reached in 1973. Monsanto phased out of the acrylic carpet fiber business in the early 1980s, and the only other supplier, BASF, exited the business in 1985. Acrylic fiber makes excellent carpets, but acrylic is at a cost disadvantage because it cannot be piece-dyed. In 1990, Monsanto reintroduced acrylic fiber into the U.S. carpet market. Several of Monsanto's carpet customers have made 100% Acrilan acrylic styles in a variety of new looks, many based upon Acrilan Plus producer dyed fiber. In addition, nylon/acrylic blends in an 88/12 ratio are the basis for Monsanto's Traffic Control Fiber System™, which identifies carpets in a Monsanto patented construction that resists matting.

Blankets is a market where the bulk and cover of acrylic fiber provide warmth with low weight in a soft fabric, an excellent match of acrylic's properties and end-use requirements. Consumption of acrylic fiber for blankets reached a peak in the 1970s as higher energy prices and energy conservation efforts led to blanket sales to offset the effect of lower thermostat settings. In the 1980s, acrylic consumption in blankets declined for several reasons:

1. Consumer focus on energy conservation waned and so did blanket sales.
2. The blanket category lost market share to alternatives like comforters and quilts.
3. Acrylic lost share to cotton as cotton blankets and afghans have grown in popularity.

Upholstery has been a growing category for acrylic fiber in the past decade. Acrylic velvets and flat wovens make excellent upholstery fabrics due to their color, feel and resistance to fading, soiling and abrasion. Growth in this market has helped to offset some of the losses in other markets. Monsanto has an extensive multifaceted program supporting the Wear Dated warranty for residential upholstery. This program has had a major influence on the growth of acrylic fiber consumption in upholstery. Without that extensive support program, the poundage would not have grown. Fiber producer programs can have a significant influence on demand, but fiber advertising, merchandising and promotion were cut back during the 1980s in response to profit pressures. This is true not just of acrylic, but other synthetic fibers as well. While synthetic fiber advertising was cut back during the 1980s, cotton advertising increased dramatically, funded by a $1 per bale levy.

Other home textiles includes markets such as drapes. This poundage has declined due to the availability of substitutes and the lack of supporting programs by the fiber producers. No published reports are available on the size of these businesses.

Industrial uses for acrylic and modacrylic fibers have included felts, battery separators, stuffing, wipe cloths, sandbags, awnings, boat covers, cement reinforcement and many others. Some of these businesses are steady and some have come and gone. The sandbag business exemplifies a business that was important for a time and then disappeared. Bag and bagging use of acrylic peaked at 27 million pounds in 1969 during the Vietnam War when bags that would not rot and deteriorate in sunlight and jungle conditions were needed in quantity. Other businesses have been steadier. Among the industrial segments that have been growing are awnings, boat covers and outdoor furniture where the excellent coloration and ultraviolet resistance of acrylic fiber are significant advantages. These advantages are maximized with pigmented fibers that offer prolonged resistance to fading. Monsanto's SEF™ and pigmented SEF Plus™ modacrylics are used in flame retardant covers and awnings where superior FR performance is needed to meet municipal building codes or enhance security.

Exports of U.S. acrylic totalled 125 million pounds in 1992 or 28% of total shipments. Many export markets and customers represent long-term relationships with U.S. acrylic fiber suppliers. Many countries, Canada for example, do not have a local acrylic fiber producer and depend on imports. Export markets are important to the U.S. acrylic fiber industry to acheive sufficient plant utilization for economic health. With nearly 500 million pounds of capacity and 1992 domestic acrylic shipments of 322 million pounds (64% of capacity), financial performance would be unsatisfactory without exports.

B. Acrylic Fiber Markets Elsewhere in the World

Students of the fiber industry in the United States are fortunate to have extensive data on end-use markets available from the Fiber Economics Bureau [6]. The FEB collects detailed data from the individual fiber producers. Individual company data is confidential, but the collective industry totals are published. In Europe, the CIRFS [7] organization collects and publishes data as does the Japan Chemical Fibres Association [8] in Japan. Statistics for the developing nations are less accessible, if they exist at all.

Today, a substantial share of the acrylic fiber produced in the developing nations of the world, especially in Asia, is used to make end products that are exported. The largest quantity of these flow to the largest markets, especially to the United States, West Europe and Japan, but trade is global and acrylic fiber goods also move from, to and within Asia, Africa, the Middle East and Latin America. This trade flow reflects the labor cost advantage of the developing countries, an advantage that is especially significant in labor intensive apparel products where acrylic fiber use is concentrated. Growth of demand for acrylic fiber in the mills of the developing nations primarily has reflected production shifting from high-cost developed countries to lower-cost developing countries.

In the future, however, an increasing share of production in the developing nations will be to satisfy a growing local demand. The basis for this expectation is empirical studies done by organizations like the World Bank and Geerdes International [9]. These studies have shown that there is a very high elasticity of demand for textile fiber products with respect to personal income. In other words, if incomes rise in developing countries, we can expect substantial increases in demand for apparel and home furnishings. Most observers predict relatively high growth rates for acrylic fiber demand in the developing regions, especially Latin America, Asia and Eastern Europe. Such rapid growth requires economic growth to drive it and will not occur otherwise.

Another significant factor driving growth of acrylic fiber demand in the developing countries that are major cotton producers, especially the People's Republic of China and India, is the need for more arable land for food production. This requires increasing sources for man-made fibers to replace cotton in the textile industry. This explains, in part, China's push to expand its acrylic fiber producing capacity.

As home market growth becomes more significant, the end-product mix of acrylic usage will shift toward local needs and preferences instead of the export market preferences that determine the mix today. One can be sure, however, that the local market acrylic fiber end products will be those that take advantage of the basic properties of acrylic: color, bulk, softness, etc. In other words, the mix of end products for local consumption in developing countries may not mirror that in Europe, Japan or the United States, but the reasons why acrylic fiber is used will be similar.

1. Western Europe

As shown in Chapter 1 (Fig. 2) Western Europe has nearly 1.8 billion pounds or about 30% of the world's acrylic fiber producing capacity, but it only needs about half of that for home markets. The balance is exported to the extent possible in order to achieve acceptable operating rates. The end-use trend in Europe is shown in the table below. Acrylic fiber mill consumption has been declining in Western Europe as acrylic has lost position in the carpet market, imports have gained share in apparel and the fashion cycle has favored other fibers and fabrications.

The following table shows the declining trend in mill consumption and also the high and increasing need for West Europe to export acrylic fiber production.

Acrylic Fiber Trend in West Europe [10] (EEC 12)
(million pounds)

	Mill consumption	Production	% of Production
1984	1016	1863	54.5
1985	1079	2041	52.9
1986	1094	2072	52.8
1987	1102	2092	52.7
1988	899	1931	46.6
1989	854	1750	48.8

By 1988 over half of acrylic fiber production in Western Europe was exported. These exports are vital to financial survival. The word survival was chosen rather than acceptable profitability, because few, if any, acrylic fiber producers worldwide had satisfactory financial performance in the late 1980s. In Sec. III, the supply side section of this chapter, we will review the industry structure changes that were triggered by the financial pressures of the time.

The end-use pattern of acrylic fiber consumption in Western Europe is similar to that in the United States, but there are market differences. To some extent these are shaped by climate and to some extent by historial raw material resources. Europe has a strong woolen industry, whereas the United States has a strong cotton industry. Differences in the end-use market distribution in the two regions reflects the fact that acrylic has a much stronger position in home textiles in Europe. In 1989, there was still a 50-million-pound carpet business and the upholstery/drapery business was much larger in Europe (125 million pounds) [11] than in the United States (25 million pounds).

Domestic Mill Consumption by Market—West Europe vs. United States, 1989 (% mill consumption)

	Western Europe	United States
Apparel	68	80
Home furnishings	27	16
Industrial	5	4
Total	100	100

2. Japan

Japan has nearly 900 million pounds of acrylic fiber producing capacity or roughly 14% of the world total. Similar to the picture drawn for Western Europe, Japan must export about half its acrylic fiber production in order to achieve acceptable operating rates. Exports have grown from 48% of production in 1985 to 55% in 1990.

Acrylic Fiber Production/Shipments Trend—Japan [12]
(million pounds)

	Production	Exports[a]	% Export
1985	844	403	48
1986	875	476	54
1987	897	472	53
1988	877	494	56
1989	809	392	52
1990	798	437	55

[a]Fiber + yarn

The end-use pattern of consumption in Japan is much more heavily weighted to home furnishings uses—blankets and carpeting—than in the United States and Western Europe. These regional differences are shown in the comparative table below.

Domestic Mill Consumption by Market—West Europe vs. United States vs. Japan 1989 (% mill consumption)

	West Europe	United States	Japan
Apparel	68	80	46
Home textiles	27	16	48
Industrial/other	5	4	6
Total	100	100	100

Sources: FEB/CIRFS/JCFA

End-use detail on acrylic fiber consumption in Japan appears in Table 2.

3. Acrylic Fiber Demand in Developing Nations

In discussing demand, it is important to distinguish between mill demand and final demand. In most of the developing nations in Asia, mill consumption of acrylic fiber is substantially above final consumption by local market consumers. Mill consumption in these nations is significantly determined by production of garments for export to the developed world. We saw earlier, for example, that more than half the acrylic sweaters purchased in the United States are imported. In the 1980s, most of these imports came from South Korea, Taiwan and Hong Kong. Their share is now shifting due to the combined effects of the antidumping margins imposed in 1990 as well as the increasing scarcity and cost of labor in these countries.

Local demand for textiles in the developing countries is low because incomes are low. The huge population in the developing world represents an enormous growth opportunity for fiber demand. As noted earlier, empirical studies by the World Bank and others have demonstrated a very high elasticity of demand for textile products as incomes rise. In the People's Republic of China, for example, fiber consumption is 10 to 11 pounds per capita while in the United States it is over 60 pounds per capita.

The opportunity for textile fiber demand does not mean that growth is inevitable. Growth in fiber demand must be preceded by personal income growth, which in turn requires successful economic development and broad participation in the value-added stream. If standards of living do rise in Asia, Africa and Latin America, the proportion of textiles production devoted to local demand in those regions will increase. This does not mean that the developing nations necessarily will reduce their exports, but it is the growth in home market

TABLE 2 Domestic End-Use Pattern of Acrylic Fibers in Japan as Percent of Domestic Use

	1985	1990
Sweaters	17	15
Jersey	11	13
Underwear	3	2
Leg knits	8	7
Hand knitting	3	1
Fake fur	6	4
Woven fabrics	3	2
Other apparel	1	1
Total apparel	52	45
Blankets	20	21
Carpets	13	17
Curtain/uphol	8	9
Total home	41	46
Other	7	9
Total domestic	100	100

Source: Japan Chemical Fibres Association.

demand that is the basis for the high acrylic fiber growth rates being forecast for the developing regions. For the near term future, however, it will continue to be mainly export-oriented production that determines mill demand in the developing world.

4. Eastern Europe

Acrylic fiber is produced in the following East European countries: Bulgaria, Hungary, Poland, Romania, Russia and Yugoslavia. Collectively, these countries (or their successors) have an estimated 600 million pounds per year of acrylic fiber capacity or about 10% of world acrylic fiber capacity. Russia accounts for 40% of the total, followed by Romania with 24%, Hungary with 13%, and Yugoslavia with 11%. Poland and Bulgaria each have 5% [13].

Much of East Europe is undergoing massive economic and political change. In this context, it is difficult to see much change in demand for acrylic fiber in the near term. When conditions become more settled, one can envision the growth of Western style retailing with more products available. If East European governments are successful in fostering economic growth, consumers will have income to buy more and this could imply more demand for sweaters and other consumer products made of acrylic fibers. To a large extent those fibers and end products will be made in East Europe, but as these economies increase their contact with the rest of the world an increasing share of those products is likely to come from the developing world, especially Pacific Rim.

C. Net World Trade in Acrylic Fiber and Products Made Thereof

1. The Relative Importance of Trade

Previous discussion has shown that roughly half the acrylic fiber that is produced in West Europe and Japan is exported. The U.S. acrylic fiber industry exported 28% of its production in 1992. While the developed world exports a substantial amount of fiber, it imports a lot of finished goods of acrylic fiber. In 1992 the United States imported an estimated 153 million pounds of acrylic fiber and fiber products [14] versus exports of 125 million pounds of fiber. The categories of imports are shown in Table 3. The United States also exports some acrylic fiber manufactures. The author has not seen good documentation of the quantity of these exports, but they are believed to be relatively small (range of 5 to 10 million pounds). On balance, acrylic product imports account for one-third of the apparent consumption of acrylic fiber in the United States.

It would be an important contribution if someone were to develop a dynamic model of apparent consumption by world region including a map of all the trade flows. Some of the data is available but much is missing. What the map would show has been suggested in the analysis above: the developed nations are exporters of large amounts of fiber. They import large amounts of manufactures and semimanufactures of acrylic, mostly garments. Conversely, the developing nations, especially in Asia, are net importers of fiber and are net exporters of manufactures.

TABLE 3 U.S. Acrylic Fiber Apparent Consumption 1992 (million pounds)

Domestic fiber shipments		322
Imported fiber	44	
Imported yarn	18	
Imported fabric	5	
Imported garments	91	
Sweater	54	
Sweatshirts	19	
Other	18	
Total imports		153
Exports garments/fabric		5
Apparent consumption		470

Source: Monsanto estimates; U.S. import statistics (IM 145).

2. Impact of Trade Policy on Acrylic Fiber Demand

Governmental trade policies such as tariffs, quotas and other regulations can have a significant effect on the worldwide movement of acrylic fiber and products made thereof. With the planned phase out of the MFA (Multi Fiber Agreement) as determined in the concluding agreements of the Uruguay Round signed in December, 1993, trade barriers will be lowered and more market access will become available to both developed and developing nations. Many other trade agreements such as GATT (General Agreement on Trade and Tariffs), NAFTA (North American Free Trade Agreement) and Europe92 impact the flow of acrylic fiber products.

The U.S. sweater market offers a good example of how trade policies can affect trade flows. There is no question that demand for acrylic fiber to produce sweaters in South Korea, Taiwan and Hong Kong was directly impacted by trade policies in the United States, especially the quotas but also tariffs and dumping margins.

Another U.S. sweater example shows that trade in products can be driven entirely by trade policy, not market demand: the growth of ramie blend sweaters. From a position of zero in the mid-1980s, ramie blend sweaters have grown to about 25% of the sweater market by 1992 [15]. Ramie is a vegetable fiber like hemp or sisal that comes primarily from China. It has been incorporated in sweaters not because it makes a good sweater fiber but because including more than 50% ramie in the blend made it possible for U.S. retailers and China PRC exporters to exploit a loophole in the United States trade laws: there is no quota on ramie sweaters, because ramie is not grown in the United States. The point is that trade policy and agreements can have a powerful impact on trade in acrylic fiber products or substitutes that impact acrylic.

3. The Role of the People's Republic of China

No discussion of world demand for acrylic fiber would be complete without mentioning the importance of China. In the last seven years, China has imported an average of 500 million pounds of acrylic fiber each year. According to trade estimates, about 80% of China's mill consumption of acrylic fiber stays in China for home market consumption. This estimate is not well documented and should be used cautiously.

In the peak year of 1987, China's acrylic imports totaled over 725 million pounds and represented about 15% of total world acrylic fiber production that year. This enormous volume of fiber has important implications for the acrylic fiber industry:

• China exerts substantial leverage on pricing and tends to set the floor on world pricing.

- China has absorbed a lot of the world's excess acrylic fiber capacity. As China builds new acrylic fiber plants in their attempt to become more self-sufficient, excess capacity elsewhere is threatened.
- Without China's purchases there would have been many more plant shut-downs than were experienced.
- The impact of China purchases varies by country but as the table below suggests is particularly important to Japan and South Korea.

China's Acrylic Fiber Imports by Country of Origin ('000 MT)		
	1992	1993
Japan	57	75
South Korea	29	52
Taiwan	39	35
Italy	79	25
Germany	25	19
United States	18	17
Mexico	23	6
Others	33	16
Total	303	245

Source: Petrochemical Consultants International, London

D. Demand Side Summary

The acrylic fiber industry is truly global. After rapid growth from the late 1950s through the mid-1970s driven by substitution demand in the economically developed nations, the industry in these regions has matured and growth is slow and in some cases is negative for reasons cited. In the future, most observers expect rapid growth in the developing nations as predicted economic growth raises incomes and boosts demand for fiber products. Growth will be high in Asia, especially China, and Latin America.

III. SUPPLY SIDE

The world acrylic fiber industry includes 55 producers who operate 64 produc-ing facilities in 29 countries. Total capacity is 6.4 billion pounds [16]. The key trend in the industry over the past three to five years has been that of rational-ization. Unprofitable operations totalling nearly 400 million pounds have been shut down. Most of the shutdowns were in West Europe and the United States. In Japan, where significant overcapacity also exists, there has not been any significant rationalization. The Japanese with their powerful trading companies are able to maintain capacity utilization by exporting a large volume of acrylic fiber, but this is probably not a firm basis for satisfactory financial performance.

A. Current Acrylic Fiber Producers

Acrylic fiber producers today tend to produce their fiber in home market factories from which they supply domestic customers and export customers. Those who had multinational production have tended to pull back. In the early 1980s, Monsanto sold its plants at Coleraine, North Ireland and Lingen, Germany. DuPont shut down its Maydown plant in the United Kingdom. This plant was subsequently sold and transferred to Magyar Viscosa, a producer in Hungary. In the 1990s, Courtaulds shut down its Calais, France plant and rationalized production at its Grimsby, U.K. plant.

The graph in Fig. 5 shows the 17 largest private sector acrylic fiber producers worldwide. The combined capacity of these companies totals (4.3 billion pounds) or 65% of world capacity. Excluded from the chart are the capacities of the centrally planned nations such as the People's Republic of China (350 million pounds capacity at year end 1992) or the former U.S.S.R. (300 million pounds of capacity at year end 1992).

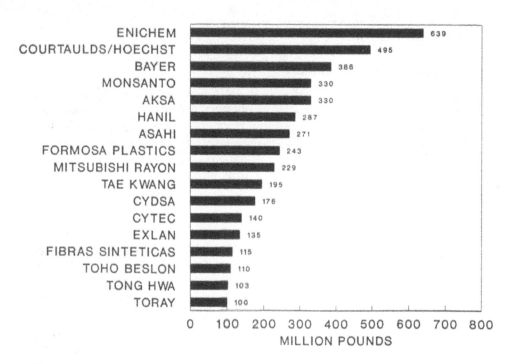

Figure 5 Top fiber sector acrylic fiber producers capacity year-end 1993.

The average size of the top 17 producers is 250 million pounds, while that of the 40 smaller producers (counting each producing location in the centrally planned countries as a producer) is under 60 million pounds. Acrylic fiber producers worldwide have been classified in Table 4 according to the following criteria: geographic location, size, solvent type and spinning method (wet or dry) [17].

B. Industry Rationalization

The acrylic fiber industry in the United States and Western Europe has been downsizing since the early 1980s as home market demand growth slowed or reversed at the same time that imports were rapidly taking greater share of the remaining demand. Indications of unsatisfactory results were apparent in several changes. Monsanto sold their European acrylic fiber plants in the early 1980s: a plant in Lingen, West Germany was sold to Bayer and a plant in Coleraine, Northern Ireland was sold to Montefibre. The Coleraine plant was later shut down. DuPont closed their Maydown, Northern Ireland plant in 1980. In 1983, Eastman Chemical shut down their Verel modacrylic plant in the United States.

1. United States

Additional rationalizations in the United States took place in 1985–1986 when all producers collectively scrapped over 200 million pounds of capacity [18]. In 1989, The Mann Group acquired the acrylic fibers business of BASF in a leveraged buyout. BASF did not want to keep this small business with weak performance and uncertain future. Buyouts rid an operation of heavy corporate overheads, but that saving can be offset by the heavy debt incurred in the buyout. In 1991, Mann Industries asked their workers to take a 15% pay cut in an effort to reduce expenses [19]. By Fall 1992 things were improving at Mann and there was talk of a turnaround [20]. But Mann discontinued operations late in 1993.

A major change in the United States was announced in 1990 when DuPont, the world's first producer of acrylic fiber, made public their decision to exit the business. Their phaseout was an orderly withdrawal that was completed in 1991. DuPont's acrylic fiber shutdown took out another 200 million pounds of capacity. The net industry change in capacity was less than that amount because Monsanto and Cytec (formerly American Cyanamid) added back some capacity.

Capacity shutdowns reflect unsatisfactory profits. Interested readers could find many producer statements in the trade press to document this assertion. Companies invest in expansions when business is profitable and growing; they downsize or divest when business is shrinking and profits are unsatisfactory.

2. Japan

In Japan, there has been limited rationalization. In 1988 Toray Industries announced that it would discontinue exports of acrylic fibers, which accounted for 17% of its sales or 13 million pounds per year. They said they would

TABLE 4 Table of World Acrylic Fiber Capacity by Producer 1993 (million pounds)

Country	Producer	Chemistry solvent	Spin method	Capacity
	USA			
USA	Cytec	NaSCN	wet	145
	Monsanto	DMAc	wet	330
Total				475
	Latin America			
Argentina	Hisisa	HNO_3	wet	35
Brazil	Celbras	DMF	dry	25
	Rhodia	DMF	wet	45
Mexico	Cydsa	DMF	wet	170
	Fibras Nationales	DMF	wet	75
	Fisisa	HNO_3	wet	115
Peru	Sudamericana de Fibras	DMF	dry	60
Total				525
	Eastern Europe			
Bulgaria	Dimitar Dimov	DMF	wet	55
Hungary	Magyar Viscosa	DMF	wet	30
	Magyar Viscosa	DMF	dry	45
Poland	ZWC Anilana	NaSCN	wet	30
Romania	Savinesti	EC	wet	140
USSR	Polymir	NaSCN	wet	150
	Navoiazot	NaSCN	wet	45
	Saratov	NaSCN	wet	45
	Mapan	DMF	wet	75
Yugoslavia	Ohis	NaSCN	wet	60
Total				675
	Western Europe			
England	Courtaulds	NaSCN	wet	200
Germany	Bayer, Domagen	DMF	dry	305
	Bayer, Lingen	DMAc	wet	80
	Hoechst, Kehlheim	DMF	dry	25
	Hoechst, Kehlheim	DMF	wet	75
	Rostextil	DMF	wet	150
Ireland	Asahi	HNO_3	wet	45
Italy	Enichem, Porto Marghera	DMAc	wet	220
	Enichem, Ottano	DMAc	wet	175
	Enichem, Porto Torres	DMF	wet	110
Portugal	Fisipe	DMF	wet	75
Spain	Courtaulds	NaSCN	wet	140
	Enichem	DMAc	wet	130
Total				1730

TABLE 4 Continued

Country	Producer	Chemistry solvent	Spin method	Capacity
Asia/Far East				
Bangladesh	Ashuganj Petro. Complex			
China	Daqing	NaSCN	wet	110
	Fushun	DMF	dry	65
	Gao Qiao	NaSCN	wet	10
	Jinshan	NaSCN	wet	150
	Lanzhou	NaSCN	wet	30
	Qinhaung Dao	DMF	dry	65
	Zibo	DMF	dry	100
India	Consolidated Fibers	NaSCN	dry	25
	India Petro. Chem.	HNO_3	wet	50
	India Acrylics Ltd.	DMF	dry	25
	J.K. Synthetics Ltd.	DMAc	wet	25
	Pasupati Acrylon	DMF	wet	40
Indonesia	Hanparan Rejecki	DMF	dry	85
Japan	Asahi	HNO_3	wet	220
	Japan Exlan	NaSCN	wet	135
	Kanebo	DMF	wet	70
	Kanegafuchi	Acetone	wet	90
	Mitsubishi Rayon	DMF	dry	10
	Mitsubishi Rayon	DMAc	wet	220
	Toho Rayon	$ZnCl_2$	wet	110
	Toray	DMSO	wet	95
North Korea	Pyong Yang	Aec	wet	10
South Korea	Hanil	HNO_3	wet	265
	Tae Kwang	NaSCN	wet	175
Taiwan	Formosa Plastic	DMF	wet	220
	Tong Hwa	NaSCN	wet	105
Thailand	Thai Acrylic Fiber		wet	45
Total				2550
Middle East / Africa				
Iran	Polyacryl Corp.	DMF	dry	50
Turkey	Aksa	DMAc	wet	330
	Yalova	DMF	wet	25
South Africa	Sasol Ltd.	NaSCN	wet	80
Total				485

"rationalize" their production equipment and despite plans to attempt to increase domestic sales believed "a reduction of business scale will probably be inevitable [21]." In 1989, Toyobo announced plans to scrap 30 million pounds per year by shutting two of the nine production lines of its Japan Exlan subsidiary. They noted they would stop producing commodity fibers but would continue to produce specialty fibers [22]. In 1991, Mitsubishi Rayon, Japan's largest acrylic fiber producer, stated plans to invest $100 million by 1994 to automate and robotize their polyester and acrylic fiber plants. A computerized central supervisory system and a robotized finishing system for acrylic staple would reduce cost and improve competitiveness [23].

3. Western Europe

Rationalization was slow to occur in Europe because governments anxious to preserve jobs were reluctant to permit closures. From the producer's standpoint, unemployment regulations that could require them to pay workers for up to two years whether they worked or not was a strong incentive to continue production.

A series of rationalization plans named for Mr. Davignon were worked out among the European producers and recommended to the European Commission. Under the second of such plans, Anic fiber in Italy announced in early 1984 the shutdown of its "aging" 90-million-pounds per year acrylic fiber plant at Pisticci [24]. That shutdown was delayed until 1986 and subsequently the plant was sold to Fisisa, and 53 million pounds of annual capacity was relocated to Mexico and started up in 1992 [25]. Another early rationalization move in Europe was Rhone Poulenc's decision to get out of the acrylic fiber business in 1984 and shut down 55 million pounds of capacity at their Colmar plant [26]. They were the only French company producing acrylic fiber, Courtaulds SA being a British parented company with a plant in France.

Additional actions came early in the 1990s. Courtaulds shut down their Calais, France, plant in 1990. That plant had a capacity of 180 million pounds per year. In addition they made some cutbacks at their Grimsby, U.K., plant believed to total 30 million pounds per year. Substantial parts of the Calais plant was subsequently sold to SASOL, which moved selected equipment to a new plant in South Africa. Enichem closed two former SNIA acrylic fiber plants: Cesano and Villacidro (each 75 million pounds per year) were shut down in 1991–1992. In 1993, Bayer announced they had shut down an acrylic fibers unit at Dormagen, affecting 230 jobs but not reducing their capacity of 385 million pounds per year. "The company's fiber operations stayed in the red [in 1992] as problems continued with acrylics." [27].

Another rationalization move in West Europe was the merger of the acrylic and viscose fibers operations of Hoechst AG of Germany and Courtaulds PLC of the United Kingdom. The joint venture will be 72.5% owned by Courtaulds and will create Europe's second largest fiber producer after Enimont of Italy [28]. Obviously the two participants expect to be able to get rid of redundant costs

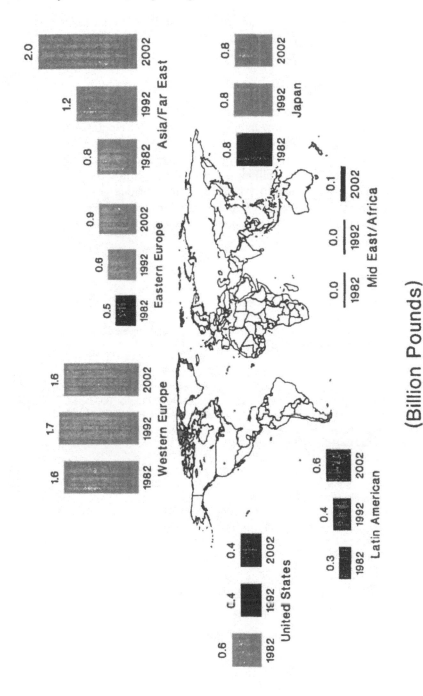

Figure 6 World acrylic fiber production: 1982, 1992, 2002

and make a stronger competitor of the new venture. The rationalization process is certainly not complete, but without insider information it is not possible to predict what specific additional moves will occur.

C. Outlook for the Future

The trends that have been detailed in this chapter indicate that most of the growth in world acrylic fiber production will be in the developing world in the decade of the 1990s. Regional trends are shown in Fig. 6. If world production grows as forecast, acrylic fiber production in the year 2002 will total 6.3 billion pounds. This is slightly over 1 billion pounds of growth in the decade ending 2002, a sustained growth rate of 1.8%. Three-fourths of that growth will be in the Far East, especially the People's Republic of China, and the balance in Latin America and Eastern Europe. The economic forces described above are the basis for the forecasts.

REFERENCES

1. Fiber Economics Bureau, *Manufactured Fiber Handbook*, Table 2, 1993, p. 10.
2. Monsanto internal estimate, unpublished.
3. MRCA Information Services, Stamford, Conn., consumer panel data, unpublished.
4. *Federal Register*, Vol. 59, No. 133, 7/13/94, p. 35750.
5. MRCA, op cit., unpublished.
6. Fiber Economics Bureau, Inc., 101 Eisenhower Parkway, Roseland, NJ, 07068.
7. Comite International De La Rayonne et Des Fibres Synthetiques, Paris, France.
8. Japan Chemical Fibres Association, No. 1-11, Chuo-ku, Tokyo 103, Japan.
9. See *World Fibers Overview*, James D. Geerdes, Presentation to Clemson University *Fiber Producer Conference 1992*, Clemson, SC, 1992.
10. C.I.R.F.S., Brussels, 1992 Fact Book.
11. C.I.R.F.S., op cit.
12. Japan Chemical Fibers Association, unpublished communication.
13. PCI—Fibers and Raw Materials; *World Acrylonitrile and Derivatives, Supply/Demand Report, 1993*, Crawley, West Sussex, England.
14. Monsanto estimates based on U.S. Government import data.
15. NPD, consumer panel data, unpublished.
16. Fiber Economics Bureau, *World Directory of Fiber Producers*, 1994-95. Also PCI —Fibres & Raw Materials, *World Acrylonitrile and Derivatives, Supply/Demand Report, 1993*.
17. Sources: Monsanto capacity estimates; Fiber Economics Bureau, *World Directory of Manufactured Fiber Producers, 1994–95;* Ciba Geigy, private communication.
18. Fiber Economics Bureau, *Manufactured Fiber Handbook, 1990–95*, p. 10.
19. *Virginia Gazette*, 12/7/91, p. 1.
20. "Mann's turnaround has 400 workers busy ... revenues have nearly doubled", *Virginia Gazette*, 10/17/92.
21. *JTN Weekly*, 9/9/88.
22. *Asian Textile Record*, 7/31/89.

23. *Asian Textile Record*, 4/22/91.
24. *European Chemical News*, 2/27/84.
25. Monsanto internal communications.
26. *Daily News Record*, 10/2/84, p. 15.
27. *Chemical* Week, April 21, 1993.
28. *Daily News Record*, April 6, 1994.

3

Polymerization

Bruce Wade

Monsanto Company, Decatur, Alabama

Raymond Knorr

Monsanto Company, Pensacola, Florida

I. INTRODUCTION

A. History

All commercial acrylonitrile polymerization processes for precursors to acrylic fibers are free radical processes. However, processes based on several other synthetic methods have been reported in the literature. Acrylonitrile (AN) can be polymerized anionically as shown below:

$$CH_2=CHCN + NaCN \rightarrow NC\text{-}CH_2CHCN^-Na^+$$
$$+ nAN \rightarrow NC\text{-}(CH_2CHCN)^-_{n+1}\, Na^+$$

The polymerization takes place at low temperature in a dry solvent such as DMF [1]. The product may have reasonable molecular weight, but usually has poor color properties owing to side reactions. The method does not introduce dye sites, and the choice of comonomers is limited; therefore no commercial process based on this chemistry has ever been developed.

A coordinate anionic process was reported by Chiang [2]. In this process, the intermediate is held in a fixed conformation while monomer units are inserted:

$$CH_2=CHCN + NaAlEt_3OEt \rightarrow NaAlEt_3CHCN-CH_2OEt$$
$$+ nAN \rightarrow NaAlEt_3(CHCN-CH_2)_{n+1}OEt$$

Solubility properties of this polymer in propylene carbonate were distinguishable from free-radical-initiated PAN, but x-ray diffraction studies failed to identify any improvement in crystalline perfection in the product. The disadvantages of comonomer limitation and lack of dye sites are also present.

A third, related, process is group transfer polymerization developed by Webster of DuPont [3]. The main focus of his work was on methacrylate monomers but some studies were also done on polymerization of acrylonitrile. Like Chiang's process, the group transfer polymerization coordinates the polymer intermediate while adding monomer units. For acrylate and methacrylate monomers, narrow molecular weight distributions were achieved, but because of the high propagation rate of acrylonitrile, broad distributions were obtained even at −50°C.

B. Properties Required in Acrylonitrile Polymers for Fiber Applications

All commercial processes are based on free radical polymerization because it gives the combination of polymerization rate, ease of control, and properties including whiteness, molecular weight, linearity, and the ability to incorporate desired comonomers and, in most cases, dye sites.

1. Molecular Weight

The molecular weight and the distribution of molecular weights are key polymer properties. Methods for molecular weight measurement in acrylonitrile polymers are discussed in Chapter 7. Typical commercial acrylic polymers for textile applications have number average molecular weights in the 40,000 to 70,000 range or roughly, 1000 repeat units. The weight average molecular weight is typically in the 90,000 to 170,000 range, with a polydispersity index between 1.5 and 3.0. The commonly measured viscosity average molecular weight is about 10% below the weight average. Higher molecular weight polymers are sometimes desirable for applications where fiber strength and modulus are important such as cement reinforcement and as a precursor for carbon fiber. In these cases, polymers having weight average molecular weights in the range of 250,000 may be employed. Further information on these applications may be found in Chapters 6 and 11.

The solubility of the polymer and rheological properties of the dissolved polymer as a dope must be precisely defined for dry or wet spinning. The polymer molecular weight is affected by all of the variables in the polymerization including initiator concentration(s), unreacted monomer concentration, comonomer type and concentration, reaction temperature and time among others. Small quantities of partially soluble gel can have a disastrous effect on dope quality. The gel may arise from very high molecular weight polymer, which may be caused by a transient control problem, presence of a divinyl monomer, or thermally induced cross-linking from unit operations such as slurry stripping or drying.

Fiber dyeability is critically dependent on the molecular weight distribution of the polymer because most acrylic fibers derive their dyeability from sulfonate and sulfate initiator fragments at the polymer chain ends. Thus, the dye site content of the fiber is inversely related to the number average molecular weight of the polymer and very sensitive to the fraction of low molecular weight polymer. For polymerization control purposes, viscosity average molecular weight— η_{sp} or η_{inh}—is usually inferred by measuring one concentration, since methods for number average are too time consuming. Thus unless the polydispersity is held constant, the dye sites may vary even with close control of the viscosity average molecular weight.

So long as the polymer is essentially linear, it may be spun over a wide range of molecular weights at the appropriate concentration considering the solvent, polymer and spinning conditions. From the standpoint of processing cost, it is desirable to have the highest possible concentration of polymer in the spin dope, because this means there is less solvent to recover per unit fiber which reduces vessel size and recovery energy. Since at a given dope viscosity, the polymer concentration is inversely related to the viscosity average molecular weight, a lower molecular weight is directionally more desirable provided that physical properties such as tenacity are not compromised. The fiber application then, will determine the polymer molecular weight. Most producers use polymers with weight average molecular weights from 90,000–170,000 for textile applications. As previously mentioned, for many fibers, dye sites come primarily from the initiator fragments on the chain ends. This is another reason to minimize molecular weight, as dyeability with cationic dyes is enhanced. With the trend to finer denier fibers where more dye is required to achieve a given color, the need for dye sites is increased. Over the years, many producers have gradually lowered the molecular weight of their polymer increase dyeability.

Polyacrylonitrile can participate in branching reactions. In a study [4] of acrylonitrile polymerization in magnesium perchlorate, branch formation occurred by a reaction in which a growing radical chain abstracts a hydrogen atom from the α-carbon, thereby starting a side chain by monomer addition. It was shown in this study that branch formation occurs when the ratio of polymer to monomer concentration is greater than one or a conversion of 80%. At this condition one branch occurs for every 2000 growth steps. Thus, at a molecular weight of 100,000 each molecule shows only one branch on the average. PAN branching by polymerization through the nitrile group has also been suggested [5]. Branch formation can also occur as a result of radical termination by disproportionation or other chain transfer reactions. Any reaction which leaves a terminal double bond can lead to long-chain branching if the double bond subsequently reacts with a growing polymer radical.

2. Comonomers

Many ethylenically unsaturated monomers have been copolymerized with acrylonitrile. The third edition of *Polymer Handbook* [6] listed 218 for which reactivity ratios have been reported in the literature! There have doubtless been many more for which this particular datum was not reported. However, only a relative handful are important with respect to the production of commercial acrylic fibers.

Nearly all acrylic fibers are made from acrylonitrile copolymers containing one or more additional monomers that modify the properties of the fiber. Thus copolymerization kinetics is a key technical area in the acrylic fiber industry. When carried out in a homogenous solution, the copolymerization of acrylonitrile follows the normal kinetic rate laws of copolymerization. The controlling variables are the relative reactivities of the monomers and the polymer radicals ending in the respective monomer units. The assumption is made that units beyond the terminal radical do not affect the reaction. In some cases, this is not true; a modified copolymerization scheme has been developed to take into account this "penultimate effect" [7]. Considering the more common case, for monomers M_1 and M_2 and polymer chain P_n:

$$P_n - M_1 \bullet + M_1 \xrightarrow{k_{11}} P_{n+1} - M_1 \bullet \tag{1}$$

$$P_n - M_1 \bullet + M_2 \xrightarrow{k_{12}} P_{n+1} - M_2 \bullet \tag{2}$$

$$P_n - M_2 \bullet + M_2 \xrightarrow{k_{22}} P_{n+1} - M_2 \bullet \tag{3}$$

$$P_n - M_2 \bullet + M_1 \xrightarrow{k_{21}} P_{n+1} - M_1 \bullet \tag{4}$$

The individual constants cannot be determined directly, but the ratios

$$R_1 = \frac{k_{11}}{k_{12}} \text{ and } R_2 = \frac{k_{22}}{k_{21}}$$

can be determined from a series of copolymerizations to low conversion. A value of R_1 much greater than 1 reveals a tendency of monomer 1 to incorporate in blocks, whereas a value of near 1 (above or below) indicates a tendency to alternate with monomer 2 along the polymer chain. A value near zero indicates an inability of the monomer to react with its own radical, in the presence of the other monomer. The same arguments of course hold for R_2. The instantaneous polymer composition may be calculated as:

$$\frac{dm_1}{dm_2} = \frac{M_1(R_1M_1 + M_2)}{M_2(R_2M_2 + M_1)} \tag{5}$$

where m_1, m_2 and M_1, M_2 are the moles of polymer and monomer respectively [6].

Comprehensive treatments of this general subject have been published [8–11]. The more specific subject of acrylonitrile copolymerization has been reviewed by Peebles [12]. The general subject of the reactivity of polymer radicals has been treated in depth [13]. Table 1 gives the reactivity ratios for acrylonitrile with comonomers of commercial importance. The ratios have mainly been developed based on a homogeneous system. In cases where the monomer or polymer is insoluble in the polymerization medium, a more complex situation exists. This will be considered further in Sec. II.C.

The monomer pair, acrylonitrile–methyl acrylate ($R_1 = 1.17$, $R_2 = 0.76$ [14]), is close to an ideal pair. Both monomers are similar in resonance, polarity, and steric characteristics. The acrylonitrile radical shows approximately equal reactivity with both monomers, and the methyl acrylate radical shows only a slight preference for reacting with acrylonitrile monomer. Many acrylonitrile monomer pairs used commercially fall into the nonideal category, e.g., acrylonitrile–vinyl acetate ($R_1 = 4.2$, $R_2 = 0.05$ [15, 16]) and acrylonitrile–vinyl chloride ($R_1 = 3.6$, $R_2 = 0.052$ [17]). In these cases incorporation of small amounts of the unreactive monomer is readily accomplished, but extreme measures must be used to make a polymer with a significant proportion of M_2. A third type of monomer pair is that which shows an alternating tendency. The tendency is related to the polarization properties of the monomer substituents [18]. Monomers that are dissimilar in polarity tend to form alternating monomer sequences in the polymer chain. An example is the monomer pair acrylonitrile-styrene. Styrene, with its pendent phenyl group, has a relatively electronegative double bond whereas acrylonitrile, with its electron-withdrawing nitrile group, tends to be electropositive. Figure 1 shows the polymer composition as a function of monomer composition for all three pairs.

Copolymer composition can be predicted for copolymerizations with more than two components, such as those employing acrylonitrile plus a neutral monomer and an ionic dye receptor. These equations are derived by assuming that the component reactions involve only the terminal monomer unit of the chain radical. The theory of multicomponent polymerization kinetics has been treated [8, 9]

Textile acrylic fibers are made from polymers of acrylonitrile, usually with at least one other comonomer, except where a homopolymer of acrylonitrile might be required for increased resistance to chemical or physical attack. Neutral comonomers—methyl acrylate (MA), methyl methacrylate (MMA), or vinyl acetate (VA)—are used to modify the solubility of the acrylic copolymers in

TABLE 1 Reactivity Ratios for Acrylonitrile Polymerizations [6]

Monomer 2	R_1	±	R_2	±	T (°C)	Reaction medium
Acrylamide	0.94	0.16	1.04	0.27	40	redox
Acrylic acid	0.35		1.15		50	peroxide/water
Allyl sulfonic	1.85	0.01	0.43	0.01	45	DMSO solution
acid, Na salt	1.00	0.01	0.38	0.02	45	DMSO solution
	1.25	0.01	0.28	0.02	45	DMSO/water 94/6 solution
1,3-butadiene	0.05	0.01	0.35	0.01	50	
1-butene	8.0		0.10		60	
2-butene *cis* or *trans*	14.0		0.0		60	
Isobutylene	1.02		0.0		60	
Isoprene	0.03	0.03	0.45	0.05	50	
Methyl acrylate	1.5	0.10	0.84	0.05	50	peroxide
	1.17		0.76		50	
	1.02		0.70		50	DMSO solution
Methacrylonitrile	0.35	0.02	2.80	0.02	50	benzene solution
Butyl methacrylate	0.31	0.05	1.08	0.40	60	peroxide

Methyl methacrylate	0.15	0.07	1.20	0.14	60	peroxide, solution
Sodium styrene sulfonate [106]	0.05	0.02	1.50	0.20	40	
Styrene	0.05	0.02	0.37	0.02	50	
	0.07	0.006	0.37	0.03	50	
Vinyl acetate	4.05	0.3	0.061	0.01	60	
	4.2		0.05		50	peroxide, DMF solution
Vinyl bromide	2.25		0.055		60	
Vinyl chloride	3.6	0.2	0.052	0.009	50	acetone solution
	3.7		0.074		50	peroxide, solution
Vinyl formate	3.0	0.05	0.04	0.005	60	
2-vinylpyridine	0.113	0.002	0.47	0.03	60	
4-vinylpyridine	0.113	0.005	0.41	0.09	60	
Vinyl stearate	4.2	0.02	0.064	0.005	60	DMF solution
Vinyl sulfonic acid	4.52		0.22		60	$ZnCl_2$/water solution
Vinylidene chloride	0.44–0.70		0.40–1.8		40	various solvents
	0.91	0.10	0.37	0.37		peroxide

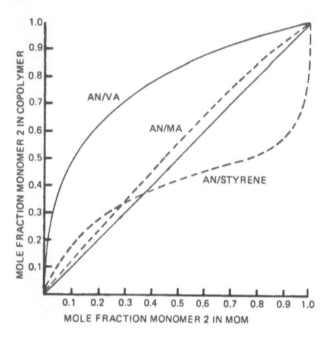

Figure 1 Incremental copolymer composition vs. unreacted monomer composition for three different reactivities: AN/VA: $R_1 = 4.05$, $R_2 = 0.06$; AN/MA: $R_1 = 1.5$, $R_2 = 0.84$; AN/styrene: $R_1 = 0.04$, $R_2 = 0.03$.

spinning solvents, to modify the acrylic fiber morphology, and to improve the rate of diffusion of dyes into the acrylic fiber. Other neutral monomers such as styrene and acrylamide have been incorporated in past commercial products. Ionic comonomers containing sulfonate groups, such as sodium *p*-styrenesulfonate (SSS), sodium methallyl sulfonate (SMS), sodium *p*-sulfophenyl methallyl ether (SPME), or sodium 2-methyl-2-acrylamidopropane sulfonate (SAMPS), may be added to provide dye sites apart from end groups and to increase hydrophilicity. Halogenated comonomers, such as vinylidene chloride, vinyl bromide, and vinyl chloride, may be incorporated to impart flame resistance to fibers used for home furnishings, awning and sleepwear markets. Modacrylic compositions (more than 15% incorporation of comonomers) are used when the end use requires high flame resistance. All of the modacrylics are flame resistant fibers with very high (35–50%) levels of halogenated comonomers. Modacrylics are discussed in more detail in Chapter 6.

Between the two most common comonomers used in commercial acrylic fibers, methyl acrylate and vinyl acetate, MA is less active in chain transfer while VA is almost as active in chain transfer as DMF ($C_m \sim 2$ at 60°C [19]).

Vinyl acetate is also known to participate in a chain transfer-to-polymer reaction [4]. This occurs primarily at high conversion, where the concentration of polymer is high and monomer is scarce. Another difficulty related to the low reactivity of VA is the precipitous change in monomer VA concentration approaching high conversion as shown in Fig. 2. Compared to MA, VA level in polymer is thus more difficult to control, being impacted by small changes in monomer to polymer conversion.

Despite its disadvantages of low reactivity, difficulty in polymer control and chain transfer in polymerization, vinyl acetate is increasingly the comonomer of choice for wet-spun acrylic fibers, primarily because of its low cost. In July 1994, the list prices for the three monomers commonly used was [20]:

Monomer	U.S.$/lb	U.S.$/kg
Vinyl acetate	0.38	0.84
Methyl acrylate	0.83	1.83
Methyl methacrylate	0.71	1.56

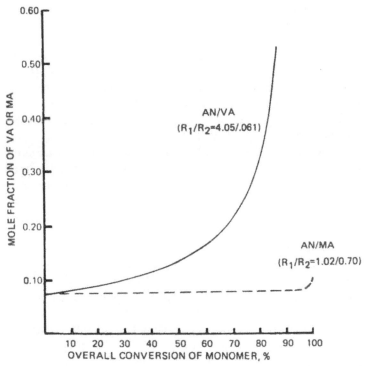

Figure 2 Drift in copolymer composition with conversion for the batch reaction of two commercially important monomer pairs.

Even though these are list prices, subject to discount, they give an idea of the cost advantage accruing to VA. Cytec (formerly American Cyanamid) switched from MMA to VA in 1994. All dry-spun acrylic fibers use MA; this is believed to be because VA is perceived to reduce the stability of the spin dope to the high temperature in the spinning tower. The amount of neutral comonomer incorporated in polymers for commercial acrylic textile fibers is in the range 3.5 to 5.5 mole% or 6 to 10 wt%. This level accomplishes the solubility, dyeability and physical property requirements with minimum reduction in hot-wet properties of the fiber.

As mentioned in the previous section, the initiator system may provide dye sites to the fiber. The total number of sulfonate and sulfate sites required to be able to dye a full range of shades with cationic dyes is 30 to 50 milliequivalents per kilogram, depending on the fiber denier and structure. Dry-spun fibers and microdenier fibers require more than 40 dye sites. Where the number provided by the end groups is inadequate, a sulfonate-containing monomer may be used to supplement [21]. In some solution polymerizations which use azo or organic peroxide initiators, all dye sites must be supplied by an added monomer. Carboxylic monomers such as itaconic acid have also been employed as dye receptors, but because they are only partially ionized at dyeing pH, they are much less efficient and therefore larger percentages are required. An old version of Courtelle is believed to have contained itaconic acid. In the early days of the acrylic fiber industry, before new classes of cationic dyes were developed by DuPont and Bayer, several acrylics were acid dyeable. These were copolymers containing vinyl pyridines, such as 2-methyl-5-vinylpyridine. These monomers were expensive, tended to impart a yellow color to the fiber and required severe conditions in dyeing; they were rapidly displaced by cationically dyeable acrylics. Several acid dyeable acrylics are still on the market, for cross-dyeing purposes; they are discussed in Chapter 6. Polymers with a high concentration (~250 milliequivalents/kg) of a sulfonate monomer are sometimes required for a specialty application such as bicomponent fiber. These are also discussed in Chapter 6.

3. Whiteness

Acrylic fibers for textile and home furnishings applications must be white in order to be used in "white" applications and also to be dyed to clear pastel colors. Some remedial work may be done in the spinning step to cover up yellowness by addition of blueing agents and optical brighteners, but a good initial polymer color is critical if a high whiteness product is to be achieved. The mechanisms of color development in acrylic polymers and fibers are discussed in detail in Chapter 8. The side reactions that make the fiber prone to color formation can be minimized by maintaining a high monomer concentration (low conversion) [22]. In aqueous processes, redox couples such as sodium nitrite with sodium bisulfite [23], metal chelates [24], and special reducing agents [25–28] have been claimed to be effective. In solution polymerization, additives such as sulfamic acid [29], thioglycerol [30], hydrazine salts [31, 32], adipic hydrazide [33] and amine salts [34–37] have been claimed effective for color control. Polymers incorporating carboxylic acids or amines as dye receptors are particularly prone to discolor as these functional groups act as initiators for cyclic imine formation. Modacrylic compositions with their high concentration of halogen, easily dehydrohalogenate during processing to form carbon-carbon double bonds which can conjugate with other chromophoric groups to lead to yellowness. They also provide Cl^- or Br^- which can initiate imine formation.

All acrylic polymers have some defects in their structure that can lead to color formation. Even in the absence of defects, the fundamental structure of the PAN molecule is relatively easily degraded to a conjugated cyclic imine chromophore. Therefore care in processing—use of stabilizers, mild temperatures and protection from oxygen—are important to achieving white fiber.

C. Polymerization Methods

Acrylonitrile and its comonomers can be polymerized by any of the well-known free-radical methods. Bulk polymerization is the most fundamental of these, but its commercial use is limited by its autocatalytic nature. Aqueous dispersion polymerization, a variant of suspension polymerization, is the most common commercial method, whereas solution polymerization is used in cases where the spinning dope can be prepared directly from the polymerization reaction product. Emulsion polymerization is used primarily for modacrylic compositions where a high level of a water-insoluble monomer is used or where the monomer mixture is relatively slow reacting. A summary of the acrylic commercial processes reported by the major acrylic fiber manufacturers is given in Table 2.

II. AQUEOUS DISPERSION POLYMERIZATION

By far the most widely used method of polymerization in the acrylic fibers industry is aqueous dispersion. Aqueous dispersion polymerization, in which water is the continuous phase, is a variation of suspension polymerization. The initiators and dispersants (if any) used in dispersion polymerization are soluble in water, while those in suspension polymerization are water insoluble (<1%) [1].

A. Initiators

The kinetics of aqueous dispersion polymerization differ very little from acrylonitrile bulk or emulsion polymerization. Redox initiation is normally used in commercial production of polymer for acrylic fibers. This type of initiator can

TABLE 2 Polymerization Processes Reported by Major Acrylic Fiber Producers

Fiber producer	Trade name	Comonomer	Solvent
	1. Fiber producers using aqueous dispersion polymerization		
Asahi	Cashmilon	MA	
Bayer	Dralon	MA	
Courtaulds	Courtelle	MA	
Cydsa	Crysel	VA	
Cytec	Creslan	VA (1994);	
(formerly American Cyanamid)		(previously MMA)	
Fabetta	Acribel	MMA	
Hoechst	Dolan	MA	
Japan Exlan	Exlan	MA	
Kanagafuchi	Kanecaron	VCl	
Mitsubishi Rayon	Vonnel	VA	
	Finel	MA	
Monsanto	Acrilan	VA	
Montefibre	Leacril	VA	
	2. Fiber producers using solution polymerization		
Toho Rayon	Beslon	MA	zinc chloride
Toray	Toraylon	MA	DMSO

generate free radicals in an aqueous medium efficiently at relatively low temperatures. The most common redox system consists of ammonium, sodium, or potassium persulfate (the oxidizer, sometimes called "catalyst"), sulfur dioxide, sodium bisulfite or metabisulfite (reducing agent, sometimes called "activator") and ferrous iron (the true catalyst). This system gives the added benefit of supplying dye sites for the fiber. This redox system works at pH 2 to 3.5 where the bisulfite ion predominates [38] and where the ferric ion generated by oxidation is sufficiently soluble. If sulfur dioxide is not used for reduction and proton generation, then another proton source such as sulfuric acid is required to maintain a pH in this range. Two reactions account for radical production: the oxidation of ferrous iron by persulfate and the reduction of ferric iron by the bisulfite ion, HSO_3^{1-}.

$$S_2O_8^{2-} + Fe^{2+} \rightarrow SO_4^{2-} + SO_4 \bullet^{1-} + Fe^{3+} \tag{6}$$

$$HSO_3^{1-} + Fe^{3+} \rightarrow HSO_3\bullet + Fe^{2+} \tag{7}$$

The sulfate and sulfonate radicals thus produced, react with monomer to initiate rapid chain growth. The iron, usually added as ferrous sulfate, is only required at a low level, less than 10 ppm based on monomer. In fact, polymerization will proceed, albeit at a lower rate, without any added iron, as small impurities in the other chemicals usually supply some. Larger iron impurities introduced through grossly impure raw materials or corrosion from improper materials of construction can be an important source of off-grade polymer in the dispersion process, since the added iron raises the rate of radical formation, thus lowering the molecular weight. Other transition metals with more than one valence state such as copper and mercury, will also serve as catalyst, but are slower than iron.

The propagation step is very fast with a rate constant (R_p) in the range 30,000 L/mol-sec [39] for dispersion polymerization and 3000 L/mol-sec [40] for solution polymerization. The half-life of a growing chain in aqueous polymerizations is reported to be about 5 min [41].

Free-radical polymer chains terminate by three mechanisms:

1. Combination
   $$P_a\bullet + P_b\bullet \rightarrow P_a - P_b$$

2. Disproportionation
   $$P_a\text{--CHX--CYZ}\bullet + P_b\text{--CHX--CYZ}\bullet \rightarrow P_a\text{--CX=CYZ} + P_b\text{--CHX--CHYZ}$$

3. Chain transfer
   $$P_a\bullet + X\text{--}Y \rightarrow P_a\text{--}Y + X\bullet$$

where Y is usually H

In AN polymerization, only reactions (1) and (3) are active. Reaction (1) also terminates the kinetic chain, whereas (3) produces a new radical.

In most commercial processes chain-transfer agents are used to control molecular weight and impart acid dye sites. Bisulfite ion, in addition to being the reducing agent is the most widely used chain-transfer agent. It apparently reacts rapidly with the growing polymer radical, $P_n\bullet$, since the bisulfite ion has a pronounced effect on polymer molecular weight with virtually no effect on the overall rate of polymerization [42].

$$HSO_3^{1-} + P_n\bullet \rightarrow SO_3\bullet^{1-} + P_nH \tag{8}$$

Reactions (7) and (8) are apparently very rapid compared with reaction (6).

The ratio of bisulfite to persulfate in the reaction mixture has a strong effect on the dye site content of the polymer [42, 43]. In the absence of chain-transfer reactions, all end-group dye sites are derived from initiator radicals. Thus if termination occurs exclusively by radical recombination, then each polymer chain contains a dye site at each end. Sulfate and sulfonate radicals are produced at equal rates, so assuming equal radical reactivity with monomer, the total dye site content should be an equal molar mixture of these two distributed among the chain ends at random. Chain transfer to bisulfite, however, terminates one chain with a hydrogen atom while starting another with a sulfonate radical. This increases the total dye site content of the polymer by reducing the polymer molecular weight, but at the same time this reaction produces chains with just one dye site. At a given molecular weight the dye site content of the polymer can, in theory, vary from two per chain at a low bisulfite level to one per chain at a very high bisulfite level. Peebles et al. [43] presented the following data:

Molar bisulfite/persulfate	12.6/1	1.4/1
Sulfates/molecule	0.16	0.24
Sulfonates/molecule	0.93	1.03
Total SAG/molecule	1.09	1.27
SAG milliequivalents/kg	21.6	31.2
M_v	1.25×10^5	1.28×10^5
M_n	5.06×10^5	4.08×10^5
M_v/M_n	2.5	3.1

In commercial practice, high reducing agent to oxidizing agent ratios are used which are equivalent to ratios of bisulfite to potassium persulfate ranging from 8 to 20 [44]. These high ratios give narrower molecular weight distributions and the combination of high activator and low initiator (to attain a specified polymer η_{inh}) gives a lower monomer to polymer conversion. Low conversion is an effective means of minimizing branching and color producing side reactions, thus producing a whiter polymer product. Control of polymerization pH is important to maintaining constant molecular weight and dye site content since the bisulfite ion is the active species in chain transfer. This is particularly true as the ratio of bisulfite to persulfate is increased thus increasing the proportion of chains terminated by transfer. At 50°C, the equilibrium constant is $10^{-2.2}$ mole/L [45] for the reaction

$$H_2SO_3 \overset{K_{eq}}{\rightleftharpoons} H^+ + HSO_3^-$$

An increase in pH from 3.0 to 3.4 increases the fraction of HSO_3^- from 0.86 to 0.94. A comprehensive review of aqueous polymerization has been published [46]. Reviews of acrylonitrile polymerization are many [47–50].

Polymerizations initiated by the persulfate-bisulfite-iron redox system may be stopped by chelating the iron using agents such as ethylene diamine tetra-acetic acid, tetrasodium salt (EDTA · 4Na) [51] or sodium oxalate [52] or by increasing the pH of the slurry (thus precipitating the insoluble $Fe(OH)_3$ using sodium bicarbonate or other bases.

Other redox systems based on sodium chlorate as oxidizing agent and bisulfite as the reducing agent, have been described in the literature [53, 54] but apparently never commercialized. The advantage of these systems is that they produce only a single radical—$HSO_3 \cdot$—avoiding the potentially hydrolyzable sulfate group. Also, the chlorine compounds directly oxidize the bisulfite, avoiding the need (and the problems caused by contamination) of iron. However, the corrosive nature of the chlorides makes equipment choice more difficult and expensive.

A side reaction of all systems using bisulfite as a reducing agent is its reaction with acrylonitrile to form the β-sulfopropionitrile [55]:

$$CH_2=CH-CN + HSO_3^- \rightarrow {}^-O_3S-CH_2-CH_2-CN$$

This reaction may occur to a small extent during the polymerization, and also continues during further processing steps, particularly if the pH is increased. The water soluble sulfonate is lost in the washing step. Bisulfite has also been shown to react with other monomers used in acrylonitrile copolymerizations [56].

Redox couples based on hydrogen peroxide and a thiol such as β-mercaptoethanol or α-thioglycerol [57] have also been used with dispersion systems to produce polymers for acid-dyeable fibers where the presence of an anionic site is detrimental.

B. Kinetics

In aqueous dispersion polymerization, up to three phases important to the reaction may be present—a continuous aqueous phase, a phase consisting of polymer particles swollen at the surface with non-water soluble monomer and in cases where the amount of monomer exceeds that which will dissolve in water or adsorb on polymer, a monomer droplet phase.

It is generally agreed that when inorganic compounds, such as persulfates, chlorates or hydrogen peroxide are used as radical generators, the initiation and primary radical growth steps occur mainly in the aqueous phase. The aqueous-phase polymer radicals may follow either of two routes shown schematically in Fig. 3. Chain growth is limited in the aqueous phase, however, because the monomer concentration is low and the polymer is insoluble in water. Nucleation occurs when aqueous chains aggregate or collapse after reaching a threshold molecular weight. If many polymer particles are present, as is the case in commercial continuous polymerizations, the aqueous radicals are likely to be captured on the particle surface by a sorption mechanism. The particle surface is swollen with monomer. Therefore, the polymerization continues in the swollen layer and the sorption becomes irreversible as the chain end grows into the particle.

In the absence of an emulsifier or other particle-stabilizing agent, an additional phenomenon may become important. As particle size and particle number continue to increase, the frequency and energy of collisions increase accordingly. Thus, the increase in collisions where particle stability is relatively low results in agglomeration of particles. Since polymer swelling is poor and the aqueous solubility of acrylonitrile is relatively high, the tendency for radical

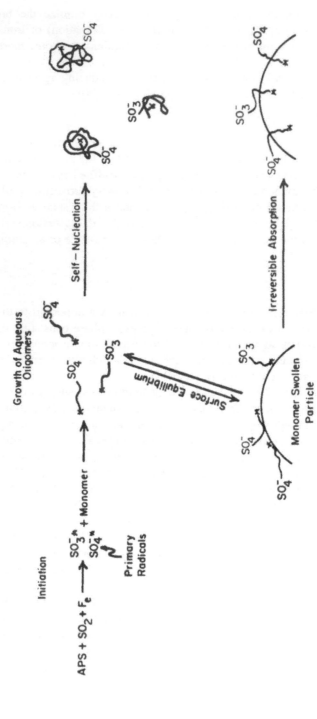

Figure 3 Schematic depiction of particle nucleation and radical absorption in dispersion polymerization of AN (APS = ammonium persulfate).

capture is very limited. Consequently, the rate of particle nucleation is high throughout the course of the polymerization and particle growth occurs predominantly by a process of agglomeration of primary particles. This can be seen in electron photomicrographs of polymer particles [58]. Unlike emulsion particles, the acrylonitrile aqueous dispersion polymer particles can be agglomerates of primary particles which are approximately 1 micron in diameter.

Lowering the water to monomer ratio in an aqueous dispersion polymerization has the effect of producing denser, more spherical particles because it favors polymer growth within particles instead of agglomeration. Because the particles are denser, the free-water/particle ratio does not decrease at the same rate as the water-to-monomer ratio. Consequently, slurry fluidity can be maintained to very low water-to-monomer ratios. Figures 4 and 5 show particles from polymers produced at 3.5 to 1 and 2.0 to 1 ratios.

Reactor agitation also has a great effect on the mean agglomerate particle size and the breadth of the particle size distribution. Generally medium-to-high shear mixing is required in aqueous dispersion polymerization of acrylic polymers due to the moderately high slurry viscosities found, usually in the range of 20 to 120 cP. Vacuum filtration is the most usual method of polymer isolation. For efficient washing and dewatering on the filter, particle sizes of 20 to 40 microns are preferred. Particles smaller than 5 microns may blind the filter cloth, preventing dewatering of the slurry cake. Manipulation of reactor agitation and baffling, plus in some cases addition of an agent such as sodium sulfate are used to ensure optimum particle size.

C. Copolymerization

When copolymer is prepared in a homogeneous solution, the kinetic expressions presented on page 41 can be used to predict copolymer composition. Dispersion polymerization and emulsion polymerization are different since the reaction medium is heterogeneous and polymerization occurs simultaneously in separate loci. In both of these systems the monomer is actually dispersed in two or three distinct phases: a continuous aqueous phase, a monomer phase consisting of polymer particles swollen at the surface with monomer, and in emulsion polymerization and some cases of dispersion polymerization, a monomer droplet phase. This affects the ultimate polymer composition because the monomers are partitioned such that the monomer mixture in the aqueous phase is richer in the more-water-soluble monomers than the two organic phases. Where polymerization occurs predominately in the organic phases, these relatively water soluble monomers may incorporate into the copolymer at lower levels than expected. For example, in studies of the emulsion copolymerization of acrylonitrile and styrene, the copolymer was richer in styrene than copolymer made by bulk polymerization, using the same initial monomer composition [59–61]. Analysis of the reaction mixtures [62] showed that nearly all of the styrene was concen-

Figure 4 Scanning electron micrograph of AN/VA polymer particle prepared by aqueous dispersion polymerization at 3.5 water-to-monomer ratio.

trated in the droplet and swollen particle phases. The acrylonitrile, on the other hand, was distributed between both the aqueous and organic phases. The monomer compositions in the droplet and particle phase were found to be essentially the same. Similar results are observed in dispersion polymerization. Vinyl acetate is incorporated more readily than would be expected based on the reactivity ratios previously quoted because it is less water soluble than acrylonitrile and thus has a higher concentration in the growing particle phase where most of the polymerization takes place.

The effect of monomer partitioning on copolymer composition is strongest with the ionic monomers since this type of monomer is soluble in water and nearly insoluble in the other monomers. It is expected that ionic monomers increase the solubility of the growing chain and thus delay precipitation to a higher degree of polymerization. Once the growing radical precipitates, however, it is unlikely to incorporate additional ionic monomer. The conclusion

Figure 5 Scanning electron micrograph of AN/VA polymer particle prepared by aqueous dispersion polymerization at 2.0 water-to-monomer ratio.

is that ionic species should be located near the initiator chain end. There will also be a greater proportion of the water-partitioned ionic monomer in the low molecular weight tail of the polymer distribution [63].

D. Polymer Processes

1. Semibatch

In the early years of acrylic fiber production, during the 1950s and early 1960s, the semibatch polymerization process was commonly used for the commercial production of acrylonitrile copolymers. In this process, the reaction vessel was either an aluminum alloy or glass-lined steel having baffles and an agitator with one or more sets of blades. The reaction was generally run at reflux (~70°C). At the end of the reaction, the small amount of excess monomer was sometimes stripped from the product. A jacket or an interior coil was used to provide heat for start-up. The vessel was charged with a portion of the reactants and the reaction was induced by using a radical initiator, such as potassium persulfate or a redox system such as potassium persulfate/sodium bisulfite/iron. Control of copolymer composition was difficult in this type of reactor because the comonomers most frequently used have vapor pressures, solubilities, and reactivities differing greatly from that of acrylonitrile. As a result, the various monomers employed in a given batch reaction mixture often reacted at widely differing rates. Often, the copolymer formed at the early stages of the reaction cycle had a much different composition than that formed at a later stage in the reaction cycle. Therefore, in some processes, monomer and initiator components were added in different proportions during the course of the reaction to maintain conditions within the reaction vessel as constant as possible. The polymer was isolated by centrifugation or filtration with washing during the separation process. Various means of drying were used including fluidized bed, kiln and belt. A laboratory scale example of such a process has been reported [64]:

One hundred and fifty-two (152) parts (2.87 moles) of acrylonitrile, 8 parts (0.093 mole) of methyl acrylate, 1.2 parts (0.033 mole) of hydrogen chloride, and 1439 parts of deionized water are charged into a round-bottomed flask. The flask is placed in a constant temperature bath, and a condenser, thermometer, stirrer, nitrogen inlet tube, and dropping funnel are attached. The monomer mixture is heated to 40°C under nitrogen for one hour. The catalyst, 0.716 part (0.00672 mole) of sodium chlorate and 5.04 part (0.04 mole) of sodium sulfite, is dissolved in 150 cc. of water into the dropping funnel. Forty (40) percent of the catalyst, 60 cc. of solution, is rapidly added to the reaction vessel. After 25 minutes, an additional 22.5 cc. of catalyst solution is added. The remaining catalyst solution is added at 25 minute intervals in volumes of 22.5, 15, 15, 7.5 and 7.5 cc. Catalyst addition is complete in 2.5 hours. The mixture is agitated 1.5 hours longer, and the polymer is collected by filtration. The pH of the effluent mother liquor is 2.5. In this example the mole ratio of sodium sulfite to monomer is 0.0135, while the mole ratio of sodium chlorate to monomer is 0.00227. conversion of monomer to polymer is 85 percent of theory. The polymer has an average molecular weight of 77,000.

2. Continuous

The most common reactor type in the 1990s is the continuous stirred tank reactor (CSTR). This reactor, constructed of an aluminum alloy, has a domed bottom, an overflow near the top of the straight section, baffles and an agitator of one or more sets of blades. For atmospheric pressure polymerizations, the overflow may be an open pipe. A jacket surrounding the reactor allows heat to be put in (steam or hot water) or taken out (chilled water). Most operate near atmospheric pressure at temperatures of 50–60°C with a condenser or an absorbing column to prevent monomer loss. In a continuous process, reactor feeds are metered in at a constant rate for the entire course of the production run. The main advantage of this process over the semibatch process is that control of molecular weight, dye site level, and polymer composition is greatly improved and production economics are better.

Aqueous dispersion processes are by far the most widely used in the acrylic fiber industry. Water acts as a convenient heat transfer and cooling medium and the polymer is very easily recovered by filtration or centrifugation. Fiber producers that use aqueous solutions of thiocyanate or zinc chloride as spinning solvent for the polymer gain an additional benefit. In such cases, wet polymer can be converted directly to dope to save the cost of polymer drying.

In an aqueous dispersion polymerization start-up, the CSTR is typically charged with a certain amount of pH adjusted water. In a more sophisticated process the start-up periods may be minimized by filling the reactor with overflow from a reactor already operating at steady state. The reactor feeds are metered in at a constant rate for the entire course of the production run, which normally continues until equipment maintenance, for example when reactor cleaning for scale buildup, is needed. A steady state is established by taking an overflow stream at the same mass flow rate as the combined feed streams.

An example of a continuous aqueous dispersion process is shown in Fig. 6 [65]. A monomer mixture composed of acrylonitrile and up to 10% of a neutral comonomer, such as methyl acrylate or vinyl acetate, is fed continuously. Polymerization is initiated by feeding aqueous solutions of potassium persulfate (oxidizer, initiator) sulfur dioxide (reducing agent, activator), ferrous iron (redox catalyst), and sodium bicarbonate (buffering agent). The aqueous and monomer feed streams may be fed at rates which give a reactor dwell time of 40 to 120 min and a feed ratio of water to monomer in the range from 3 to 5. The reaction mass is held at constant temperature (usually 45–60°C) by circulating chilled water in the reactor jacket. The product stream, an aqueous slurry of polymer particles, is mixed with tetrasodium ethylenediaminetetraacetic acid (iron chelating agent) or sodium oxalate to stop redox radical initiation and thus shortstop the polymerization. This slurry is then fed to the top section of a baffled monomer stripping column. The separation of unreacted monomer can be effected by contacting the slurry with a countercurrent flow of steam introduced at the bottom of the column. The monomers are condensed from the overhead stream and separated from the resulting water mixture for reuse using a decanter. The stripped slurry is taken from the column bottoms stream and separated from the water using a continuous vacuum filter. After filtration and washing, the polymer is pelletized to improve drying efficiency in belt drying, dried, ground, and then stored for later spinning.

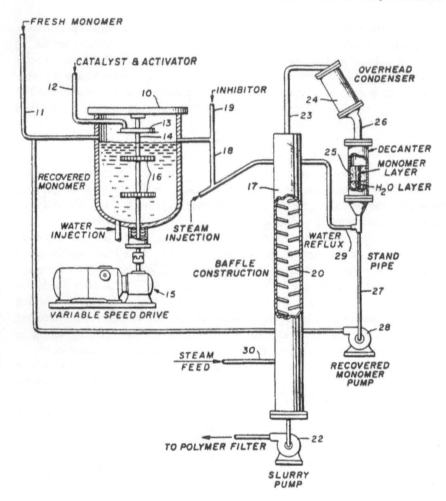

Figure 6 Schematic diagram of aqueous dispersion polymerization reactor and slurry stripping column [65].

The monomer recovery process may vary in commercial practice. A less desirable sequence is to filter or centrifuge the shortstopped slurry with concurrent washing to recover the polymer and unreacted monomers. The combined filtrate and washings are then distilled to separate the unreacted monomers from the water and salts; the monomers then may be reused in the polymerization. In this case, monomer fumes are present in the separation and drying steps which can cause both industrial hygiene and environmental problems and also a yield loss. Further yield loss results from the longer contact time of the acrylonitrile monomer with sodium bisulfite to form the addition product sodium β-sulfopropionitrile.

If no chelating agent is added, the need for monomer recovery may be minimized by using two-stage filtration with filtrate recycle to the reactor after the first stage. Nonvolatile monomers, such as sodium styrene sulfonate, or sodium p-sulfophenyl methallyl ether can be partially recovered in this manner. This often makes process control more difficult because some reaction by-products or other contaminants can affect the rate of polymerization and often the composition may be variable. When recycle is used it is often done to control discharges into the environment rather than to reduce monomer losses. Other methods of recovery of nonvolatile monomers, such as adsorption on activated carbon [66] have been reported.

Cost reduction has been a focus of fiber producers since the overall market for acrylic fibers has not grown in the developed countries. A significant cost reduction is realized by operating continuous aqueous dispersion processes at very low water-to-monomer ratios. Mitsubishi Rayon, for example, has reported [67–69] ratios as low as 1.75. This compares to ratios of 4 to 5 widely used in the 1970s. The low water-to-monomer ratios produce a change in the nucleation and particle growth mechanisms which yields denser polymer particles. The cost reduction comes in the drying step. While conventional water-to-monomer ratios give wet cake moisture levels of 200% (dry basis) the modified process yields wet cake moisture levels of 100% or less. Thus a savings in drying cost is realized. The low water-to-monomer process has the added advantage of increased reactor productivity. Most major producers have now adopted a low (≈2.1) water-to-monomer process.

III. SOLUTION POLYMERIZATION

The only other major method of commercial polymerization used for acrylic fibers is solution polymerization. The main advantages, as previously mentioned, are that the separate operations of polymer separation, drying and solutioning are eliminated. However, on the minus side, cationic dye sites are often difficult to incorporate, comonomers such as vinyl acetate and vinyl chloride cannot practically be incorporated, reaction rates are low, and polymer color is inferior. No new manufacturing plants based on solution polymerization have been built for many years.

In theory, any of the spinning solvents for acrylic fiber can be used in solution polymerization. Some, however, such as dimethylacetamide, have such high-chain-transfer constants ($C_s = 2.8 \times 10^4$) [70] that polymer of molecular weight suitable for textile end uses cannot be made at reasonable reaction rates. DMF also has a high rate of chain transfer ($C_s = 2.8 \times 10^4$) [70], but several commercial processes have used it. The reported chain-transfer constants for acrylonitrile polymerization solvents are listed in Table 3.

TABLE 3 Rate Constants for Chain Transfer to Solvent in Acrylonitrile Polymerization

Solvent	$T\,(^\circ C)$	$C_s \times 10^4$
γ-butyrolactone	50	0.658
		0.74
N,N-dimethylacetamide	50	4.495
		5.05
N,N-dimethylformamide	20	1.4
	25	4.97
	40	3.24
	50	1.0 (2)
		2.7
		2.78
		2.8
		2.83
		10
	60	2.412 (2)
		4.494
		5.0

Source: From Ref. 70.

Factors other than chain-transfer constant are important in the choice of solvent—among these are the quality of the fiber that can be spun from the solvent, the initiator systems that will function in the solvent and the comonomers that can be used. If the fiber is to be dry-spun, then the choice is limited to DMF as the other organic solvents have boiling points too high for practical application. Use of the sodium thiocyanate system precludes the use of redox initiators due to reaction with the solvent. Monomers such as vinyl acetate and vinyl chloride are too unreactive to be suitable in solution polymerization; the sodium or potassium salts of the sulfonate monomers have limited solubility in some solvents and need to be converted to the amine salt. A list of producers using solution polymerization and what has been revealed about their processes has been shown in Table 2.

Compared to aqueous dispersion polymerization, the kinetics of solution polymerization are straightforward, as there is only a single phase to be considered. The conventional kinetic scheme developed for vinyl monomers [71, 72] represents the situation for AN polymers:

Radical formation (azo or peroxide initiator)

$$I \xrightarrow{k_d} 2R\cdot$$

Chain initiation

$$R\bullet + AN \xrightarrow{k_i} P_1\bullet$$

Chain propagation

$$P_n\bullet + AN \xrightarrow{k_p} P_{n+1}\bullet$$

Radical transfer to monomer, solvent, additives

$$P_n\bullet + X{-}Y \xrightarrow{k_{ct}} P_n{-}Y + X\bullet$$

Termination by radical recombination

$$P_n\bullet + P_m\bullet \xrightarrow{k_{tr}} P_{n+m}$$

Termination by radical disproportionation

$$2P_n{-}CH_2{-}CHCN\bullet \xrightarrow{k_{td}} P_n{-}CH{=}CHCN + P_n{-}CH_2{-}CH_2CN$$

Termination by metal ion (such as ferric)

$$P_n\bullet + FeCl_3 \xrightarrow{k_{tm}} P_n{-}Cl + FeCl_2$$

The rate of polymerization is

$$R_p = k_p\,[AN]^m (k_d f[I]/K_{tc})^n$$

where f is the initiator efficiency, and m and n are exponential factors that characterize the rate dependence on the monomer and initiator concentrations respectively. The term for degree of polymerization is written [1]

$$X_n = \frac{K_p[AN]}{(fK_dK_{tc}[I])^{0.5}}$$

At monomer concentrations above 2.0 to 2.5 molar, the reaction order has been found to be m (monomer) = 1.0 and n (initiator) 0.5 [73, 74]. In DMF however, a monomer reaction order greater than expected was explained by chain transfer to solvent followed by slow reinitiation by the DMF radical [75]. At high monomer concentrations, the acrylonitrile can act as a nonsolvent and cause the reaction orders to deviate. The reaction has been found to be heterogeneous above 4 molar in DMF and 6 molar in ethylene carbonate [73]. In ethylene carbonate, the rate and molecular weight went through a maximum when the concentration exceeded 6 molar; in DMF the molecular weight increased monotonically at acrylonitrile concentrations above 4 molar. The reaction order may also deviate from unity at low AN concentration, presumably due to the significance of side reactions such as nitrile polymerization [74].

Solution polymerization can be implemented by feeding the monomers to a continuous mixing tank along with a solvent for the polymer. The overflow stream from this tank together with the initiator solution(s) in the same solvent, is then routed to a continuous stirred tank reactor where the polymerization is carried out in homogeneous solution. Typically, longer residence times are required in solution polymerization than in aqueous dispersion polymerization due to the lower radical and monomer concentrations. Jacketed, stainless steel reactors with agitators to ensure good mixing and heat transfer are employed. Monomer plus some solvent are removed from the product stream using some form of evaporator, and the resulting polymeric solution is used directly for spinning. Following is an example of solution polymerization from the patent literature [75]:

> 1.5 parts of cupric chloride ($CuCl_2 \cdot 2H_2O$) and 0.07 part of sodium chlorate ($NaClO_3$) were dissolved in 850 parts of 60% aqueous zinc chloride solution and then 100 parts of acrylonitrile were added to this solution to form a homogeneous solution. To the resulting solution, a solution of 1.68 parts of sodium hydrosulfite in 50 parts of 60% aqueous zinc chloride was added at 60°C. with stirring, whereupon the polymerization is started at once. After three hours a transparent, colorless and fine polymer solution was obtained. The final conversion was 98% and the average molecular weight of polyacrylonitrile produced was 79,000.

IV. EMULSION POLYMERIZATION

Although many commercial free radical polymer processes are based on emulsion polymerization, it has apparently disappeared as a method for producing acrylonitrile polymers for fiber end use. Emulsion polymerization has the advantage of segregating radicals from one another by trapping them in monomer micelles. Normally only one radical populates a micelle, therefore recombination is not possible and higher molecular weights and rates of polymerization may be achieved. In the past, polymers such as the acrylonitrile/2-methyl-5-vinyl pyridine copolymer used in Acrilan by Chemstrand (now Monsanto), the acrylonitrile/vinyl chloride copolymer in Union Carbide's process for Dynel and the process for the acrylonitrile/quaternary ammonium monomer copolymer used in Cytec's acid-dyeable carpet fiber [76] were produced by emulsion processes.

In emulsion polymerization, a small amount of the added monomer is dispersed by the emulsifier (a soap or detergent) as aggregates of 50–100 molecules, ca. 50 Å diameter, called micelles. The remaining monomer is in droplet form (10,000 Å diameter) and serves as a reservoir to replenish the micelles. The polymerization is initiated by water soluble agents such as potassium persulfate/sodium bisulfite. The radicals formed in the aqueous phase are rapidly captured by the micelles [77].

Polymerization proceeds rapidly within the micelle, with additional monomer being supplied by the droplets. The number of micelles is fixed by the amount of emulsifier used and is normally in the region of 10^{14} to 10^{20} per liter. In an ideal case, polymerization is initiated within a micelle when the first radical enters and is terminated by recombination on entry of a second radical. It resumes when a third radical enters, etc. Thus at a given time, one-half of the micelles contain a growing chain (those which have had an odd number of radicals enter). Acrylonitrile, as well as some other monomers [78–80], deviates from ideality, because the growing chains may desorb from the micelle [81, 82]. Factors favoring desorption include small particle size, high AN water solubility, high-chain-transfer activity—especially with a water-soluble agent—and low particle swelling by AN monomer. Under these conditions, the radical population may be as low as one per 1000 particles [83–91].

The main examples of emulsion polymerization of acrylonitrile polymers for fiber end use has been in the production of flame-retardant modacrylics based on vinyl chloride. Dynel, a product formerly made by Union Carbide Corporation, was a copolymer made up of 40% acrylonitrile and 60% vinyl chloride. The reactivity ratios for this pair are $r_1 = 3.7$, $r_2 = 0.074$ (see Table 1). To make the composition containing 60% vinyl chloride, a monomer composition of 82% VCl/18% AN must be maintained. At 18% AN, the reaction rate is low, and chain-transfer reactions can reduce molecular weight to an unacceptable level. The emulsion process isolates the radicals from each other and from other potential transfer agents in the aqueous phase and thus allows production of polymer with adequate molecular weight for fiber applications.

There are many examples of commercial emulsion polymerization in the literature [92–95]. The Dynel process [96, 97] used a semibatch technique wherein the acrylonitrile was added in increments to keep the concentration in free monomer at 18% throughout the course of the polymerization; this was accomplished by monitoring the reactor pressure: 75–76 p.s.i.g. (5.27–5.34 kg/cm^2) = 18% AN/82% VCl at 40°C. The process required 19 additions of AN over a 77-hr time span! The copolymer yield was 66%. As only a single addition of initiator was used, the molecular weight distribution was likely broad.

V. BULK POLYMERIZATION

In theory, bulk polymerization processes have definite advantages over water-based processes in that the separation of unreacted monomer from water is avoided as is the disposal of the waste water. However, due to the highly exothermic nature of acrylonitrile polymerization and the insolubility of polymer in monomer, no bulk process for acrylonitrile polymers has ever been used commercially.

The difficulty arises because the polymer precipitates from the reaction mixture barely swollen by its monomer. The heterogeneity has led to kinetics that deviate from the normal. When initiator is first added, the reaction medium remains clear while particles 10 to 20 nm in diameter are formed. As the reaction proceeds the particle size increases, giving the reaction medium a white milky appearance. When a thermal initiator, such as azo-bis-isobutyronitrile (AIBN) or benzoyl peroxide, is used, the reaction is autocatalytic. This contrasts sharply with normal homogeneous polymerizations in which the rate of polymerization decreases monotonically with time. Studies show that three propagation reactions occur simultaneously to account for the anomalous autoacceleration [98–101]. These are chain growth in the continuous monomer phase; chain growth of radicals that have precipitated from solution onto the particle surface; and chain growth of radicals within the polymer particles [102, 103].

Montefibre, however, reports [104, 105] getting around these difficulties with a commercially feasible process for bulk polymerization in a continuous stirred tank reactor. The heat of reaction is controlled by operating at relatively low (<50%) conversion levels of monomer. Operational problems with thermal stability are controlled by using a free-radical redox initiator with an extremely high decomposition rate constant. Since the initiator decomposes almost completely in the reactor, the polymerization rate is insensitive to temperature and can be controlled by means of the initiator feed rate. Polymer molecular weight and dye site content are controlled by using mercapto compounds and oxidizable sulfoxy compounds. They report, for example, that by using 0.16% (based on monomer) cumene hydroperoxide and 0.12% sodium methyl sulfite, which has a half life ($t_{1/2}$) at 50° of 0.29 hr, a stable, fluid slurry can be maintained at 48% conversion. The reaction could also be maintained at higher and lower temperatures with similar conversions and no decrease in stability. In contrast, polymerization initiated by AIBN, was characterized by a low rate and poor control at 40°C and 50°C and out of control autocatalytic polymerization at 60°C.

REFERENCES

1. G. Odian, *Principles of Polymerization, 2nd ed.*(Wiley, New York, 1981), p. 288.
2. R. Chiang, J. H. Rhodes, and R. A. Evans, *J. Polym. Sci.*, 4A: 3089 (1966).
3. D. V. Sogah, W. R. Hertler, O. W. Webster, and G. M. Cohen, *Macromolecules*, 20: 1473 (1987).
4. J. Ulbricht, *Z. Phys. Chem. (Leipzig)*, 5/6: 346 (1962).
5. L. H. Peebles, Jr., *J. Am. Chem. Soc.*, 80: 5603 (1958).

6. R. Z. Greenley, in *Polymer Handbook, 3rd ed.* (Wiley, New York, 1989), Sec. II, p. 165–171.
7. G. E. Ham, *J. Polym. Sci., 45*: 169 (1960).
8. G. E. Ham, in *Copolymerization* (Wiley-Interscience, New York, 1964).
9. A. Valvassori and G. Sartori, *J. Polym. Sci., 5*: 28 (1967).
10. T. Alfrey Jr., J. J. Bohrer, and H. Mark, *Copolymerization* (Wiley-Interscience, New York, 1952).
11. G. E. Ham, in *Encyclopedia of Polymer Science and Technology, 2nd ed.* Vol. 4 (Wiley, New York, 1966), p. 165.
12. L. H. Peebles, in *Copolymerization*, (Wiley-Interscience, New York, 1964).
13. A. D. Jenkins and A. Ledwith (Eds.), *Reactivity, Mechanism and Structure in Polymer Chemistry* (Wiley-Interscience, New York 1974).
14. S. Iwatsuki, M. Shin, and Y. Yamashita, *Makromol. Chem., 102*: 232 (1967).
15. I. S. Dorokhina, A. D. Abkin, and V. S. Klimenkov, *Khim. Volkna, 1962*: 1/49 (1962); from C.A. *59*: 2956H (1963).
16. I. S. Dorokhina, A. D. Abkin, and V. S. Klimenkov, *Vysok. Soedin., 5*: 385 (1963).
17. B. R. Thompson, and R. H. Raines, *J. Polym. Sci., 41*: 265 (1959).
18. F. R. Mayo and C. Walling, *Chem. Rev., 46*: 191 (1950).
19. K. C. Berger and G. Brandrup, in *Polymer Handbook, 3rd ed.* (Wiley, New York, 1989), Sec. II, p. 81–142.
20. *Chemical Marketing Reporter,* July 25, 1994.
21. R. J. Anders and W. Sweeny, U.S. Pat. 2,837,500 (June 3, 1958) to DuPont.
22. U.S. Pat. 3,813,372 (July 27, 1970) to Mitsubishi Rayon.
23. Japan Pat. Appl. 64,949 (Dec. 28, 1973) to Mitsubishi Rayon.
24. N. Yakouchi, T. Kawamura, and T. Tokitada, Japan Pat. 38,561 (Nov. 13, 1971) to Mitsubishi Rayon.
25. J. R. Kirby, U.S. Pat. 3,784,511 (March 28, 1972) to Monsanto.
26. British Pat. 1,360,669 (Dec. 5, 1970) to Bayer.
27. U.S. Pat. 3,681,275 (Jan. 27, 1969) to Japan Exlan.
28. British Pat. 1,339,669 (Sept. 28, 1970) to Mitsubishi Rayon.
29. S. Matsumura and C. Kanemitsu, Japan Pat. 26,333 (July 30, 1971) to Teijin.
30. S. Nakao, N. Numato, and T. Yamamoto, Japan Pat. 29,874.
31. H. Sakai, Z. Izumi, H. Kitagawa, S. Hamada, and M. Hoshira, British Pat. 1,295,529 (Aug. 8, 1972) to Kanebo.
32. H. Sakai, M. Hoshima, S. Hamada, Z. Izumi, and H. Kitagawa, Japan Pat. 37,553 (Nov. 28, 1970) to Toray.
33. S. Nakao, N. Numato, and T. Yamamoto, Japan Pat. 29,872 (Sept. 13, 1973).
34. S. Nakao, N. Numato, and T. Yamamoto, Japan Pat. 29,873 (Sept. 13, 1973).
35. I. Ito, Z. Izumi, and H. Kitagawa, Japan Pat. 28,114 (Dec. 3, 1968) to Toyo Rayon.
36. H. Kitagawa and T. Aragane, Japan Pat. 28,470 (Dec. 6, 1968) to Toyo Rayon.
37. H. Sakai, Z. Izumi, and H. Kitagawa, Japan Pat. 28,466 (Dec. 6, 1968) to Toyo Rayon.
38. W. Pasiuk-Bronikowska, J. Ziajka, and T. Bronikowska, *Autoxidation of Sulfur Compounds* (Ellis Horwood, New York, 1992), pp. 9–22.
39. F. S. Dainton and R. S. Eaton, *J. Polym. Sci., 39*: 313 (1959).
40. J. Ulbricht, *J. Polym. Sci., Part C, 16*: 3747 (1968).
41. F. S. Dainton and D. G. L. James, *J. Polym. Sci.,* 299 (1959).
42. L. H. Peebles, Jr., *J. Appl. Polym. Sci., 17*: 113 (1973).

43. L. H. Peebles, Jr., R. B. Thompson, Jr., J. R. Kirby, and M. Gibson, *J. Appl. Polym. Sci.*, *16*: 3341 (1972).
44. British Pat. 837,041 (filed Aug. 26, 1957) to DuPont.
45. W. Pasiuk-Bronikowska, J. Ziajka, and T. Bronikowska, *Autoxidation of Sulfur Compounds* (Ellis Horwood, New York, 1992), p. 10.
46. S. R. Palit, T. Guha, R. Das, and R. S. Konar, in *Encyclopedia of Polymer Science and Technology*, Vol. 2 (Wiley, New York, 1965), p. 229.
47. K. Stueben, in *Vinyl and Diene Monomers, Part 1, High Polymers Series*, Vol. 24 (Wiley-Interscience, New York, 1970) p. 181.
48. W. M. Thomas, *Adv. Polym. Sci.*, *2*: 401 (1961).
49. Anon., *The Chemistry of Acrylonitrile, 2nd ed.* (American Cyanamid, New York, 1959).
50. A. D. Jenkins, in *Vinyl Polymerization, Part I* (Marcel Dekker, New York, 1967), pp. 369–400.
51. W. Pasiuk-Bronikowska, J. Ziajka, and T. Bronikowska, *Autoxidation of Sulfur Compounds* (Ellis Horwood, New York, 1992), p. 32.
52. R. B. Thompson, Jr. and W. K. Wilson, U.S. Pat. 3,153,024 (Oct. 13, 1964) to Monsanto.
53. A. Hill, U.S. Pat. 2,673,192 (March 23, 1954) to Diamond Alkali.
54. M. Wishman and W. R. Kocay, U.S. Pat. 3,012,998 (Dec. 12, 1961) to American Cyanamid.
55. R. T. E. Schneck and I. Danishefsky, *J. Org. Chem.*, *16*: 1683 (1961).
56. G. Wentworth, *J. Org. Chem.*, *41*: 2647 (1976).
57. G. N. Milford and W. K. Wilkinson, U.S. Pat. 3,065,212 (Nov. 20, 1962) to DuPont.
58. B. G. Frushour and R. S. Knorr, in *Handbook of Fiber Science and Technology*, Vol. IV (Marcel Dekker, New York, 1985), pp. 190–191.
59. R. G. Fordyce and E. C. Chapin, *J. Am. Chem. Soc.*, *69*: 581 (1947).
60. R. G. Fordyce, *J. Am. Chem. Soc.*, *69*: 1903 (1947).
61. R. G. Fordyce, G. E. Ham, *J. Polymer Sci.*, *3*: 891 (1948).
62. W. V. Smith, *J. Am. Chem. Soc.*, *70*: 2177 (1948).
63. D. J. Stookey, unpublished, Monsanto.
64. W. R. Kocay and M. Wishman, U.S. Pat. 3,012,997 (Dec. 12, 1961) to American Cyanamid.
65. D. W. Cheape and W. R. Eberhardt, U.S. Pat. 3,454,542 (July 8, 1969) to Monsanto.
66. G. Wentworth, U.S. Pat. 3,970,604 (July 20, 1976) to Monsanto.
67. S. Ito, Y. Kawai, and T. Oshita, *Kobunshi Ronbunshu, 43*: 645 (1986).
68. S. Ito, *Sen'i Gakkaishi, 43*: 236 (1987).
69. S. Ito and C. Okada, *Sen'i Gakkaishi, 4.23*: T618–T625 (1986).
70. K. C. Berger and G. Brandrup in *Polymer Handbook, 3rd ed.* (Wiley, New York, 1989), Sec. II, pp. 94–133.
71. J. C. Bevington, *Radical Polymerization* (Academic Press, New York, 1961).
72. C. H. Bamford, W. C. Barb, A. D. Jenkins, and P. F. Onyon, *The Kinetics of Vinyl Polymerization by Radical Mechanism* (Academic Press, New York, 1958), pp. 1–97.
73. G. Vidotto, A. Grossatto-Arnaldi, and G. Talamini, *Die Makromol. Chem.*, *122*: 91 (1969).
74. G. Vidotto, A. Grossatto-Arnaldi, and G. Talamini, *Die Makromol. Chem.*, *140*: 263 (1970).

75. M. Taniyama, I. Kanda, M. Nakajima, and M. Amaya, U.S. Pat. 3,287,307 (Nov. 22, 1966) to Toho Rayon.
76. D. Coleman, S. African Pat. Appl. 686,986 (Aug. 30, 1968) to American Cyanamid.
77. R. M. Fitch and Lih-bin Shih, *Progr. Col. Polym. Sci.*, *56*: 1 (1975).
78. D. Gershberg, *J. Chem Eng. Symp. Ser. Inst. Chem. Eng.*, *N3*: 3 (1965).
79. Z. Izumi, H. Kiuchi, and M. Watanabe, *J. Polym. Sci.*, *A1*: 455, 469 (1967).
80. T. O'Neill and V. Stannett, *J. Macromol. Sci.–Chem.*, *A8*: 949 (1974).
81. N. Nomura, *Emulsion Polymerization* (Academic Press, New York, 1982), pp. 191–219.
82. R. W. Thompson and J. D. Stevens, *Chem Eng. Sci.*, *30*: 663 (1975).
83. J. Ugelstad, *Pure Appl. Chem.*, *53*: 323 (1981).
84. J. Ugelstad and P. C. Mork, *Br. Polym. J.*, 2: 31 (1970).
85. J. Ugelstad and F. K. Hansen, *Rubber Chem. Technol.*, *49*: 536 (1976).
86. J. Ugelstad, P. C. Mork, and J. O. Aasen, *J. Polym. Sci.*, *A5:* 2281 (1967).
87. J. Ugelstad, P. C. Mork, P. Dahl, and P. Rangnes, *J. Polym. Sci.*, *C27*: 49 (1969).
88. J. T. O'Toole, *J. Polym. Sci.*, *9*: 1281 (1965).
89. M. Nomura, K. Yamamoto, I. Horie, K. Fujita, and H. Harada, *J. Appl. Polym. Sci.*, 27: 2483 (1982).
90. M. Nomura, M. Kubo, and K. Fujita, *J. Appl. Polym. Sci.*, *28*: 2767 (1983).
91. M. J. Ballard, D. H. Napper, and R. G. Gilbert, *J. Polym. Sci. Polym. Chem Ed.*, *19*: 939 (1981).
92. H. Saka, S. Hamada, H. Inove, and S. Hosaka, Japan Pat. 90,380 (Nov. 26, 1973) to Toray.
93. I. Mata, I. Sakai, and M. Tomita, Japan Pat. 13,645 (Jan. 31, 1975) to Unitika.
94. A. F. Lupo *et al.*, Rom. Pat. 55,795 (Sept. 9, 1973) to Institute de Cerecetour Chimice.
95. M. R. Bechtold, U.S. Pat. 2,972,511 (Feb. 21, 1961) to DuPont.
96. E. W. Rugeley, T. A. Field, Jr., and G. H. Fremon, *Ind. Eng. Chem.*, *40*: 1724 (1948).
97. L. C. Shriver and G. H. Fremon, U.S. Pat. 2,420,330 (May 13, 1947) to Union Carbide.
98. C. H. Bamford and A. D. Jenkins, *Proc. R. Soc. London*, *A226*: 216 (1953).
99. C. H. Bamford and A. D. Jenkins, *Proc. R. Soc. London*, *A228:* 220 (1955).
100. C. H. Bamford, A. D. Jenkins, M. C. R. Symons, and M. G. Townsend, *J. Polym. Sci.*, *34*: 181 (1959).
101. A. D. Jenkins, in *Vinyl Polymerization, Part I* (Marcel Dekker, New York, 1967), Chap. 6.
102. K. E. J. Barret and H. R. Thomas, *Dispersion Polymerization in Organic Media* (Wiley, New York, 1958).
103. L. H. Peebles, in *Copolymerization* (Wiley-Interscience, New York, 1964), Chap. 9.
104. P. Melacini, L. Patron, A. Moretti, and R. Tedesco, U.S. Pat. 3,878,365 (Jan. 22, 1974) to Montefibre.
105. P. Melacini, L. Patron, A. Moretti, and R. Tedesco, U.S. Pat. 3,839,288 (Oct. 1, 1974) to Montefibre.
106. C. E. Grabiel and P. L. Decker, *J. Polymer Sci.*, *59*, 425 (1962).

4

Wet-Spinning Technology

G. J. Capone

Monsanto Company
Decatur, Alabama

I. INTRODUCTION: SOLUTION SPINNING

Wet-spinning acrylic fiber had its origins in the mid-1850s with the wet spinning of cellulosics by Chardonnet. Raw cotton was nitrated with a sulfuric acid/nitric acid solution. The resultant cellulose nitrate was then dissolved in a 40/60 ethyl ether/ethyl alcohol solvent to make a viscous "dope." This dope was forced through a spinnerette to form fibers of cellulose nitrate. After washing and stretching to orient the polymer chains and thus improve strength, the fiber was denitrified in an ammonium hydrosulfide bath. Further treatments were required to remove impurities and bleach the product. This first commercial process for synthetic fiber which was practiced into the late 1940s is shown schematically in Figure 1. The Chardonnet process exemplifies important principles used in synthetic fibers—the use of a soluble intermediate to effect the conversion to fiber, phase separation in the formation of a fibrous network, and aftertreatment to generate the desired fiber structure. Elements of these solution-spinning processes are still practiced today in the manufacture of rayon fibers.

Two other general classes of solution spinning are utilized to form fibers—those which are reacted after spinning and those which are not changed chemically (Table 1). In the case of a reactive process, if the reaction had taken place before spinning, the resultant product would be unprocessable. In contrast to the previous rayon process, the conversion to a soluble intermediate is unnec-

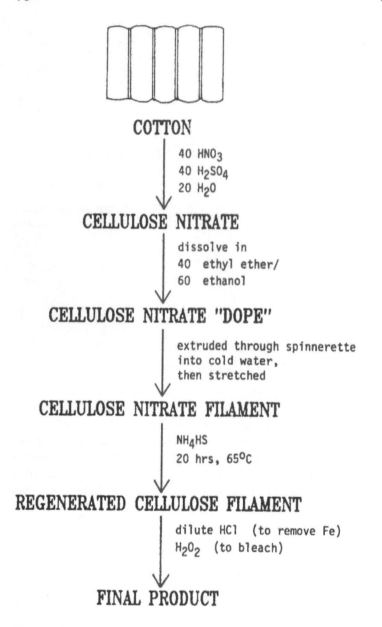

Figure 1   Chardonnet rayon flowchart.

Table 1 Solution Shinning Process

Polymer	Intermediate	Fiber	Example
Insoluble	Soluble	Insoluble	Viscose rayon Cuprammonium rayon
Soluble	Soluble	Insoluble	Polyvinyl alcohol Polybenzoxazole
Soluble	Soluble	Soluble	Polyacrylonitrile Polyvinyl chloride Poly(m-phenylene-terephthalamide)

essary. An example of the reacted fiber structure is the system for spinning polyvinyl alcohol. For the group of solution spun fibers that are not changed chemically because the starting polymer is soluble. and the polymer makes a useful product, solution spun polyacrylonitrile is the most significant commercial process used today. As shown in Table 2, solvents for wet spinning cover a wide range, from reactive systems for cellulose to organics such as acetone, salt solutions such as aqueous sodium thiocyanate, and to inorganic acids. One characteristic all these solvents have in common is that they are miscible with water. This is a practical necessity for purifying the fiber by removal of solvent in the aftertreatment processes.

II. GENERAL DISCUSSION

A. Acrylic Spinning

Solution spinning of polyacrylonitrile may be subdivided into two major processes: dry spinning and wet spinning [1–9]. In dry spinning the heated

Table 2 Solution Spinning Solvents

Solvent	Polymer example
Aq. NAOH + CS_2	Cellulose
Sulfuric acid	Poly(p-phenylene-terephthalamide)
Acetone	Cellulose acetate
N,N-Dimethylformamide	Spandex
Aq. NaSCN	Polyacrylonitrile
Aq. NaOH	Polyvinyl alcohol

polymer solution is pumped through a spinnerette into a tower with cocurrent flow of heated inert gas. Most of the solvent is evaporated, causing gelation of the solution. A bundle of filaments emerges from the bottom of the tower (Figure 2) [1, 4] and is subjected to various aftertreatments. This process is the subject of Chapter 4. In wet spinning, the polymer solution is pumped through a spinnerette that is submerged in a liquid bath containing a coagulant (Figure 3) [1, 4]. The spin-bath composition is usually the spin solvent plus water, although other nonsolvents are sometimes used. Solvent diffuses out of the polymer solution and the nonsolvent diffuses in causing a phase change to a solid polymer plus a liquid (solvent + nonsolvent). When a nonpenetrating bath liquid is used, such as polyethylene glycol, only outward diffusion takes place and gelation may occur in wet spinning. Nonsolvent or coagulant is continuously metered into the spin bath to maintain constant concentration. The formed filaments emerge from the bath and are subjected to successive aftertreatments.

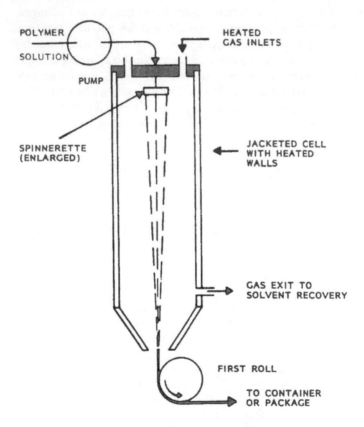

Figure 2   Schematic of dry-spinning tower.

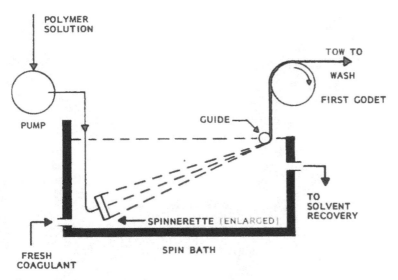

Figure 3 Schematic of wet-spun coagulation bath.

These two spinning processes for producing acrylic fiber account for 2.7 million metric tons of acrylic fiber production around the world [2, 10–14] with wet-spinning systems representing 85% of this production (Table 3) [14]. Wet-spinning systems are attractive because of the variety of solvents that can be used. The solvent choices are made based on raw material sourcing, solvent manufacturing economics, polymer solubility parameters, and fiber aftertreatment processes. Wet spinning is also characterized by high production capability per unit manufacturing area and low potential for human exposure to solvents. Finally, wet-spinning systems generate a porous gel structure conducive to in-line dyeing.

B. Solvents

A summary of solvent types for dry spinning and wet spinning are shown in Table 4. Commercial dry spinning is limited to dimethylformide because of the need for high vapor pressure to evaporate the solvent in the spinning tower. The major commercial solvent types for wet spinning are dimethylformamide (DMF), dimethylacetamide (DMAc), sodium thiocyanate, and nitric acid having 24%, 24%, 20%, and 10% of the production capacity, respectively. Each solvent type generates characteristic fiber structures that are related to the interactions between the solvent, the polymer composition and coagulant. The choice of solvent/polymer/coagulant determines the options manufacturers can employ in the aftertreatment steps such as drying and relaxation. Typical examples of solvent/polymer spinning solution compositions are given in Table 4.

Table 3 Acrylic Fiber Production

Country	Producer	Solvent	Spin-ning	Capacity MT/yr. (1993)	Gel dyed	Pigment ed	Technology/Comments
North and South America							
United States	Cytec (American Cyanamid)	NaSCN	Wet	60,000	x		
	Mann Group	ZnCl$_2$	Wet		x	x	Dow-Badische—Shutdown 1993
	Monsanto	DMAc	Wet	136,000	x	x	
Total				196,000			
Latin America							
Argentina	Hisisa	HNO$_3$	Wet		x	x	Asahi—Shutdown 1992
Brazil	Celbras	DMF	Dry	19,000			Mitsubishi Rayon
	Rhodia	DMF	Wet	18,000			Rhone-Poulenc
Mexico	Cydsa	DMF	Wet	75,000	x	x	Rhone-Poulenc
	Fibras Nationales de Acrilica	DMF	Wet	34,000	p		Snia
	Fibra Sinteticas	HNO$_3$	Wet	30,000			Asahi
Peru	s.d.F.	DMF	Dry	31,000			Bayer
Total				207,000			
Western Europe							
Germany	Bayer, Dormagen	DMF	Dry	120,000		x	
	Bayer, Lingen	DMAc	Wet	50,000			Monsanto
	Hoechst, Kehlheim	DMF	Dry	8,000			
	Hoechst, Kehlheim	DMF	Wet	35,000		x	
	Markische Faser	DMF	Wet	48,000	x	x	
Great Britain	Courtaulds	NaSCN	Wet	80,000	x	x	

Table 3 Continued

Country	Producer	Solvent	Spin-ning	Capacity MT/yr. (1993)	Gel dyed	Pigment ed	Technology/ Comments
Greece	Vomvicryl	DMF	Wet				Snia— Shutdown 1990
Italy	Enichem Fibre	DMAc	Wet	140,000	x		
	Montefibre	DMAc	Wet	120,000	x	x	Monsanto
	Snia	DMF	Wet		x		Shutdown
Ireland	Asahi Cashmilon	HNO$_3$	Wet	25,000			
Spain	Courtaulds	NaSCN	Wet	62,000	x		
	Montefibre España	DMAc	Wet	70,000	x		
Portugal	Fisipe	DMAc	Wet	36,000			Mitsubishi Rayon
Total				794,000			

Eastern Europe

Country	Producer	Solvent	Spin-ning	Capacity MT/yr. (1993)	Gel dyed	Pigment ed	Technology/ Comments
Bulgaria	Bulana Burgas	DMF	Wet	15,000	p	x	Snia
Hungary	Magyarviscosa	DMF	Wet	11,000	x		Snia
	Magyarviscosa	DMF	Dry	20,000			DuPont/ Simon-Carves
Poland	ZWS Anilina	NaSCN	Wet	15,000	x	x	Courtaulds
Romania	Savinesti	Ethylene carbonate	Wet	65,000	p	p	
Russia	Novopolatsk	NaSCN	Wet	10,000	x	x	Modacarylic Kanegafuchi
	Navoiazot	NaSCN	Wet	40,000			Courtaulds
	Saratov	NaSCN	Wet	40,000			Courtaulds
	Mapan Novopolatsk	DMF	Wet	35,000	x	x	Snia
Yugo-slavia	Ohis	NaSCN	Wet	30,000	x	x	Courtaulds
Total				281,000			

Middle East

Country	Producer	Solvent	Spin-ning	Capacity MT/yr. (1993)	Gel dyed	Pigment ed	Technology/ Comments
India	Ind. Petro Chem	HNO$_3$	Wet	24,000			Chemtex
	Ind. Synth. Co.	DMAc	Wet				Planned Montefibre

Table 3 Continued

Country	Producer	Solvent	Spin-ning	Capacity MT/yr. (1993)	Gel dyed	Pigment ed	Technology/ Comments
	Ind. Acrylic Ltd.	DMF	Dry	12,000			DuPont/ Chemtex
	Reliance Text.	DMF	Dry				Planned DuPont
	Consolidated	NaSCN	Wet	Under constr.			Exlan 15,000 T/A
	J. K. Synthetics	DMAc	Wet	24,000			Montefibre
	Pasupati	DMF	Wet	15,000			Snia
Iran	Polyacryl Corp.	DMF	Dry	30,000			DuPont
Turkey	Aksa	DMAc	Wet	150,000	x		Montefibre
	Yalova	DMF	Wet	15,000	x		Snia
Total				270,000			

Far East

Country	Producer	Solvent	Spin-ning	Capacity MT/yr. (1993)	Gel dyed	Pigment ed	Technology/ Comments
China	Daqing	NaSCN	Wet	50,000	p		Cytec
	Anqing	NaSCN	Wet	Under constr.			Cytec 50,000 T/A
	Qinhuangdao	DMF	Dry	30,000			DuPont/ Chemtex
	Gao Qiao	NaSCN	Wet	2,300	x		
	Ningbo	DMF	Dry	Under constr.			DuPont/ Chemtex 30,000 T/A
	Jin Shan	NaSCN	Wet	52,000	x		
	Jin Yang	NaSCN	Wet	20,000			
	Mao Ming	DMF	Dry	Under constr.			DuPont/ Chemtex 30,000 T/A
	Lanzhou	NaSCN	Wet	14,000	p		Courtaulds
	Fushun	DMF	Dry	30,000			DuPont/ Chemtex
	Zibo	DMF	Dry	45,000			DuPont/ Chemtex
Total China				243,300			
Indonesia	Hamparan Rajecki	DMF	Dry	40,000			DuPont/ Chemtex
Japan	Asahi	HNO$_3$	Wet	101,000			3,000 T/A filament
	Kanebo	DMF	Wet	29,000		x	

Table 3 Continued

Country	Producer	Solvent	Spin-ning	Capacity MT/yr. (1993)	Gel dyed	Pigment ed	Technology/ Comments
	Kanegafuchi	Acetone	Wet	41,000		x	Modacrylic
	Mitsubishi Rayon	DMF	Dry	5,000			
	Mitsubishi Rayon	DMAc	Wet	98,000			Monsanto
	Nippon Exlan	NaSCN	Wet	60,000			
	Toho Beslon	ZnCl$_2$	Wet	43,000			
	Toray	DMSO	Wet	43,000			
Total Japan				420,000			
North Korea	Anilon	Ethylene carbonate	Wet	4,000			
South Korea	Hanil	HNO$_3$	Wet	118,000	x		
	Taekwang	NaSCN	Wet	79,000	p		
Taiwan	Formosa Plastic	DMF	Wet	102,000			
	Tong Hua	NaSCN	Wet	55,000			
Thailand	Thai Acrylic Fibers	NaSCN	Wet	17,000			
Total				1,741,600			

C. Polymer

The limits of polymer concentration in the spinning solvents are determined by both physical and practical considerations. At the high polymer concentrations, both the polymer solubility and solution spinning pressure limitations are controlling. The polymer solubility limits are determined by solvent strength to maintain a spinning solution without gelation. For example, polymer compositions greater than 30% in DMF and DMAc will gel rapidly, particularly in large commercial equipment where conditions enhance gelation as opposed to laboratory conditions where elevated and controlled temperatures minimize gelation. Another limitation on high polymer solution concentration is the design pressure for the wet-spinning systems. In wet spinning acrylic fiber the polymer weight average molecular weight is typically 100,000–120,000 and solution viscosities are in the range of 500 poise. In practice, wet-spinning system spinnerettes contain large numbers of capillaries and are thin-plate designs operating at low linear speeds and at the temperatures of the spin-bath coagulant system. Therefore, both polymer molecular weight and polymer solution concentration are also limited based on equipment design. The lower limits on polymer solution

Table 4 Solvents for Acrylic Fiber Spinning

Solvent	% Polymer
Dimethylformamide	28–32
Dimethylacetamide	22–27
Aq. NaSCN	10–15
Aq. ZnCl$_2$	8–12
Dimethylsulfoxide	20–25
Nitric acid	8–12
Ethylene carbonate	15–18

concentration are also governed by fiber property and practical limitations. As the polymer concentration is lowered, significant fiber structure changes can occur to the point where coagulation is not possible as there is not enough polymer present to form a fiber network. In addition, low polymer concentration is more costly because of higher solvent recovery requirements.

Typical examples of acrylic polymer compositions for commercial acrylic fibers are shown in Table 5. The acrylic polymer can be 100% polyacrylonitrile. However, for most commercial applications neutral comonomers are incorporated in the acrylonitrile polymer to improve fiber physical properties and increase the rate of dyeing by increasing fiber amorphous regions. Acid comonomers are sometimes employed to increase the sites for cationic dyes beyond those provided by the polymer end groups. Other comonomers are sometimes employed for special purposes such as fire retardancy or imparting differential shrinkage. The neutral and acid comonomer units interact with the solvent and coagulant to alter the rate of diffusion and phase separation and ultimately influence the fiber structure.

Table 5 Polymer Composition of Acrylic Fibers

Acrylonitrile	Neutral comonomer	Acid comonomer
90–94%	6–9%	0–1%
	Methyl acrylate	Sodium styrene sulfonate
	Vinyl acetate	Sodium methallyl sulfonate
	Methyl methacrylate	Sodium sulfophenyl methallyl Ether
		Itaconic acid

D. Coagulation

The initial fiber formation as the spinning solution exits the spinnerette capillaries is the most critical element in a solution-spinning process. In general, wet-spinning acrylic fiber is accomplished with very large numbers of capillaries in the spinnerette at low linear speeds. Wet spinning coagulation systems use 20,000 to 100,000 capillaries in each spinnerette with diameters in the range of 0.05 to 0.25 mm. Typically, the linear speeds in the coagulation bath are 3 to 16 m/min. The fiber denier per filament (weight in grams per 9000 m) is calculated by mass balance and is given by the equation [43]

$$\text{denier/filament} = \frac{9.672 \times 10^3 \, dWQ}{SV} \tag{4.1}$$

where

d = spinning dope density in grams per liter
W = polymer concentration in the dope expressed as the weight fraction
Q = volumetric flow rate per spinnerette hole in liters per second
V = first roll speed expressed as the linear speed in meters per second at the exit of the coagulation
S = overall stretch ratio including any relaxation

As in any physiochemical process of forming a shaped article, the precise control of polymer content, flow rate, and temperature are critical to the ultimate fiber properties and set the stage for the complex transformations from a dope to a fiber. Certainly, in comparing solution-spinning processes such as wet and dry spinning, the major differences in fiber characteristics are a direct result of this initial fiber-forming step. Likewise, within a given wet-spinning system, the process parameters for polymer solution behavior at the exit of the spinnerette capillaries are key to final fiber properties.

As shown schematically in Figure 3, a typical wet-spinning system for an acrylic polymer consists of the spinnerette submerged in a solution of the solvent and a nonsolvent. If, for instance, the spinning solvent were DMF, the spin bath might be 60% DMF, 40% water at 45°. The nonsolvent, water, in the spinning solution surrounding the spinnerette causes phase separation and coagulation of extruded filaments. Within a very short distance of the face of the spinnerette, the fundamental mechanisms of heat transfer, mass transfer, fluid mechanics, and solution thermodynamics interact on the dope stream to form a precursor fiber structure. All of the physical parameters that one expects to govern the processes of heat transfer, mass transfer, and solution thermodynamics play a role in this most critical period of a few seconds during the filament exposure to the spinning media. The temperatures of the polymer solution and spin bath fluid are important to the rate of heat transfer and, subsequently, to the rate of diffusion and phase separation. The polymer composition and content in the solution and the relative concentrations of the solvent and nonsolvent in the spinning media also determine the diffusion rates of the solvent and nonsolvent by setting the driving force for phase separation. The underlying principle for all the processes and the determination of the fiber structure is the control of the relative diffusion rates of solvent and nonsolvent and the phase separation characteristics of the polymer/solvent/nonsolvent system.

There has been a great deal of fundamental work on modeling the diffusion of solvents and nonsolvents for wet spinning [15–28]. The modeling efforts were primarily focused on the determination of the time and distance from the capillary for complete structure formation. Generally, each successive model has been a refinement of previous models to remove constraints on boundary conditions. The diffusion models do not define or predict fiber structure. The area of least study is phase behavior for acrylic polymer/solvent/nonsolvent systems. It is likely that incorporation of phase separation phenomena into the diffusion models would be more predictive of fiber properties and offer more areas of future investigation.

While the acrylic wet-spinning system is complex with multiple variables to consider and control, these same variables can be utilized to alter fiber structure over a wide range. The system can readily produce fibers of different cross-sectional shapes, surface textures, lusters, and physical properties. A variation of wet spinning is known as dry-jet wet spinning or air-gap spinning. In this process, the spinnerette is located a short distance (<1 cm) above the spin bath with the filaments extruded vertically into the fluid. This has the advantage of allowing a dope temperature which is independent of the spin-bath temperature and also avoids the high stress on the protofilaments at the jet face which occurs in wet spinning. As a consequence, dry-jet wet spinning is suited to processes which require higher than normal solution solids and high linear speed. Thus it is used for the production of specialty filament products where spinning speed compensates for the small number of holes per spinnerette. The dry-jet spun filament gels before reaching the spin bath and thus the structure resembles that produced by dry spinning. Dry-jet wet spinning has not been successfully used with spinnerettes as large as those commonly employed in wet spinning.

E. Aftertreatments

All acrylic fiber solution spinning processes utilize some type of washing and orientation process to remove solvent remaining from the initial fiber formation step and to develop the internal fiber morphology for greater strength [62, 63]. In wet-spinning systems the washing and orientation steps are a continuous part of the spinning system as shown schematically in Figure 4.

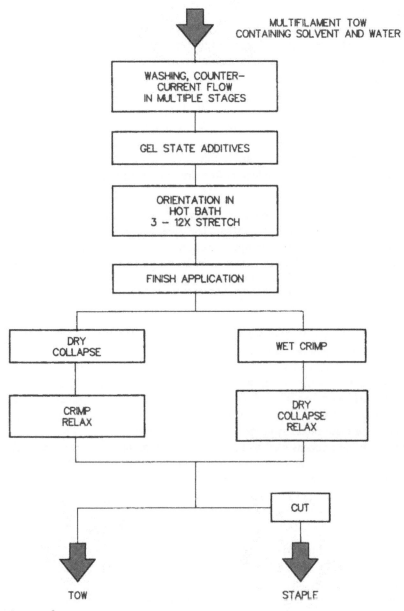

Figure 4   Processing of acrylic fiber tows.

1. Washing

As with the initial fiber formation step, fiber washing is a diffusional process that is driven by temperature and wash liquid concentration. Commercial acrylic fiber systems use water as the wash media with a wash temperature limited to the boiling point of the water at the steady-state concentration of extracted solvent. As the fiber structure is an easily accessible gel, the optimum washing temperature may be well below 100°. Washing may be conducted mainly preceding, or concurrent with, stretching. A small amount of residual solvent is usually purposefully left in fiber to optimize processing behavior. The fibrillar network remains a porous structure at this point in the process and gel state additives, such as dyes, can be added before drawing or stretching the fiber.

2. Orientation

The orientation or stretching process is very important in altering the fiber structure and enhancing fiber properties. The fiber temperature is increased to above the wet glass transition temperature, and the fiber is stretched 3 to 12 times between roll sets with hot water as the heating medium. In this stretching process the orientation of the fibrillar network formed in the spin bath is increased and the fiber strength is increased.

3. Finish

At some point subsequent to the washing step, a chemical treatment is applied to the fiber. The chemical or "finish" is normally an aqueous solution or emulsion and has both lubriants and antistatic components that facilitate fiber processing in the subsequent collapse/drying steps and transforming the fiber into yarns and fabrics. The finish types and amount used in the fiber are critical to good performance in the textile mills and often the finish compositions are changed for specific yarn and fabric systems. If the finish is applied before fiber collapse and drying, the interior, as well as the surface, will contain finish components. Depending on end use, finish quantity based on fiber may be from 0.1 to 0.7%. Additional applications of finish are sometimes made after the drying, collapsing, and relaxing steps discussed in the next section. These finish treatments also facilitate fiber handling and can be very effective because most of the finish remains on the surface of the fiber where it is fully utilized.

4. Drying, Collapsing, Relaxing

The drying, collapsing, and relaxation steps in processing acrylic fibers can take different forms and can occur in different sequences. The process sequences are usually dictated by the initial fiber structure generated in the spin bath. For those processes generating an open gel structure with pore sizes greater than 0.3 micron, a drying/collapsing process is employed by passing the fiber over heated zones and/or hot rolls. The drying portion of the process is the removal of both external and internal fiber water from the washing and orientation steps. While drying the fiber is important, a more significant process of fiber collapse is also occurring as the water is driven from the internal fibrillar network. The fiber collapse is the closure of the oriented open pore structure that had its origin in the spin bath. If collapse is not complete, the fiber will be brittle and luster differences between collapsed and uncollapsed regions will be evident. The collapsed fiber is then crimped and relaxed. The crimping operation is a mechanical treatment to add cohesion and bulk to a multifilament bundle of fibers and is very important to processing steps in conversion to yarns and fabrics. Stuffer box crimpers are almost always employed.

The relaxation process is the final step in changing fiber morphology and significantly alters the fiber stress-strain behavior. The relaxation of acrylic fiber requires energy input by means of a hot, wet environment; both batch and continuous relaxation systems are utilized.

For those acrylic fiber processes that develop a gel network with small (<0.3 micron) internal fiber pores, a different route to fiber drying/collapsing/relaxing can be used. As shown in Figure 4, the acrylic tow can be wet crimped to impart cohesion for textile processing and subsequently followed by a one-step dry/collapse/relax treatment. The water from the orientation step is trapped within the small pores in the fiber. By supplying heat energy, the fiber temperature is increased above the wet glass transition temperature of the fiber before the water can escape the pores. Fiber collapse occurs, aided by the surface tension of the liquid water, as the water is being driven from the fiber and fiber relaxation is achieved. The choice of drying/collapsing/relaxing steps for acrylic fiber for commercial processes has significant cost implications. These process options are driven mainly by the initial fiber structure formed in the spin bath, which is in turn a function of the polymer/solvent/nonsolvent system.

Wet-spinning processes employ jets of up to 100,000 holes and linear speeds of up to 100 m/min. In addition, jets can be ganged in a spin bath and several spin baths can be arranged end to end. Given this flexibility, the limitations on the productivity of the process are the capability of the downstream equipment to process the bundle and, for a product sold as tow, the availability of stretch-breaking equipment (Chapter 10) able to utilize it. Spinning machine productivities of up to 2 metric tons per hour are known, although producers with older equipment, may be limited by washing efficiency to rates of less than 0.5 ton/hr.

Acrylic fiber from wet spinning systems is sold mainly into the apparel and home furnishing markets as staple or tow. The denier range is from 0.1 (for synthetic suede) to 40 (for wigs). Typical physical properties are between 2 and 4 g/d tenacity with elongation between 30% and 60%.

III. DIFFUSION

One of the fundamental processes governing the formation of wet-spun acrylic fiber is diffusion. The fiber structure is controlled by the diffusion of solvent from the polymer solution as the solution exits the spinnerette capillary and by the counterdiffusion of the nonsolvent into the polymer solution. These diffusional processes are described by concentration differences along and within the fiber as coagulation proceeds at the exit of the capillary. The relative rates of solvent to nonsolvent diffusion set the driving force for phase separation and the rate of phase separation. Like any diffusion process, the independent variables for wet spinning are rate, concentration, and temperature.

The rate term includes the overall spinning rate and the degree of stretch that is taken within the spin bath. The overall spin rate influences the shear rate at the capillary wall and degree of dope swell at the exit of the capillary and thus alters the rates of diffusion. The jet stretch, defined as the ratio of the first roll take-up velocity to the dope extrusion velocity, behaves as a pump to draw coagulant into the fiber and express solvent from the polymer solution. Therefore, increasing stretch ratio in the coagulation bath increases the rate of diffusion, but the relative diffusion of solvent out to nonsolvent in remains constant [15].

Both the concentrations of polymer in the spinning solution and the relative concentrations of the solvent/nonsolvent pair in the spin bath impact the diffusion rates during coagulation. Increased polymer concentrations in the spin solution increase the resistance of the boundary layer within the filament which limits diffusion of both solvent and nonsolvent. Also, the relative diffusion rate of solvent out to nonsolvent in increases.

The solvent in the spin bath could theoretically vary from 0% to 100%. At 0% the spin bath contains only nonsolvent and very rapid coagulation occurs. For most wet-spinning systems this is not a practical operating condition as the amount of nonsolvent which must be processed to recover the spinning solvent is very large. If the bath contains all solvent, there is no coagulation and no process. However, within the boundaries of 0% and 100% solvent there exist infinite solvent/nonsolvent ratios for changing the diffusion rate. As the solvent-to-nonsolvent bath ratio increases, the diffusion rate of both solvent and nonsolvent decreases and the relative diffusion rate of solvent to nonsolvent increases [19].

Temperature is a key variable controlling the diffusion of solvent and nonsolvent. As in most diffusion processes, an increase in temperature increases the diffusion of solvent and nonsolvent whether the temperature change is in the polymer-spinning solution or in the coagulation bath. However, again the diffusion rate for the solvent increases faster than the rate for the nonsolvent [19]. For acrylic-fiber-spinning systems, the coagulation baths are normally within a temperature range of 0 to 50° and the polymer spin solution is between 25 and 120° C.

Given the main variables of polymer-spinning solution and coagulation bath, many investigators [15–28] have derived mathematical models for the time to coagulate an extruded filament, diffusion rates of solvent and nonsolvent, and estimated the diffusion coefficients for a variety of spinning systems. The work of Paul [19], using the gelation of large cylindrical rods, showed the solvent diffusion was a function of the square root of the gelation time. The mathematical arguments presented by Paul are for three physical models for diffusion. These models are (1) an equal-flux model, (2) constant-flux ratio model, and (3) variable-flux ratio model. The equal-flux model assigns equal flux for coagulant inward and solvent outward. The solution to Fick's law yields a boundary movement within the coagulated filament proportional to the square root of the time. However, the model prediction deviates from experimental data as the coagulation time increases. The variable-flux model assumes the flux ratio of solvent to coagulant is constant at the coagulation boundary and changes from the coagulated filament surface to the center of the filament. The mathematical equations became a trial-and-error procedure and did not represent the experimental results for concentrations at the boundaries. The constant-flux ratio model, in which the solvent to nonsolvent flux ratio was assumed constant at the boundary and within the filament, was reported by Paul to be most representative of the experimental data and is a reasonable physical depiction of the diffusion process. A graph from Paul's model of the diffusion coefficient as a function of spin bath concentration is shown as Figure 5. The experimental data are for dimethylacetamide (DMAc) solvent and water as the nonsolvent. The author acknowledged that all the models are very complex and require approximations and curve fitting. The experimentation is made particularly difficult because of phase change.

More recent investigations of modeling diffusion can be seen in work by Baojun, Ding, and Zhenqiou [15]. The rate of diffusion is modeled from cylindrical coordinates based again on Fick's law and shown below:

$$\frac{dC}{dt} = \frac{1}{r}\left[\frac{\delta}{\delta r}\left(rD\frac{\delta C}{\delta r}\right)\right] = D\left(\frac{\delta^2 C}{\delta r^2} + \frac{1}{r}\frac{\delta C}{\delta r}\right) \tag{4.2}$$

The solution to the diffusion equation is:

$$\frac{M_t}{M_0} = 4\sum_{n=1}^{\infty}\frac{1}{(\lambda_n)^2}e^{-(\lambda_n)^2 D_s t/R^2} \tag{4.3}$$

where

C = solvent concentration within the filament
r = radial position at any location with the filament
M = mass of the solvent within the filament as a function of time
λ_n = positive root satisfying the zero order Bessel function
D = diffusion coefficient

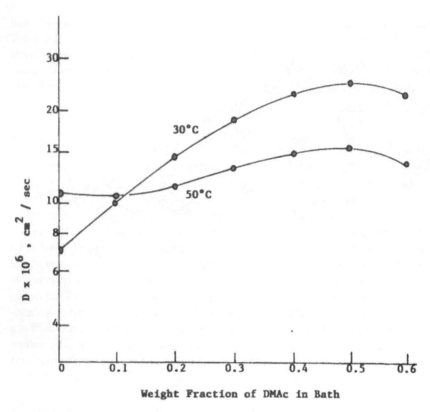

Figure 5   Diffusion coefficient—DNAc/H_2O [19].

R = radius of the filament
t = coagulation time

Baojun and co-workers used laboratory spinning with the dimethylfor-
mamide/water system to capture filaments and analyzed the composition of the
filaments. Correlations are presented for diffusion coefficients and flux ratios as
functions of jet stretch, polymer solution concentration, and coagulation
temperature. The flux ratios they reported are similar to those reported in Paul's
data, 20 years earlier. Also, the diffusion coefficients are in the same range of 4
to 10×10^{-6} cm²/sec that Paul found for DMAc/H_2O systems.

Other investigators, such as Jian and co-workers [20] and Terada [23], studied
the diffusion relationships in DMF, DMSO, and NaSCN solvents with the
nonsolvent water. Similar models were used; and the reported data show diffusion coefficients again in the same range as reported by Baojun. If the diffusion
coefficients are similar for different solvents, then what, if any, differences are

there in the properties of the fibers generated from the different solvents? Grobe and Heyer [28] alluded to this in a discussion of differences noted with hydrophilic polymers. Their analysis was that, while the thickness at the boundary is predicted by diffusion models, the pore size of the boundary is different and is the controlling mechanism for solvent diffusion. The pore size of the fibrillar network is a function of the polymer/solvent/nonsolvent interaction and phase behavior. The phase behavior in wet spinning is the same as other ternary systems that are controlled by temperature, concentration, and bonding energy between solvents, nonsolvents, and polymers. While the diffusion mechanism is very important, phase separation is also important, and incorporation of phase behavior leads to mechanisms for fiber structure development in wet-spinning acrylics.

IV. PHASE SEPARATION

In wet-spinning acrylic fiber there is a transformation of a polymer solution to a solid phase or porous gel network by the action of a nonsolvent. The polymer/solvent/nonsolvent interaction and the rate at which the changes occur determine the fibrillar network and the ultimate fiber properties [29–42].

Cohen, Tanny, and Prager [31] describe the porous membrane formation in cellulose acetate-acetone-water, cellulose acetate-acetic acid-water, and polystyrene-toluene-ethanol systems. A schematic ternary diagram is shown in Figure 6. In the analysis of interactions between the components there are described porous structures of differing degrees; the degree depends on the position within the phase diagram and the driving forces for separation of a polymer phase. If the composition remains above the binodal line, a one-phase system will exist. If the composition is between the binodal and spinodal lines, the single phase is metastable and a dense phase will form. If the composition is in the spinodal region, the system is unstable and a porous membrane forms. The phase diagrams are developed from binary interaction parameters and molar volumes. The interaction parameters and molar volumes define the molar chemical potentials. The overall flux (J_i) for the diffusion of the components is expressed as a product of the diffusion coefficient and gradients of chemical potential:

$$J_i = -(RT)^{-1} D_i(\phi_i, \phi_j)\phi_i \frac{\delta\mu_i}{\delta m} \tag{4.4}$$

where

D_i = diffusion coefficient of component i

ϕ_s = volume fraction of components i and j

μ = chemical potential of component

m = mass of component

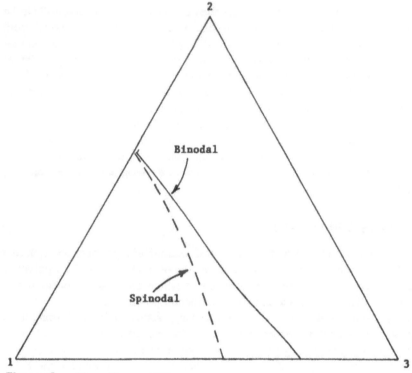

Figure 6   Phase diagram [31].

T = temperature
R = gas constant

By applying phase separation to the mechanism of acrylic fiber wet-spinning gelation, the fundamental properties of molar volumes and chemical interaction parameters become a part of the overall diffusion mechanism and a defining mechanism for the fiber structure. Of interest in the phase separation mechanism is the idea of spinodal decomposition. The spinodal decomposition in an analysis of Cahn [32] describes the unstable regime of the phase diagram as the region where the second derivative of the Gibbs free energy with respect to component, $\delta^2 G/\delta c^2$, is equal to zero. In the metastable region, this free-energy function is greater than zero or the free energy is increasing with increasing component concentration. Within the metastable region a finite fluctuation is needed to cause the solution to enter the unstable or spinodal region. The fluctuation is called the nucleus. The work to form the nucleus is a measure of the metastability and is zero at the spinodal boundary. Changes in temperature or composition can cause the nucleus to form. The composition changes within the system are a function of diffusion and driven by the chemical potential differences between the species. Both theoretical models and laboratory measurements have been reported by other investigators [34–36, 38–40] concerning the theories of binodal and spinodal phase boundaries. These data are primarily for binary systems involving a polymer-solvent pair with polymer concentration and solution temperature as the control variables. Similarly, a polymer rich phase and solvent/nonsolvent phase are formed in the wet spinning coagulation process. In the orientation process a gel network is formed with an interconnected, fibrillar structure. The voids within the network are the result of the solvent/nonsolvent phase separation and diffusion processes.

Ziabicki [30] presented a qualitative picture of wet-spinning systems using ternary phase diagrams and phase separation models. He concluded that thermodynamic and kinetic aspects of phase separation are controlling the wet-spinning process. Rende [16] developed an acrylic fiber wet-spinning model that includes a phase equilibrium model. With the assumption that the chemical potentials of the species are equal at equilibrium, an equation was derived for the relationship between the molar volumes and the interaction parameters for each species. The equation correlates the time for coagulation with the diffusion coefficient and species concentrations in terms of molar volumes. The fiber radius was found to vary as the square root of the coagulation time, which is the same functional relationship reported for the experimental work of Paul [19].

Given the phase separation and diffusion models for wet-spinning acrylic fiber, there are fiber structure implications that dictate the physical properties of wet-spun acrylics and determine fiber property differences brought about by different polymer solvent/nonsolvent spinning systems. The coagulation bath produces a porous gel network formed by separation of a polymer solution into a polymer-rich phase and a solvent-rich phase. The polymer-rich region phase is composed of an interconnecting network of bonded polymer chains. The pore size within the polymer network for wet-spun polyacrylonitrile polymers is large compared to the single gel phase generated by a dry-spinning mechanism. This is a result of the relatively high rate of phase separation with systems using nonsolvents and having a counterdiffusion mechanism.

V. STRUCTURE

A. Coagulation Bath

The fiber structure generated in the wet-spinning coagulation bath is a result of the counterdiffusion of solvent and nonsolvent and the phase separation of the polyacrylonitrile polymer. The structure is described as a gel network of interconnected polymer fibrils separated by voids [13, 41]. The polymer chains are laterally bonded by intermolecular forces to form fibrils that are connected by extended polymer chains. The void size is determined by the rate of diffusion and phase separation. The void size increases as the rates of diffusion and phase separation increase. As a consequence of the increasing void size, the number of voids decreases. The size and number of voids impact the fiber physical properties and the type of aftertreatments used to dry, collapse, and relax the final product. A wet-spun acrylic fiber void size is approximately 0.3 micron or greater, as contrasted with dry-spun acrylic fiber void size of approximately 0.1 micron. Very large voids on the order of 10 microns can be formed as shown in Figure 7. These macrovoids are in addition to the "normal" voids in the 0.3-micron range.

The experimental work on wet-spinning coagulation extends to all the process variables, including solvents, polymers, nonsolvents, temperatures, and flows [15, 17, 21, 22, 42, 45, 48, 60, 64, 65]. An example of the impact of solvent on the initial fiber structure is shown in Table 6 as reported by Jenny [13]. The data on fiber-specific surfaces show an increase from a low value of 90 m^2/g for DMF solvent to a high of 204 m^2/g for HNO_3. The larger surface area values have smaller void size and represent fibers produced under slower coagulation conditions. Examples of the large voids or macrovoids formed in wet-spinning DMF is in the work of Takahashi [45]. Large macrovoids are formed at coagulation bath solvent concentrations from 20 wt.% to 70 wt.% Only at DMF bath concentrations greater than 75% or, when the diffusion rate was reduced, did the macrovoids disappear. It is believed that the fiber void sizes from the different solvents are predictable based on the molar volumes and interaction parameters discussed in Section IV on phase separation. Acrylic

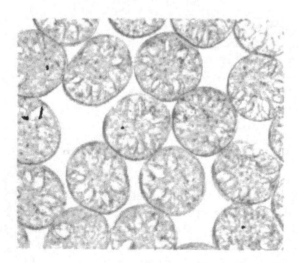

Figure 7   Large macrovoid photo—DMAc/H_2O.

Table 6 Jenny Surface Area Comparisons

Type of fiber	Specific surface in m^2/g	Relative S
NaSCN·A	160	2.5
HNO$_3$	204	1.5
NaSCN·B	140	2.6
DMF	90	1.1
DMAc	114	1.2

fiber containing small voids can be dried, collapsed, and relaxed using milder conditions than those conditions employed to treat fiber containing large voids. These treatment steps impact the cost of producing acrylic fiber.

Another interesting aspect of wet-spinning acrylic fiber is the concept of maximum spin bath stretch as a function of solvent concentration. As mentioned previously, the spin-bath stretch or jet stretch is defined as the ratio of the take-up velocity to the theoretical polymer solution velocity at the exit of the capillary. The maximum jet stretch is the highest ratio that is attainable without filament breakage in the coagulation bath. The maximum jet stretch data shown in Figure 8 [66] are for wet-spinning systems with DMF or DMAc solvents and water nonsolvent. Both solvents' responses go through a minimum as the spin-bath solvent concentration increases. This behavior is not explained by diffusion considerations alone, but can be more fully understood by solvent/nonsolvent interactions. The jet-stretch minimum is where the spin bath molar ratio of water to solvent is 2 to 1 for both DMF and DMAc. At the 2/1 molar ratio it is known that the solution viscosity goes through a maximum, indicating an association of water with solvent. Above the 2/1 mole ratio, there is an excess of water in the system, coagulation occurs rapidly, and a porous structure is formed. This porous structure can be drawn and supports high stretch tensions. As the 2/1 molar ratio is approached the coagulation slows, the associated solvent/nonsolvent pair behaves as a coagulant, and the fiber structural differences from the coagulated outer radius to the fluid center of filament cannot support the higher stress. Also, as mentioned above, the solvent/nonsolvent viscosity increases sharply within this concentration range, and this increased viscosity contributes more drag on the filaments. As the solvent level is increased further, there is excess solvent relative to the water, less driving force for phase separation, the radial structure differences within the fiber are less with a thinner skin, and therefore the maximum jet stretch ratio increases dramatically. Takahashi [45] reported data for DMF solvent at a bath temperature of 30°C. The maximum stretch ratio shows a curve similar to Figure 8 with a minimum at 75% DMF. Paul [19] indirectly measured the degree of swelling for fiber coagulated in a DMAc–water bath at 30 and 50°C as a function of spin-bath concentration

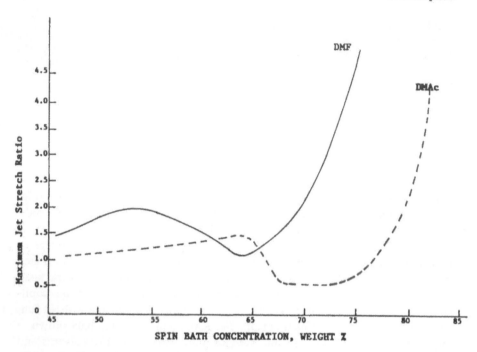

Figure 8 Comparison of DMF and DMAc solvents at the maximum jet-stretch ratios.

(Figure 9). The shape of the curves are the same as that for jet stretch and going through a minimum at 70 to 80 wt.% solvent. As the degree of swell decreases, the ability to draw the fiber also decreases and limits the jet stretch.

Wet-spinning systems using other nonsolvents are represented in Figure 10 [42]. The DMF solvent was spun in conjunction with the nonsolvents water, butanol, ethylene glycol, xylene, and carbon tetrachloride. All of the maximum jet stretch curves go through a minimum as a function of the DMF coagulation bath concentration. The same mechanism of coagulation is occurring with different nonsolvents and, as discussed in the section on phase equilibrium, the interaction parameters of the solvent and nonsolvent should be considered to fully understand the coagulation process. Interestingly, the phase behavior of the systems, presented [42] in the form of cloud point determinations, correlates with the definition of the nonsolvent concentration required to initiate the phase change.

The effects of jet stretch on fiber properties are primarily exhibited in fiber luster and the dry/collapse/relax processes. Low-jet-stretch conditions produce small void structures which are more transparent to light and have a lustrous

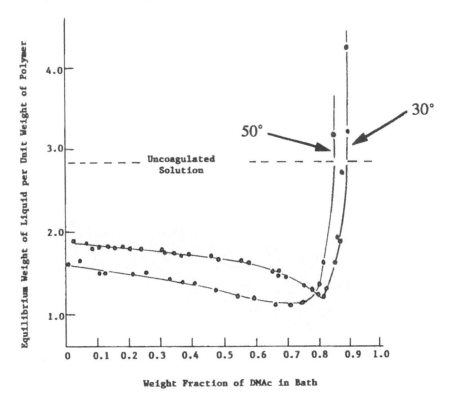

Figure 9   Degree of swell (Paul [19]).

appearance. Also, small void structures (low jet stretch) dry, collapse, and relax under milder conditions than larger void structrues (high jet stretch). These structures ultimately determine the process equipment and conditions for producing acrylic fiber. However, after the acrylic fiber is stretched, dried, collapsed, and relaxed, the jet stretch does not significantly change acrylic fiber tensile properties.

Other coagulation bath variables also have significant impact on the pore size of the fiber generated in the spin bath. These independent variables are polymer concentration in the spin solution and coagulation bath temperature. Examples of the effects and experimental data for DMAc/water systems are given in the work of Knudsen [44]. Increasing the percent polymer in the spinning solution or decreasing the spin-bath temperature decrease the pore size. The mechanism is again the reduction in the rates of diffusion and phase separation. However, at the present time there is no correlating expression to predict the pore size of wet-spun acrylic fiber.

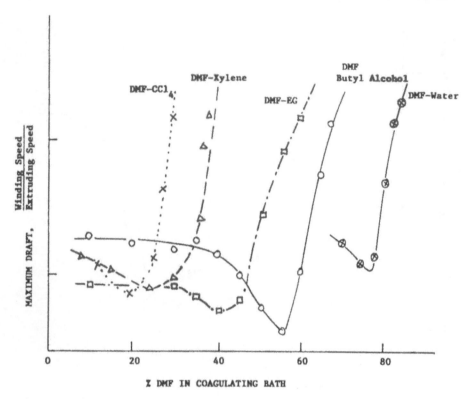

Figure 10   Maximum jet stretch—other nonsolvent [42].

The wet-spinning process has the flexibility to produce a wide variety of fiber cross-sectional shapes (Figure 11). Fiber cross section can be altered by changing the coagulation conditions for a given polymer/solvent/nonsolvent [6, 58, 60]. In general, using round capillaries, the cross sections will be different "bean" and round shapes. The bean-shaped fibers are produced by formation of a weak skin during coagulation; the subsequent volume reduction by solvent diffusion pulls the outer skin toward the center of the fiber. Round fiber can be produced by (1) increasing the coagulation rate and building a thick outer structure that remains round as diffusion proceeds or by (2) decreasing the coagulation rate and producing a thin outer structure that uniformly coagulates with the interior of the fiber. Modified shapes, such as "dogbone," elliptical, ribbons, and others, are produced with shaped spinnerettes and coagulation conditions selected to retain the desired cross section.

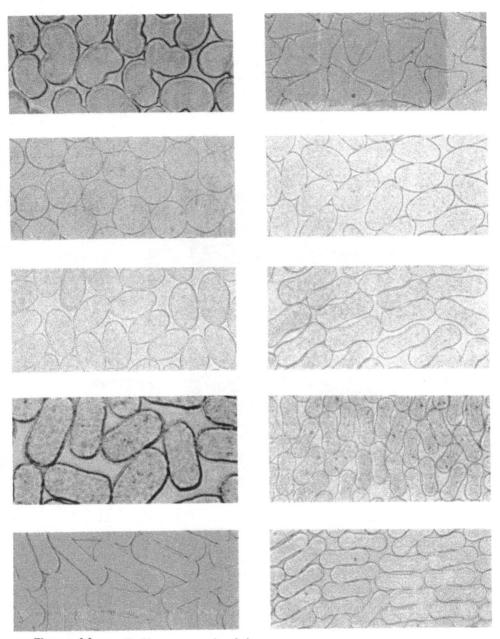

Figure 11 Acrylic fiber cross-sectional shapes.

B. Orientation

At the exit of the coagulation bath the fiber structure is characterized by an interconnected fibrillar network that has a measurable void volume. The void size is a function of the polymer solvent/nonsolvent system and the variables used to control diffusion and phase separation. This initial structure impacts the aftertreatment processes that are used to increase fiber strength and toughness.

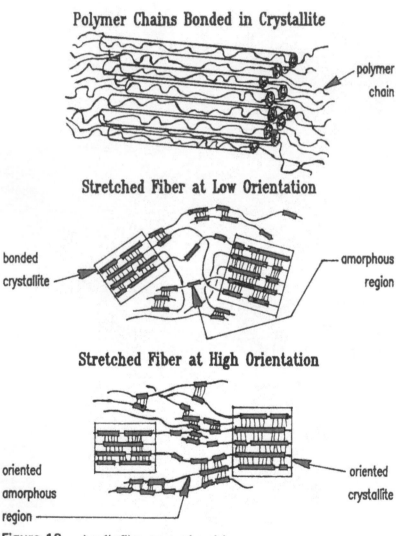

Figure 12   Acrylic fiber structural model.

A model for the fibrillar network is shown in Figure 12. The polymer chains are depicted as rodlike cylinders that are bonded in crystallite bundles. The crystallites are laterally joined with polymer chains forming interconnecting bridges to other bonded crystallites. The coagulated fiber at the exit of the coagulation bath has low orientation [55] in the direction of the fiber axis. The tangled polymer chains extending beyond the bonded structures forms an amorphous region that readily accepts gel additives, such as organic dyes. As mentioned in Section II E and shown in Figure 4, the gel fiber is washed by countercurrent flow of wash medium, normally water. The amount of wash and therefore the residual solvent remaining in the fiber are controlled to maintain constant fiber properties such as dyeability and shrinkage.

The rate of solvent removal is a mass transfer process and is dependent on wash temperature and wash rate. As the wash temperatures increase and the wash solution solvent concentration decreases (high wash rate), the rate of solvent diffusion increases. At high solvent diffusion rates, the gel fiber can swell and increase fiber void size. Subsequently, the final acrylic fiber tensile properties are reduced. For these reasons, the wash systems for commercial spinning systems are chosen to provide controlled washing rates and temperatures. These systems take many forms, but most are countercurrent and segmented to impose a gradient of solvent concentration and temperature as the fiber is sequentially washed. The wash segments nearest the coagulation bath with high solvent concentration have the lowest wash temperatures. As the solvent concentration is decreased in subsequent wash zones, the temperature is increased. The fiber is stretched or oriented within a range of 3 to 12 × by passing over driven rolls. In order to stretch the fiber between the driven rolls, the fiber temperature is increased above the wet glass transition temperature (T_g) of about 65°C. In stretching the fiber, both the amorphous and crystalline regions are oriented in the direction of the fiber axis and the void regions are elongated with water trapped within the voids. The elongated void regions with entrapped water are depicted in Figure 13.

Wash water rates for commercial processes will vary from about 3 to 8 kg. of water per kilogram of fiber. Economically there is no incentive to reduce the water ratio beyond that required for dilution of the spin bath. Water employed for washing is recycled from the solvent recovery process and may contain a small amount of base to ensure fiber neutrality.

C. Drying, Collapsing, Relaxing

The drying and collapsing steps are normally achieved simultaneously, but the method of achieving the processes may vary, depending on the void structure. For collapse to occur, the fiber temperature is increased to the wet glass transition temperature [57]. The collapse is dependent on time, temperature, and the mobility of the chain segments. As the fiber collapses, the diameter decreases. In

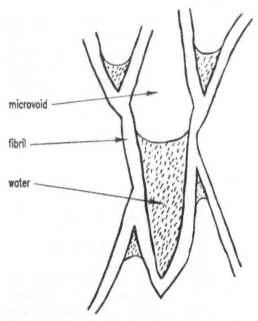

Figure 13   Acrylic fiber structure—drying and collapsing fibrillar network.

addition, there is a consolidation of fibrils and possibly bonding between fibrils. This collapse mechanism determines the stability of the fiber to reswelling upon exposure to hot wet conditions. If the void structure reopens, the fiber loses strength and luster changes are observed.

This collapse mechanism may be accompanied by relaxation if the fiber is not under tension. However, fiber containing large voids will not completely collapse in the relaxed state because the large voids allow the water to vaporize too quickly and the fiber temperature drops below the wet T_g necessary for proper collapse. In those wet-spinning processes that generate gel fibers with large voids, on the order 0.3 micron or greater, it is necessary to maintain tension on the fiber to increase the pressure within the voids and achieve collapse. For those wet-spinning systems that use dry/collapse processes under tension, a separate relaxation step is needed to achieve desirable fiber properties, such as tenacity, elongation, and dye rate. The polymer chain orientation in

Stretched Fiber Before Relaxation

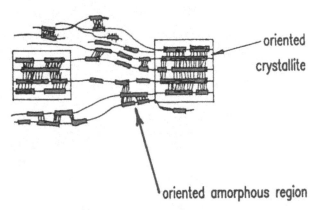

oriented
crystallite

oriented amorphous region

Fiber Network After Relaxation

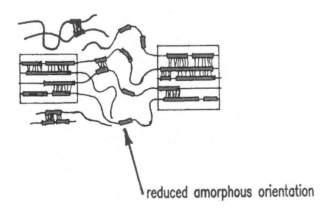

reduced amorphous orientation

Figure 14   Acrylic fiber structure—stretched fiber before relaxation, fiber network after relaxation.

the final collapsed and relaxed fiber is shown in Figure 14. The fiber, which had increased orientation in the amorphous and crystalline regions before collapsing and relaxation, now has reduced amorphous orientation with the crystalline region maintaining an oriented configuration. Typical fiber tenacity and elongation values for wet-spun acrylics for apparel use are 2 to 4 g/d and 30% to 60%, respectively. The wide range in elongation is a result of the relaxation conditions. Of course, higher-tenacity fibers [43] can and are produced for end uses requiring this property.

REFERENCES

1. G. J., Capone and J. C. Masson, 3rd Chemical Congress of North America & 195th American Chemical Society National Meeting, Toronto, Canada, 1988.
2. P. Bajaj and M. S. Kumari, Developments in acrylic fibers: an overview, *Man-Made Textiles in India*, 211–223 (May 1987).
3. P. Bajaj and S. Kumari, Modification of acrylic fibers: an overview, *JMS-Rev. Macromol. Chem. Phys.*, C27(2), 181–217 (1987).
4. J. C. Masson, and A. L. McPeters, Acrylic fiber technology: a 1984 perspective, *Fiber Producer*, 34–42 (June 1984).
5. R. Wiedermann, Spinning acrylic fibres, *Chemiefasern/Textilindustrie*, *31/83*, 481–484 (1983).
6. W. Carl, Acrylic fibres and their familiar modifications, *Chemiefasern/Textilindustrie*, *30/82*, 518520 (1980).
7. W. Carl, Acrylic fibres and their familiar modifications, *Chemiefasern/Textilindustrie*, 30/82, 568570 (1980).
8. G. Prasad, Wet spinning of acrylic fiber and effects of spinning variables on fiber formation, *Synthetic Fibers*, 616 (Jan. and March 1985).
9. G. Prasad and A. Vaidya, Acrylic fiber production: developments in wet and dry spinning processes, *Textile Mag.*, 4056 (April 1986).
10. M. Gulrajani, New vistas of growth for synthetics III, *Indian Textile J.*, 22–31 (Sept. 1984).
11. R. Ramah, Acrylic fiber in India, *Synthetic Fibers*, 13–17 (April/June 1990).
12. Current market trends of acrylic fibers, *Chemiefasern Textilindustrie*, *41/93*, 1350 (Dec. 1991).
13. R. Jenny, Influential factors and selection criteria for dyestuffs for gel-dyeing of wet spun acrylic fibers, *Textile Chem., Art Dyeing*, no. 9, 75–81 (Sept. 1990).
14. R. Jenny, Ciba-Geigy, private communication.
15. Q. Baojun, P. Ding, and W. Zhenqiou, The mechanism and characteristics of dry-jet wet-spinning of acrylic fibers, *Adv. Polym. Technol.*, 6, 509529 (1986).
16. A. Rende, A new approach to coagulation phenomena in wet spinning, *J. Appl. Polym. Sci.*, *16*, 585594 (1972).
17. Q. Baojin, Q. Jian, and Z. Zhenlong, Void formation in acrylic fibers, I, *Textile Asia*, 40 (April 1989).
18. Q. Baojin, Q. Jian, and Z. Zhenlong, Void formation in acrylic fibers. II, *Textile Asia*, 30 (May 1989).
19. D. Paul, Diffusion during the coagulation step of wet spinning, *J. Appl. Polym. Sci.*, *12*: 383–402 (1968).
20. Q. Jean, L. Zhaofeng, and P. Ding, Diffusion and microvoid structure in blend acrylic fibers during wet spinning process", *J. China Textile Univ.*, no. 2, 15–31 (1986).
21. A. Savitskii, and I. Gorshkova, Quantitative analysis of the precipitation media in fiber spinning from polymer solutions, *Khem. Volokna*, *15*, 4–6 (1973).
22. K. Terada, Diffusion during the coagulation process of wet spinning of acrylic fibers. I: mathematical approach, *Sen'i Gakkaishi*, *29*, no. 8 (1973).
23. K. Terada, Diffusion during the coagulation process of wet spinning of acrylic fibers II; relation between the mutual diffusion in filament and the formed structure, *Sen'i Gakkaishi*, *29*, no. 8 (1973).

24. K. Terada, Diffusion during the coagulation process of wet spinning of acrylic fibers. III: simulation of diffusion phenomena when fiber radius and diffusion coefficient chanqe simultaneously, *Sen'i Gakkaishi*, *29*, no. 12 (1973).

25. E. Alieva, Y. Kozhevnikov, V. Nedvedev, and A. Serkov, Diffusion processes in spinning polyacrylonitrile fibres, *Khimicheskie Volokna*, no. 5: 6–8 (Sept. 1990).

26. V. Grobe and H-J. Heyer, Investigation on diffusion processes in wet spinning polyacrylonitrile fibers 1st communication: principles of diffusion measurements, *Faserforsch. Textiltechn.*, *18* (1967).

27. V. Grobe and H-J. Heyer, Investigations of diffusion in wet spinning of polyacrylonitrile fibers second report: experimental testing of the boundary conditions, *Faserforsch. Textiletechn.*, *19*, 33–36 (1968).

28. V. Grobe and H-J. Heyer, Investigations of diffusing processes in wet spinning of polyacrylonitrile fibers, 3rd communication: results obtained with aqueous coagulation baths, *Faserforsch. Textiltech.*, *19*: 313–318 (1968).

29. A. Vaidya and D. Gupta, Influence of spinning dope additives and spin bath temperature on the structure and physical properties of acrylic fibers, *Textile Res. J.* 601–608 (Oct. 1989).

30. A. Ziabicki, The role of phase and structural transitions in fiber spinning processes, Society of Fiber Science and Technology, Japan 25th Anniversary, Tokyo-Osaka (1969).

31. C. Cohen, G. Tanny, and S. Prager, Diffusion controlled formation of porous structures in ternary polymer systems, *J. Poly. Sci. Polym. Phys. Ed.*, *17*, 477–489 (1979).

32. J. Cahn, Phase separation by spinodal decomposition in isotropic systems, *J. Chem. Phys.*, *42*, 93–99 (1965).

33. N. Palsara and E. Mauman, Spinodal decomposition in polymer-polymer-solvent systems induced by continuous solvent removal, Dept. of Chem. Eng., Rensselaer Polytechnic Institute, Troy, NY.

34. K. Kawanishi, M. Komatsu, and T. Inoue, Thermodynamic considerations of the sol-gel transition in polymer solutions, *Polymer*, *28* 980–984 (1987).

35. J. van Aartsen, Theoretical observations on spinodal decomposition of polymer solutions, *Eur. Polym. J.*, *6*, 919–924 (1970).

36. J. van Aartsen and C. Smolders, Light scattering of polymer solutions during liquid-liquid phase separation, *Eur. Polym. J.*, *6*, 1105–1112 (1970).

37. F. Nimts, A. Geller, B. Geller, and B. Yunusov, Structure features of polyacrylonitrile gel fibers", *Khemicheskie Volokna*, no. 3, 11–13 (May 1986).

38. G. Caneba and D. Soong, Polymer membrane formation through the thermal inversion process. I: experimental study of membrane structure formation, *Macromolecules*, *18*, 2538–2545 (1985).

39. G. Caneba and D. Soong, Polymer membrane formation through the thermal inversion process. II: mathematical modeling of membrane structure formation, *Macromolecules*, *18*, 2545–2555 (1985).

40. M. Komatsu, T. Inoue, and K. Miyasaka, Light scattering studies on the sol-gel transition in aqueous solutions of polyvinyl alcohol, *J. Polym. Sci.*, *Polym. Phys. Ed.*, *24*, 303–311 (1986).

41. V. Grobe and G. Mann, Structure formation of polyacrylonitrile solutions into aqueous spinning baths, *Faserforsch. Textiletech.*, *19*, 49–55 (1968).

42. M. Takahashi and M. Watanabe, Publication no. 21: information concerning the yarn spinnability in a yarn spinning bath liquid at the critical concentration, *J. Soc. Textile Cellulose Ind., Jpn., 17*, 249–252 (1961).

43. B. Frushour and R. Knorr, Fiber chemistry, in *Handbook of Fiber Science and Technology, IV*, pp. 171–369.

44. J. Knudsen, The influence of coagulation variables on the structure and physical properties of an acrylic fiber, *Textile Res. J., 33*, (Nov. 1963).

45. M. Takahashi and M. Watanabe, Studies on acrylic fiber, publication 21, filament formation in a solvent-water type coagulation bath, *J. Soc. Textile Cellulose Ind., Jpn., 17*, 243–248 (1961).

46. M. Sotton, A. Vialard, and C. Rabourdin, Influence of wet and dry thermal treatments on the structure of polyacrylonitrile fibers—contribution to the study of the relations between structure and dyeing properties, part 1, *Bull. Sci ITF, 2*, 173–203 (1973).

47. M. Takahashi, Studies on acrylic fiber. VII: relation between coagulating forces of spinning baths and properties of fibers, *Sen'i Gakkaishi, 15*, 959–959 (1959).

48. B. Catoire, P. Bouriot, R. Hagege, M. Sutton, and J. Menault, Acrylic spinning formations, *Textile Asia*, 73–80 (Feb. 1989).

49. A. Stoyanov, Influence of the contents of polymer with different molecular weights in spinning solutions on properties of acrylic fiber, *J. Appl. Polym. Sci., 27*, 235–238 (1982).

50. A. Stoyanov and V. Dpestev, Influence of drawing and molecular weight on structural mechanical properties of acrylic fibers, *J. Appl. Polym. Sci., 24*, 583–588 (1979).

51. A. McPeters and D. Paul, Stresses and molecular orientation generated during wet spinning of acrylic fibers, *Appl. Polym. Symp.*, no. 25, 159–178 (1973).

52. Z. Wu, J. Qin, and B. Qian, The development of texture in acrylic fibers, in *Internationale Chemiefasertagung International Man-Made Fibers Congress*, Dornbirn, Austria, (September 1988).

53. M. Sutton, J. Jacquemart, and R. Monroiq, Influence of Dry and Wet Thermal Treatments Upon the Structure and Dyeing Properties II, Relations Between Structure and Dyeing Properties, *Bull. Sci. IFT, 2 (8)*: 247–261 (1973).

54. D. Paul and A. Armstrong, The elastic stresses generated during fiber formation by wet spinning", *J. Appl. Polym. Sci., 17*, 1261–1282 (1973).

55. D. Paul, Spin orientation during acrylic fiber formation by wet spinning, *J. Appl. Polym. Sci., 13*, 813–826 (1969).

56. A. Stoyanov and V. Krustev, Influence of structure on the changes during heat treatment and properties of acrylic fibers, in *Morphology of Polymers, Process, Europhys. Conf.*, Macromol Phys. 17th Meeting, 1985, pp. 639–644.

57. J. Dumbleton and J. Bell, The collapse process in acrylic fibers, *J. Appl. Polym. Sci., 14*, 2402–2406 (1970).

58. G. East, J. McIntyre, and G. Patel, The dry-jet wet-spinning of an acrylic fiber yarn, *J. Text. Inst., 75*, 196–200 (1984).

59. S. Ogbolue, Structure/properties relationships in textile fibres, *Textile Inst., Textile Progr., 20* (1990).

60. C. Han, and J. Park, A study of shaped fiber formation, *J. Appl. Polym. Sci., 17*: 187–200 (1973).

61. N. Beder, D. Kabanova, and I. Dvoeglazova, A method of investigating the structure formation process of unoriented polyacrylonitrile specimens", *Khimicheskie Volokna*, no. 2, 31–32 (1986).
62. E. Wittorf, "New Concepts for Acrylic Fiber After Treatment Lines for Large Tow Deniers", *Chemiefasern Textilindustrie, Man-Made Fiber Year Book*: (1989).
63. W. Wagner, Comparison of further development of wet and dry spinning processes for acrylic fibre, in *2nd International Conference on Man-Made Fibers*, Beijing, China, November 1987.
64. P. Eberhard, P. Anneliese, and S. Hartig, A process for the manufacture of void free and constant cross sectional shaped polyacrylonitrile fibers with high packing densities, East German Patent No. 78624, December 20, 1970.
65. D. Paul, A study of spinnability in the wet spinning of acrylic fibers, *J. Appl. Polym. Sci., 12*, 2273–2298 (1968).
66. G. J. Capone, unpublished Monsanto Company data.

5

Dry-Spinning Technology

B. von Falkai

Bayer AG
Dormagen, Germany

I. INTRODUCTION

The melting point of polyacrylonitrile is higher than the temperature at which it decomposes. Consequently, the polymer can only be spun into filaments and fibers from solution (solvent spinning) or from a plasticized gel. In the solvent spinning process, the liquid filaments/fibers formed as the solution is pressed through the spinnerets are solidified either by coagulation in suitable baths (wet-spinning process, see Chapter 4) or by evaporation of the solvent in a spinning tube or cell. This method is known as dry spinning.

The technical process involved in dry spinning is relatively simple: the solvents in the spinning solution are evaporated by hot inert gases or hot air as it passes out of the spinnerets. In order to ensure that the solvent evaporates quickly, the temperature of the gas used for this purpose is generally above the normal boiling point of the solvent. The evaporated solvent is removed from the system with the gas and recovered. The fibers solidify in the vertical spinning tube as the solvent is removed.

The first synthetic filament yarn, Chardonnet silk, was discovered in 1884. It was based on nitrocellulose and produced by a dry-spinning process. This spinning process is still used today in the manufacture of acrylonitrile homo- and copolymers, various manmade fibers such as acetate silk, polyvinyl alcohol fibers (PVA), and some aromatic polyamides.

Solvent spinning relies on the use of a suitable solvent. The use of dimethylformamide for the production of polyacrylonitrile fibers was discovered by researchers at Du Pont in the United States in 1942 [1], paving the way for the first industrial-scale production of polyacrylonitrile fibers by the Du Pont company at Waynesboro in 1948 [2].

The same discovery was made independently by the I. G. Farben AG in Germany, also in 1942. In fact, their discovery predates Du Pont's by two months [3]. However, World War II and its aftermath meant that Bayer was only able to start production of polyacrylonitrile fiber at its plant in Dormagen, Germany, in 1954 [2].

The production process was patented by Du Pont and Bayer, thus effectively preventing other companies producing dry-spun polyacrylonitrile for many years. This patent barrier was one reason why fiber manufacturers wishing to produce polyacrylonitrile fibers in the early 1950s concentrated their efforts on the development of wet-spinning technology.

For successful dry spinning, the solvent must fulfill the following preconditions. The polymers must have adequate stability at the solvent boiling point. The solvent should also have good dissolving properties, not react with the dissolved polymers and have a low boiling point. Similarly, its heat of vaporization must not be too high, it must have sufficient thermal resistance, low toxicity, a very low tendency to produce static charges, low risk of explosion and be relatively easy to recover. And the price should, of course, be as low as possible. Taking these criteria as a basis, dimethylformamide is the most suitable solvent for dry-spinning polyacrylonitrile.

Following these general remarks on the dry-spinning process, let us turn our attention to the basic principles of the production process. We will then look at the cost effectiveness of the classic two-step production process and the quality of the fibers produced compared with the new integrated one-step process (continuous process).

II. STAGES IN FIBER PRODUCTION

The main stages in fiber production are

Polymerization
Production of the spinning solution
Spinning
Aftertreatment

A. Production of the Spinning Solution

Polymerization of acrylonitrile can be by suspension, emulsion, mass, or solvent polymerization. Spinning solutions produced by solvent polymerization are only suitable for wet spinning, because they have a relatively low polymer content. The polymerizate used for dry spinning is therefore produced by continuous suspension polymerization.

1. The Dissolving Process

As mentioned, dimethylformamide has become the universal solvent for dry spinning of polyacrylonitrile fibers. It has the following physical properties:

Boiling point (1012 mbar)	153°C
Heat of vaporization	579 J/g
Flammability limits in air	
lower:	2.2 vol.%
upper:	15.2 vol.%
Explosive limits in air	
lower:	50–55 g/m^3
upper:	200–250 g/m^3
Ignition temperature	445°C

For production of high-quality fibers, dimethylformamide should have the following purity levels:

Water	<0.3%
Amines (dimethylamine)	<0.1%
Formic acid	50 mg/L
Iron	<0.05 ppm
pH (20% aqueous solution)	6.5–9.0

Many attempts have been made to use other solvents in the dry-spinning process as they became known as solvents for polyacrylonitrile. However, the only other organic solvent that is of any interest is dimethylacetamide.

Under normal dissolving conditions (approx. 80°C) and polymer concentrations of 25–28%, the viscosity obtained with dimethylacetamide is double that obtained with dimethylformamide. However, its main disadvantage is that its ignition point is over 100°C lower than that of dimethylformamide, at around 320°C. Consequently, about 30% less energy can be used in the spinning tube, thus reducing production capacity by around 30%.

2. Dissolving Techniques

When producing the spinning solution, the powder polymer and the solvent should be pasted up as smoothly as possible prior to dissolving to cut down the dissolving time. This can be achieved by

Producing compact microgranules of the polymer

Mixing the polymer and solvent streams thoroughly as they are added, e.g., in mixing screws, turbomixers, or paddle mixers

Maintaining the solvent at a low temperature, which allows formation of a suspension in which the polymer is dispersed

The spinning solution can be produced either batchwise or continuously.

One example of a batch process involves dissolving the polyacrylonitrile powder by stirring it into the dimethylformamide at 70–90°C. It is then filtered and spun into filaments. For technical reasons, it is generally necessary to use large vessels although this can impair the color of the material if spinning times are long.

In the case of polymers that are difficult to dissolve, e.g., polymers with a high molecular weight and acrylonitrile homopolymer, the dissolving temperature has to be raised substantially to produce spinnable solutions of stable viscosity with no clumps of undissolved particles. This causes even more problems as regards the color consistency of the material over long spinning periods.

In another batch process (Figure 1), the polyacrylonitrile powder is mixed in a pasting-up unit and the resultant suspension is then drained into an intermediate vessel, heated to 70–90°C in a dissolving vessel with heat exchanger to dissolve any remaining particles and then pressed into a deaeration vessel. The spinning solution is then pumped into a spinning vessel from which it passes into the spinning unit via filter [4]. The whole process takes 8–10 hr from the introduction of the suspension into the solvent system to completion of spinning. The relatively long reaction time may impair the color of the solution during spinning, as with the process described above.

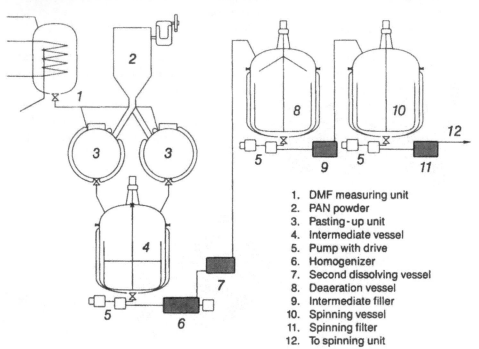

1. DMF measuring unit
2. PAN powder
3. Pasting-up unit
4. Intermediate vessel
5. Pump with drive
6. Homogenizer
7. Second dissolving vessel
8. Deaeration vessel
9. Intermediate filler
10. Spinning vessel
11. Spinning filter
12. To spinning unit

Figure 1 Schematic representation of a PAN dissolving unit.

Another variant is the thermal shock process. In a first stage, a high-viscosity polyacrylonitrile dimethylformamide suspension is produced in a turbomixer at room temperature in a moderate vacuum. The suspension is then heated to 130–150°C in 3–5 min (shock process). The clear but inhomogeneous spinning solution produced is transferred to a high-performance homogenizer, e.g., a static mixer, without intermediate cooling, homogenized, and conveyed immediately to the spinning machine (Figure 2). This process produces fiber material with a solids content of 25–30%.

To produce a stable-viscosity spinning solution the solution has to be treated for a given length of time at a minimum temperature before spinning. The temperature and time depend on the concentration of the solution and the molecular weight of the polymer.

If up to approximately 33% by weight polyacrylonitrile is mixed with dimethylformamide at temperatures of up to 50°C, a high-viscosity paste is obtained. This can be converted into a stable-viscosity spinning solution by heating it to 130–140°C in 3–4 min. Very high spinning drafts can be produced with such spinning solutions. These can be used in the production of microfibers [5] (see Section IIB5).

By blending the dimethylformamide with nonsolvents which cause differing degrees of precipitation, modified polyacrylonitrile fibers can be produced. In

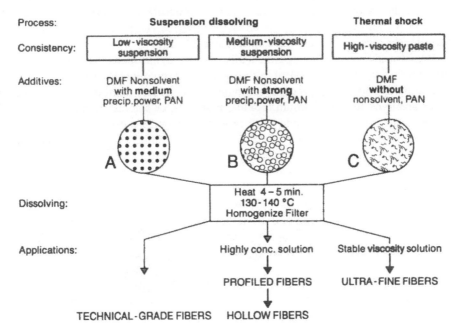

Figure 2 Thermal shock and suspension dissolving processes.

order to produce hydrophilic fibers, the dimethylformamide is blended with nonsolvents with moderate precipitating power for the polymer, for example glycols or polyvalent alcohols, and then mixed with the polyacrylonitrile. This suspension is then converted to a stable-viscosity spinning solution by heating it to 130–140°C in 3–4 min, then homogenized, filtered, and spun [6].

If a nonsolvent with a strong tendency to precipitate polyacrylonitrile, e.g., water, is added to the spinning solvent, more viscous but easily transportable suspensions are produced, depending on the polymer concentration, the proportion of nonsolvent and the mixing temperature. The maximum permissible solids content is about 40% by weight. For viscosity reasons the temperature of the suspensions should not exceed 20°C. The suspensions are converted to stable-viscosity solutions by heating them to 130–140°C in 3–4 min. Fibers with a variety of different cross sections [7] and hollow fibers [8] can be produced from these concentrated spinning solutions (see Section IIB5).

Figure 2 contains a schematic representation of the thermal shock and suspension dissolving methods. To minimize yellowing and to ensure that the color of the spinning solution remains fairly constant throughout the dissolving stage, it is advisable to carry out each stage of the process in an inert atmosphere. It is also possible to add chelating agents or reducing stabilizers such as ethylenediamine-tetraacetic acid (EDTA) to prevent discoloration by traces of heavy metals or oxidizing impurities [9].

A stable-viscosity solution is the main prerequisite for a smooth spinning process. Various additives such as thiosemicarbazide can be used to lower and stabilize the viscosity of the spinning solution [10, 11].

It is also advisable to incorporate additives such as phosphonocarboxylic acid that prevent any residual catalyst salting out during spinning [12, 13].

During the dissolving process, or immediately before spinning, if preferred, other additives such as titanium dioxide, pigments, or soluble dyestuffs can be incorporated into the solution to produce low-luster or dope-dyed fibers.

The polymer content of the spinning solution is limited by the solubility and rheological properties of the polymer (fiber-forming capacity) and by any tendency it may have to gel. In practice, as we have already seen, concentrations of 25–35% are commonly used for dry-spinning processes, depending on the molecular weight of the polymer. The temperature of the spinning solution also has to be kept within certain limits up to the end of the spinning process [14]. For example, solutions must not be kept at temperatures far below 70°C or above 100°C for prolonged periods; otherwise they tend to gel.

3. Storage Stability of the Spinning Solution

Over the past few decades, various research teams have studied the storage stability of concentrated polyacrylonitrile solutions at below 70°C and above 100°C.

Jost [15] was the first to store polyacrylonitrile solutions at below 70°C. He investigated the solutions with x-ray structural analysis. He found reflections that intensified as the storage period increased. These findings suggest that crosslinking occurs in the solution. This crosslinking is generated by local structures and increases with the length of storage. This suggests that the crosslinking points in the gels are a sign of emergent paracrystalline regions [16].

However, Labudzinska and Ziabicki [17] found that the gelling of polyacrylonitrile and other polymer solutions is not necessarily a preliminary stage in crystallization. According to them, crystallization and phase separation may occur as a result of gelling but are not characteristic of gelling. Ziabicki proposed a model of "pure gelling" for gel formation during storage at temperatures below 70°C. According to this model, the structure of the solution is maintained when it gels and the polymer chains are only linked by the force of the secondary valences in the dipolar bond between the nitrile groups [17].

Figure 3 is an Arrhenius diagram showing the stages in the gelling of a polyacrylonitrile homopolymer MW= 150,000 at various concentrations at temperatures below 70°C [18]. The activation energy calculated from the gradients of the curves shows that these values increase as the concentration rises.

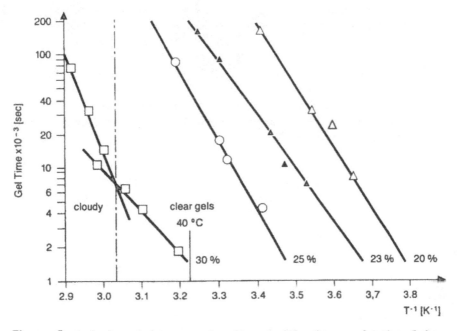

Figure 3 Arrhenius gelation curves: logarithm of gelling time as a function of absolute reciprocal gelling temperature of PAN solution. Parameter: concentration (after [18]).

If a 30% solution is stored at 60–70°C for a prolonged period, it also becomes cloudy. While the mean activation energy of 20% and 25% solutions at E_a= 96 kJ/mol, this value rises by over 70% to E_a = 170 kJ/mol for 30% solutions. This suggests a higher density of physical crosslinking points, which manifests itself as a cloudy appearance at visible wavelengths.

In spinning solutions stored at over 100°C for prolonged periods, gelling occurs through crosslinking reactions. The tendency to irreversible gel increases with the viscosity of the solution and the molecular weight of the polymer and decreases with the polydispersity of the product [19].

The problem of gelling during storage of highly concentrated spinning solutions must not be underestimated when considering the design and construction of industrial plants.

Here are some aspects that have to be considered in order to produce usable spinning solutions:

1. Air bubbles incorporated in the solution at the dissolving stage cause problems during spinning, so the solution has to be thoroughly deaerated after dissolving. This is generally carried out in a mixer vessel under a moderate vacuum.
2. In order to ensure problem-free spinning, the spinning solution must not contain any undissolved or swollen polymer particles or impurities. It is therefore passed through a filter as it leaves the deaeration vessel. The filters are comprised of densely woven filter cloths or metal sieves. The efficiency of the filter (amount of polymer passed through per m^2 filter surface area before pressure drop becomes excessive) can be taken as a measure of the quality of the spinning solution (freedom from swollen particles, etc.).
3. After filtration the spinning solution is pumped to the spinning unit. The deaerator, filters, pumps, and pipes must be kept at a constant temperature in order to enable good spinning.

B. The Spinning Process

1. Rheology of the Spinning Solution

The spinning stage is primarily characterized by various flow processes and is therefore influenced to a large extent by the rheological properties of the substances used. The polyacrylonitrile solutions used in the production of fibers have non-Newtonian rheological properties. Their flow behavior thus differs from Newtonian fluids in two ways: first, they are pseudoplastic, in other words, their viscosity decreases as the shearing force increases, and, second, they are viscoelastic and are thus able to store elastic energy, especially during short periods of deformation, for example extrusion through the spinneret capillary. The temperature dependence of the elastic modulus suggests that they have an entropy-elastic mechanism.

These phenomena are connected with the macromolecular structure of polyacrylonitrile and in qualitative terms they can be explained by the fact that during the flow processes the molecules are "hooked" together by entanglements of the chains. Shearing forces release those links to some extent, leading to flow resistance and a reduction in viscosity.

At the same time, the hooks act as temporary physical crosslinking points, which dissolve and re-form constantly during the deformation process, so the liquid can be compared to rubber, where crosslinking points are not fixed but vary in site and time.

2. Internal and External Spinning Processes

The production of the fiber comprises the following basic stages:

Flow of the spinning solution in the pipework
Processes in the capillary tube
Behavior of the material as it leaves the capillary tube
Behavior of the liquid material during deformation
Solidification of the filament.

The first two processes are "internal" (i.e., inside the spinning unit), while the other three are "external" (outside the spinning unit).

The internal spinning processes are characterized by the rheology of the spinning solution in the pipework and spinnerets. As the high-viscosity solution flows through the pipes a virtually parabolic velocity profile can be traced, which means that some of the solution remains in the pipework for longer than the rest. In order to counteract this, static mixers are built into the pipes. These comprise lamellar arrangements that divide the solution into a number of streams, alter their course and mix them with one another. As the spinning solution flows through the pipes leading to the spinneret, the arrangement of the chain molecules is disordered or at most displays some sort of order in small regions, but without any preorientation.

The spinneret at the end of this part of the process is particularly important since it is responsible for distributing the spinning solution pumped to it precisely and evenly. It is at this stage that the spinning solution receives the desired cross section (see also Section IIB5). Since the spinning solution only remains in the capillary tubes for a very short time (10^{-4}–10^{-2} sec), the rate at which it enters them is of some significance.

The viscosity of the spinning solution may be as much as several thousand poise at room temperature. It is generally highly pseudoplastic so the shearing viscosity in the spinneret only accounts for a small proportion of the total viscosity.

The spinning solution undergoes a process of deformation as it flows into the capillary tube. Like most fiber-forming polymers, polyacrylonitrile spinning solutions are entropy-elastic liquids, so elastic deformation always involves a certain degree of molecular orientation. The elastic energy stored during its passage through the capillary tube manifests itself to some extent in the broadening of the stream of liquid as it leaves the spinneret. As a result of the viscoelasticity of the spinning solution, the smaller the ratio of contact time in the capillary to the relaxation time of the spinning solution, the less the reduction in pressure at the spinneret aperture conforms to the Hagen-Poiseuille equation.

The external spinning processes start as the solution leaves the spinneret. In viscoelastic spinning solutions, normal stresses that have not been reduced during the brief period in the capillary are released as the filament flows out of the spinneret. Consequently, the stream of liquid bulges (Figure 4). The greater the relaxation period and the shorter the contact time in the capillary tube, the more the solution bulges on leaving the spinneret. This is known as the Barus effect.

This effect increases with the viscoelasticity and viscosity of the solution and the spinning speed. Similarly, it decreases as the temperature of the spinning solution and spinneret increase and increases with the speed at which the material is drawn off. As a result of plastic deformation, the filament becomes continuously thinner after the bulge until it solidifies.

The viscosity and the modulus of elasticity of the spinning material increase as the solvent is evaporated. This occurs particularly quickly up to the point where the solution is converted to a rubberlike state due to the increasing concentration of the polymer.

Figure 5 is an overview of the changes in the various parameters during the spinning process. The titer (denier) drops constantly up to the solidification point. The speed and tension of the material increase by the same amount. The rate of solidification accelerates towards the solidification point while the relaxation rate decreases continuously up to this point. The tension, solidification and orientation gradients reach their maximum points while the filament is in the rubber-elastic state.

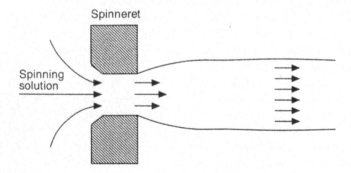

Figure 4   Bulge in spinning solution as it leaves the spinneret.

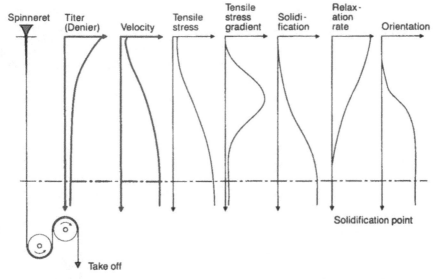

Figure 5 Processes occurring during spinning (direction of spinning downwards, diagram shortened).

The spinnability of the material is restricted on the one hand by filament breakages if the material is spun at too low a viscosity and, in particular, if it is spun at unsuitably high temperatures. However, on the other hand, filament breakages also occur if the viscosity of the spinning solution and the take-off speed are too high.

The solvent evaporates as the spinning solution leaves the spinneret and enters the hot atmosphere of the spinning tube. In unfavorable conditions this can occur very quickly as a reaction to the sudden reduction in pressure immediately after the spinneret outlet, resulting in vapor bubbles in the filaments. These can impair the tensile properties of the fiber.

Further evaporation of the solvent during spinning primarily depends on the diffusion of the solvent in the spinning solution. The evaporation process slows down as the filaments solidify. The material, concentration and temperature-related diffusion coefficients remain constant throughout the evaporation stage, as does the titer. The amount of residual solvent in the fiber material depends on the time it is held under evaporation conditions, the take-off speed, and the length of the spinning tube. It naturally also depends on the temperature of the heated jacket and the gas temperature.

In dry-spinning processes the spinning draft (ratio of velocity at take-off to theoretical velocity of solution at spinneret exit) is 1–20, far lower than in melt spinning. The draft does not normally have much impact on the properties of the spinning material and the end product.

Although the general laws governing the forces in the filament during spinning apply, the dry-spinning process is even more difficult to describe in terms of mathematical formulae than the melt-spinning process since the changes in the parameters depend to a considerable extent on the evaporation conditions which cannot be quantified accurately.

We have recently gained an insight into the fiber-forming mechanism of the dry-spinning process through extensive theoretical work based on a mathematical model. The melt-spinning process can be described fairly easily in mathematical terms, using a separate formula for equilibrium of the polymer melt, equilibrium of the various forces, and thermal equilibrium. For dry spinning, however, the presence of the solvent makes it necessary to include transport of the material in solution, diffusion of the solvent through the surface of the fiber after solidification, and concentration-related viscosity [20].

3. Theory of Fiber Cross Section Formation

The diffusion and evaporation stages that comprise the solidification of the fiber in dry spinning generally induce radial inhomogeneity of the filaments. The outer sections solidify faster than the inner sections, producing a core-sheath structure. Further diffusion of the solvent from the core through the solidified sheath reduces the mass of the core. As a result, the sheath collapses inward. The faster the evaporation rate compared with the diffusion rate, the more likely the cross section of the fiber is to change from circular through bean-shaped to a dog-bone (dumbbell) shape.

Figure 6 shows these interrelationships in schematic form. If the rate of evaporation and the rate of diffusion of the solvent are roughly equal, a circular cross section is formed (a). However, if the rate of evaporation is far higher than the rate of diffusion, the cross section collapses to a dog-bone shape (b) [21].

Two parameters which can be further used to describe the formation of the cross section are the rate of evaporation of the dimethylformamide and the dwell

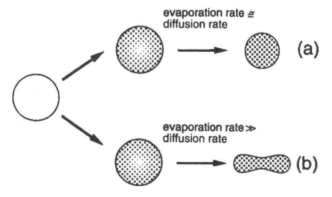

Figure 6 Cross section formation during solidification (a) and (b) after dry spinning.

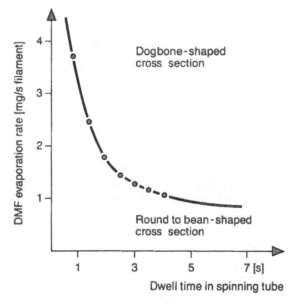

Figure 7 Influence of thermal spinning parameters on fiber cross section (after [22]).

time of the filaments in the spinning tube. The rate at which the solvent evaporates can be calculated from the amount of solvent used, the residual solvent content of the filaments, and the dwell time in the spinning tube [22].

Figure 7 shows the evaporation rate of dimethylformamide in mg/s filament as a function of the dwell time in the spinning tube. The curve obtained can be divided into dog-bone and other cross sections.

Since the cross section is important for the textile properties of the fiber, the fiber technology can be regulated by selecting suitable spinning parameters, as described above.

4. Dry-Spinning Technology

The most important stage in the dry-spinning process is the evaporation of the solvent after spinning. This determines the construction of the spinning machines.

As described earlier, the spinning solution, which is kept at a constant temperature of around 90°C, is conveyed to the spinning pump after a variety of preliminary operations such as central filtration, deaeration, etc.

Spinning is normally carried out at a pressure of 8–15 bar. In order to facilitate evaporation of the solvent, the solution is heated to 130–140°C as it passes through the candle filter and then transported to a spinning manifold with built-in spinnerets.

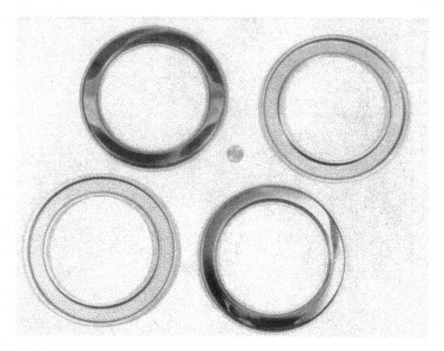

Figure 8 Ring spinnerets for dry spinning; 50-Pfennig piece in center (2-cm diameter) shows relative size of spinnerets.

The spinnerets are made of stainless steel alloy. The diameter of the capillary tubes is 0.1–0.3 mm. Bowl/hat-shaped spinnerets are used for spinning filaments, while ring or bowl spinnerets with up to 2800 apertures are normally used for spinning fibers. The ring spinnerets are mounted in the spinning head from inside and outside (Figure 8). The main advantages of ring spinnerets are process related. The gas used to dry the fibers can be introduced both centrally through the inner mount and outside around the spinneret.

As it leaves the spinneret, the solution flows into the spinning tube in which the solvent is evaporated. The spinning tube, which is also known as a spinning cell, is 6–10 m long and 250–450 mm in diameter. The tube is kept at the requisite temperature by a jacket heated by a fluid such as diphyl (26.5% diphenyl, 73.5% diphenyloxide) (Figure 9).

It is often advantageous to divide the heating into sections and to intensify the heat energy in the zone requiring the most heat, i.e. near the spinneret [23]. The drying gas, heated to 300–350°C, flows co- or countercurrent to the material as it leaves the spinnerets. In the cocurrent process, the gas enters the tube through filters positioned close to the spinneret and is removed at the other end by suction. In the countercurrent process, it enters at the bottom of the tube and is removed above the spinning head. Both systems have advantages and

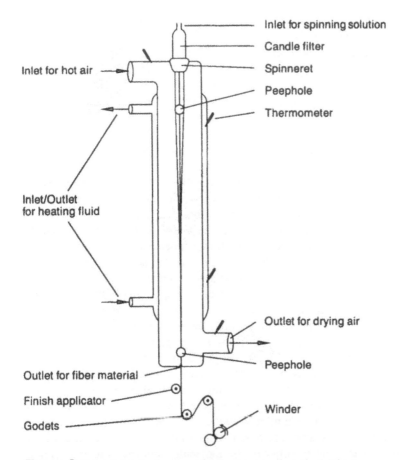

Figure 9 Spinning tube for dry-spinning process (length: 6–10 m).

disadvantages. A partial vacuum prevents the solvent-laden gas from leaving the spinning tube through the fiber exit opening.

The flow rate of the gas depends on the throughput rate and solvent content of the spinning solution, the desired residual solvent content and the physical properties of the solvent, e.g., evaporation ratio and explosive limits.

Excessive flow rates and turbulence must be avoided to prevent damp filaments coming into contact with each other and sticking together. The gas generally flows through the spinning tube at a rate of 1–2 m/sec. Air, an inert gas, or saturated steam can be used for this purpose.

If saturated steam is used, the same volume as air introduces more energy. The absence of oxygen in the spinning tube allows the use of higher temperatures. For example the temperature of the steam can be above 400°C, while the spinning tube can be heated to over 240°C. This can be exploited to increase the output of the spinning tube [24].

In the dry spinning of polyacrylonitrile fibers, output is normally 8–15 kg polyacrylonitrile solids per spinning tube per hour. Outputs of over 30 kg/hr are possible, but special spinning heads are needed and the temperature of the spinning solution and the flow of the gas around the spinneret have to be altered [25]. If air is used to evaporate the solvent, the solvent content must be kept below the explosion threshold at all times.

Some processes use an inert gas, e.g., pure nitrogen or air doped with nitrogen instead of air. These systems are usually closed circuits. The oxygen content of air doped with nitrogen is normally reduced to 6–8%. This has a favorable effect on the explosion threshold of the dimethylformamide and reduces yellowing of the fiber material. The hot gas provides most of the heat required for evaporation of the solvent. However, the spinning tube is also heated so that the gas does not cool to a temperature which allows condensation of the solvent.

It is never possible (or desirable) to remove all the solvent from the fiber material in dry-spinning processes. There is always a residual solvent content of 5–25%, depending on the exact process used. This aids orientation of the filaments in the subsequent drawing process.

The residual solvent content and the high temperature in the spinning tube (about 200°C) inhibit molecular orientation, so comparatively little orientation occurs during the spinning process. The degree of orientation occurring during spinning increases with the viscosity of the spinning solution, the spinning speed and also to a slight extent, with the spinning draft. It also increases with the speed at which the material solidifies, e.g., fine titers and microfibers, which have a very large surface area in relation to their volume.

The temperature of the spinning tube is both a favorable and an unfavorable influence. It increases the mobility of the molecules and thus hinders orientation. Higher orientation during spinning reduces the drawability of the fibers or increases the tenacity of the final product if constant drawing ratios are used. However, the temperature of the spinning tube has far less impact than the residual solvent content of the fibers.

When spinning filament yarns, an oil-based finish is applied to the filaments of the bottom of the spinning tube. They are then drawn off via godet rollers and wound at 200–500 m/min. In the production of tow, the finish (lubricant) is applied as the filaments leave the spinning tubes. The filaments are then passed round a guide roller and combined to form a single tow. Each spinning machine comprises 10–70 spinning tubes. Tows of up to 500,000 dtex are drawn off by special take-off rollers at speeds of 250–450 m/min and coiled in spin cans (Figure 10).

If desired, the finish can be applied in the spinning tube (Figure 11). This guarantees extremely uniform distribution of the lubricant film on the surface of

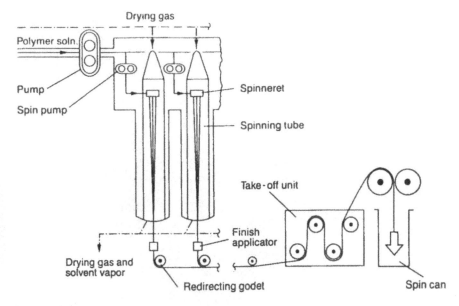

Figure 10 Dry-spinning unit for production of staple fibers.

the individual filaments [26]. An overall view of a dry-spinning machine is shown in Figure 12.

In dry spinning there are four elements requiring a constant temperature: the spinning solution, the spinning head, the drying gas, and the sides of the spinning tube. This requires a considerable amount of measuring and regulation technology. The entire spinning machine also needs special protection against the risk of explosion. Even when it has been coiled in the cans, the spinning material may release solvent vapor.

In order to make sure that the TLVs (threshold limit values) are met in production premises, all working areas should have adequate ventilation and air extractors.

5. Modified Spinning for Specialty Fibers

Let us look briefly at how slight modification of the spinning solution and/or spinning conditions can be used to produce interesting fibers. Many of these are already being produced commercially.

The types of fiber we shall consider here are

Microfibers
Profiled
Hollow
Absorbent

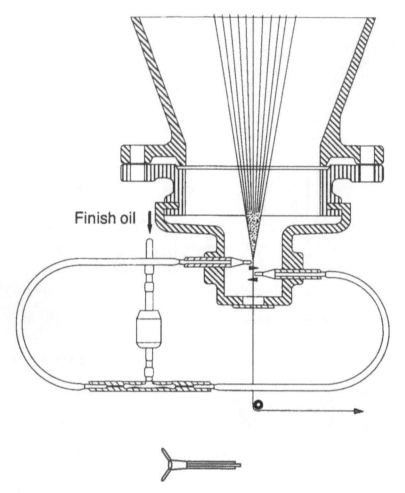

Figure 11 Finish applicator (after [26]).

Other specially modified fiber types (many of which can be produced by dry spinning) are mentioned in Chapter 6.

Microfibers. Recently the chemical industry has renewed its efforts to produce fibers and filaments with titers of less than 1 dtex. These are known as microfibers. They have a range of advantages over conventional synthetic fibers, e.g., polyester and polyacrylic fibers with titers of over 1 dtex. These include brightness, an attractive luster, a soft hand, and comfortable wear properties.

There are various methods of producing microfibers [27]. The amorphous-amorphous structural change that polyacrylonitrile microfibers undergo in the

Figure 12 View of PAN dry-spinning plant (Bayer AG, Dormagen, Germany).

drawing process means that a high drawing ratio, e.g., up to 1:20 can be used. The disadvantage of this method is that it requires particularly strict adherence to all drawing parameters, e.g., speed, temperature, etc.

As mentioned previously (page 109), a more reliable method of manufacturing microfibers is described in U.S. patent 4,497,868 [5]. The following parameters have to be observed. One of the main preconditions for producing microfibers with the desired textile properties is that the viscosity of the spinning-solution should be stable prior to spinning. In order to ensure this, when working with pure acrylonitrile polymers with a mean K value of 80 ($\bar{M}_n$ = 20,000; $\bar{M}_w$ = 138,000) and concentrations of up to about 35% in dimethylformamide, the polymers should first be pasted up with the solvent at low temperatures. The highly viscous paste produced must be treated at a minimum of 140°C for at least 4 min before proceeding.

When using acrylonitrile copolymers, as is normally the case in industrial production, the high-viscosity paste can be pretreated at slightly lower temperatures (approx. 125–130°C) for the necessary period in order to ensure that the viscosity is stable.

The viscosity of the solution is considered to be stable if it remains constant for several hours (measured in seconds in a dropping-ball viscometer). Such solutions can be spun into fine-titer fibers and filaments in high drafts (30–500) under mild thermal conditions (slow evaporation of the solvent). The temperature of the spinning solution should be 150°C. The temperature in the spinning tube should not exceed 200°C, while the temperature of the gas/air used to evaporate the solvent should not be over 400°C.

The properties of a 0.6 dtex fiber produced under these conditions are listed below. (The properties of a 1.3 dtex fiber are included as a comparison [22].)

Titer		0.6 dtex	1.3 dtex
Cross section		Round	Dog-bone
Density		1.181	1.175
Brightness		7.1	4.8
Tenacity	cN/tex	30–32	28–30
	grams/denier	3.4–3.6	3.2–3.4
Breaking elongation %		23–25	30–35
Modulus of elasticity			
	cN/tex	458	327
	grams/denier	52	37
Spinning limit Nm*		200	100
Yarn strength			
	cN/tex	17.2	14.9
	grams/denier	1.95	1.69

*Nm = metric number, 10,000 Nm = 1 dtex

Under even milder thermal conditions extremely high drafts can be used, producing extremely fine titers. However, such high drafts require solutions of extremely stable viscosity as the spinning solution gels if the temperature is too low. Using the method outlined in [51], filaments with a titer of 0.2 dtex were produced from a stable-viscosity 29.5% polyacrylic spinning solution using a draft of 457. After stretching in boiling water in a ratio of 1:3.6, filaments with a final titer of 0.07 dtex were obtained.

The solvent evaporates more slowly under these milder spinning conditions. As a result, the fiber shrinks more uniformly over the entire cross section. As Figure 13 shows, the fiber cross section therefore remains round after solidification [21]. Moreover, more uniform solidification of the fiber structure in the spinning tube alters the morphology of the fiber.

Figure 14 shows that the surface of the filament becomes smoother as the fineness of the titer increases. The fibrillar surface structure observed at coarser titers virtually disappears. The structure of the fiber is maintained even after drawing. The smooth surface structure is responsible for the increased brightness of fine-titer fibers.

0.6 dtex/round   1.3 dtex/Dralon

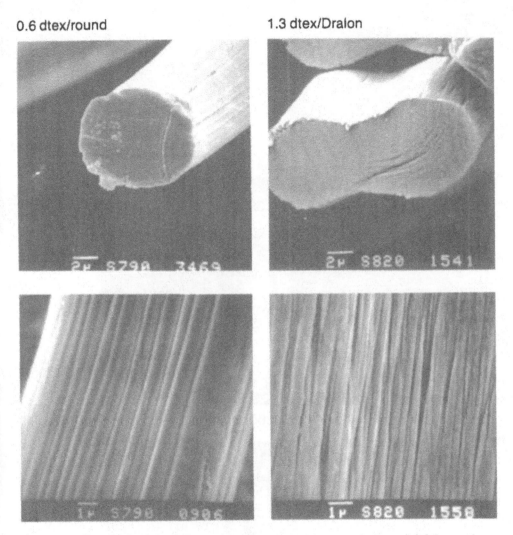

Figure 13 Comparison of the cross sections and surface morphology of 0.6 dtex and 1.3 dtex fibers.

Profiled Fibers. Production of synthetic fibers with modified cross sections by melt spinning or wet spinning has been state-of-the-art for many years. For example, melt-spun polyamide and polyester fibers are often extruded through special profiled spinnerets to obtain a specific level of brightness or produce a particular hand, luster, or general appearance.

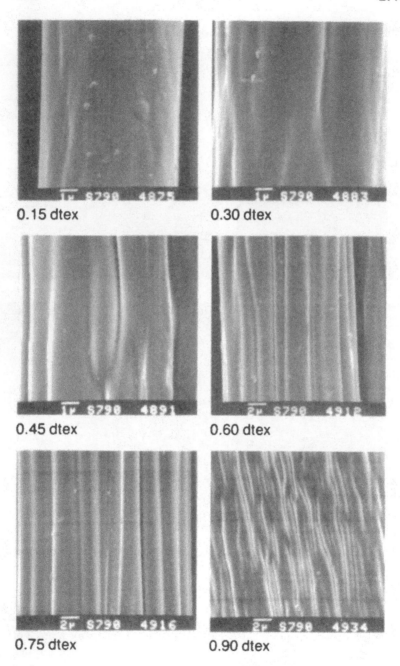

0.15 dtex

0.30 dtex

0.45 dtex

0.60 dtex

0.75 dtex

0.90 dtex

Figure 14 Fiber surface morphology as a function of titer.

Profiled polyacrylonitrile fibers cannot be produced by dry-spinning solutions of normal concentration. However, using the method outlined in U.S. patent USP 4,810.448 [7], dry-spun fibers can be produced with any desired cross section. As outlined in Section IIA2 the following procedure is used:

1. The spinning solvent, e.g., dimethylformamide, is first mixed with a nonsolvent with strong precipitation properties, e.g., water. The proportion of nonsolvent should be 2–8% by weight calculated on the total suspension.
2. The dimethylformamide/water mixture is then blended with the polymer (acrylonitrile homo- or copolymer) to produce a suspension. The polyacrylonitrile should have a K value of around 80, and the solids content can be up to 40%.
3. Before spinning, the suspension is converted into a stable-viscosity spinning solution of constant viscosity by heating it briefly to just below the solvent boiling point.
4. The spinneret apertures can be shaped as required, but the surface area should not exceed 0.2 mm^2.

Mild thermal conditions are needed to spin profiled fibers. The temperature of the spinning tube should be around 160°C, and the air/gas should be about 150°C. Figure 15 shows the cross sections of profiled fibers produced by this method.

Figure 15 Dry-spun profiled fibers.

Hollow Fibers. For a number of years now, melt-spinning and wet-spinning processes have been used to produce hollow fibers. As the fiber-formation process in dry spinning is somewhat complicated, not all production processes used for wet spinning can be transferred to dry-spinning technology.

In order to produce hollow fibers by the dry-spinning method, the viscosity of the spinning solution must be at least 120 sec (dropping-ball viscometer[*]) at 80°C. The ring in ring spinnerets should comprise three segments of equal size, positioned 0.2 mm from one another. The apertures should be less than 0.2 mm^2, and their arms should not be more than 0.1 mm wide. The air should be introduced in the center of the spinneret and flow counter to the direction of spinning.

The steps following spinning are the same as for other dry-spun acrylic fibers [8]. Figure 16 is a photo of dry-spun hollow fibers.

Absorbent Fibers. In an effort to improve the wear comfort of synthetic fibers, producers have accorded particular attention to absorbent fibers. These fibers absorb moisture from the air and from perspiration, either because they have good swelling properties (e.g., cotton) or because they have a special porous structure.

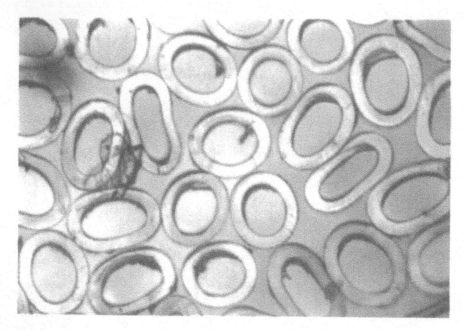

Figure 16 Dry-spun hollow fibers.

[*]Jost, K., *Rheol. Acta, 1*, 303–308 (1958). Ball diameter 0.125 inch, stainless ball, density 7.85, distance of drop 10 cm.

Fibers with good swelling properties can be produced by incorporating comonomers such as acrylic acid, methacrylic acid or dimethylaminoethyl methacrylate [28]. However, when swollen, such fibers do not transport the moisture quickly enough. One successful method of producing highly absorbent acrylic fibers is by creating a pore system inside the fiber. Here it is important that the porous core be protected by a sheath of suitable thickness to prevent problems during processing.

The sheath must contain a large number of tiny channels that transport the water into the porous core of the fiber. Absorbent fibers incorporating these features have high absorbency, do not swell, do not easily feel wet, transport moisture rapidly, and allow it to evaporate quickly [29]. Figure 17 shows scanning electron microscope images of an absorbent core-sheath fiber.

As mentioned in Section IIA2, when producing hydrophilic fibers the spinning solvent, e.g., dimethylformamide, must first be mixed with a nonsolvent of medium precipitation power. One can use 5–50% by weight nonsolvent (calculated on solvent and solids content). The greater the proportion of nonsolvent used, the more hydrophilic the fiber. Water retention values of 100–150% can be achieved (measured in accordance with DIN 53814) [6].

The solvent/nonsolvent mixture is then normally blended at low temperatures with acrylonitrile copolymer to form a suspension. The K value of the polymer should be about 80 and the concentration of the suspension should ideally be between 28% and 33% by weight.

The suspension is then converted into a stable-viscosity spinning solution by heating it briefly (3–5 min) at temperatures 30–60°C above the temperature at which the suspension looks homogeneous.

Suitable nonsolvents with moderate precipitating power include both solids and liquids, for example, single and multiple-substituted alkyl ethers and alkyl esters and polyvalent alcohols such as glycerin, diethylene glycol, triethylene glycol, and tetraethylene glycol. High-boiling alcohols such as 2-ethylcyclohexanol can also be used [6].

Absorbent acrylic fibers with a core-sheath structure and pore system are already available commercially. Bayer's Dunova is an example.

6. Special Spinning Methods

The spinning processes described in this section are of great interest in theory but have not yet achieved any technical significance. We will look in more detail at two special spinning processes:

Melt spinning
Gel spinning

Melt Spinning

Polyacrylonitrile. The melting point of the polyacrylonitrile monocrystal is around 320°C [30]. However, chemical changes start to occur below this

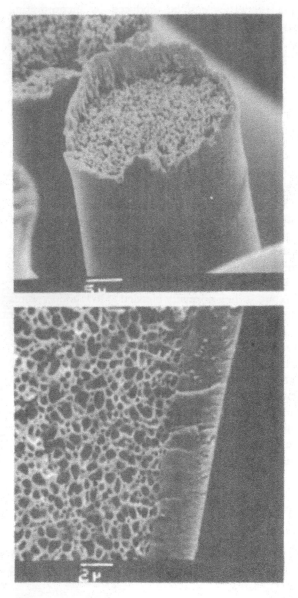

Figure 17 Absorbent core-sheath fiber as seen under a scanning electron microscope.

temperature, from around 200°C, in particular cyclization in air. In order to spin polyacrylonitrile from the melt, its melting point would have to be lowered by about 120°C.

Various attempts have been made to lower the melting point and/or melting viscosity of the polymer in order to develop a more economical solvent-free melt-spinning process for high-quality polyacrylonitrile fibers. The main reason for the high melting point of this polymer is its low entropy of fusion ΔS_m ($T_m = \Delta H_m/\Delta S_m$). The polyacrylonitrile molecule becomes helical as a result of strong electrical interaction between neighboring CN dipoles (see Chapter 7). The helixes are not destroyed in the melt but the intramolecular interaction becomes ineffective. The increase in entropy ΔS_m during melting is thus small. To bring the melting point of the polyacrylonitrile down to below 200°C the intramolecular interaction of the CN dipoles must be weakened. Chemically this can be achieved by copolymerization of acrylonitrile with other comonomers. The distance between the CN groups is thus increased.

Another way of weakening the interaction of the CN dipoles is by surrounding the CN groups with other dipoles. The H_2O molecule which forms hydrogen bridges to the nitrogen is the most mobile, takes up little space, and diffuses through the melt or solid relatively quickly. Apart from water, other suitable additives which form H bridges to the nitriles via their OH groups are glycols and glycerin. The reduction in dipole interaction depends on the dipole moment. One attraction of this method of weakening the CN dipoles is that the substance used, e.g., water, can be removed from the fiber afterwards. The properties with which the fiber is endowed during production are thus largely maintained.

Melt-spinnable polyacrylonitrile solutions can also be produced by incorporating, e.g., 5–20% low-molecular polyacrylonitrile by weight or an admixture containing an incompatible component, e.g., a polyolefin or polyamide (PA 12).

Copolymers. The melting point of the polymerizate can be lowered by incorporating a statistically distributed comonomer. This avenue was first explored in the early 1950s but did not lead to a breakthrough in this technology. Recently this work has been taken up again, primarily by Japanese companies.

Acrylic esters or methacrylic esters can be used as the comonomers. Emulsion polymerization can produce a latex which is easy to process. For example, 85% acrylonitrile and 15% methacrylic ester gives a polymerizate with reduced viscosity of 0.62. If 11% of the polymer is extracted with acetone, η_{red} increases to 1. The melt can be spun at 1500 m/min at 200°C. After hot drawing at a ratio of 4.5, a 1.8 denier fiber with a tenacity of 4.5 g/den is obtained [31].

It is well known that the heat shrinkage of drawn copolymerizates increases as the comonomer content increases. This is due to the weakening of the CN interaction by statistical distribution of comonomers. This reduces the glass transition point. However, this will only be viable on an industrial scale if the

copolymer molecules can be partially crystallized to counteract their tendency
to shrink.

Polyacrylonitrile Hydrates. The production of melt-spun fibers from a poly-
acrylonitrile/water system was first described in 1948 [32]. This led to the
production of fibers for paper production by a melt-spinning technique using a
polyacrylonitrile/water/carboxymethyl cellulose system [33].

In the early 1970s fiber manufacturers refocused their attention on these
developments following new findings on the theoretical side [34–36].

The phase diagram shown in Figure 18 shows that, depending on the temperature and water content, a one-phase melt can be produced from polyacrylonitrile hydrate. The minimum water content needed for an extrudable hydrate melt
can be determined with the aid of DTA. This shows the water content at which
all CN groups become uncoupled. To be on the safe side, an additional 7% water
should be added per polymer unit [33]. Figure 19 shows a plant developed for
melt spinning of polyacrylonitrile hydrate [37]. Excess water can be removed
from the polymerizate under pressure after washing. Additional drying of the
polymerizate then becomes unnecessary.

A pressure between 30 and 70 bar is needed for extrusion of the hydrate melt.
Pressure-resistant spinnerets with 60 μm apertures and an L/D ratio of 2 should
be used for spinning. In zone II the polyacrylonitrile hydrate melt is only stable

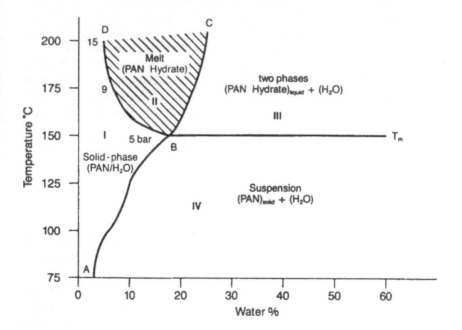

Figure 18 Phase diagram polyacrylonitrile/ water.

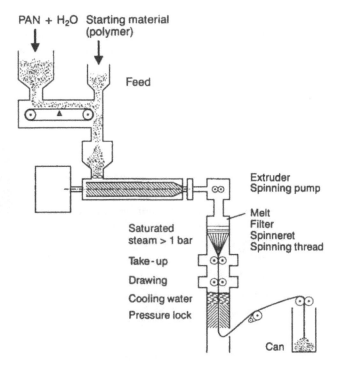

Figure 19 Melt spinning of a PAN hydrate (after [37]).

under pressure (see Figure 18). At the spinning stage it is important to release the water content (between 15% and 25% by weight) without destroying the filament. The spinning tube should therefore be under pressure to provide the atmosphere needed to evaporate the water. The difference between the pressure inside and outside the fiber should be positive but as low as possible. The evaporation rate should not exceed the rate at which the water diffuses in the filament.

The maximum obtainable spinning speed in a spinning tube of a given length is thus a function of the rate of diffusion of the water in the polymer. Drawing of the tow in the spinning tube immediately after spinning, if necessary in a two-step process, promotes evaporation of the water. In this way the fiber cross section is reduced by up to 100:1 to give finer deniers. After cooling to the temperature of the saturated steam (150°C), the tow should be passed through a cooling water zone before leaving the pressurized spinning tube through a pressure lock. The pressure lock may be created with running water. If necessary, the tow can be drawn before being coiled in the cans. It is then lubricated, crimped, and dried without tension at approx. 85°C.

Fibers produced in this way have a tenacity of about 4.5 g/den and a breaking elongation of 40%. The cross section is round with a core-sheath structure and the surface has a corrugated appearance. Fabrics produced from these fibers have a somewhat harsher, crisper hand.

In order to establish the economic and technical viability of this production process, the following factors would have to be investigated prior to scaling up for industrial plant:

How to overcome pressure problems
Maximum output per spinning position
Energy consumption
Probable quality of the fibers

An attempt has been made to lower the melting point of a polyacrylonitrile hydrate still further through copolymerization (10% of the monomers should have hydrophilic properties) and by restricting the degree of polymerization (6,000–16,000). Statistically distributed polar comonomers, e.g., methacrylic ester, block the nitrile groups so less water in needed to free the CN dipoles.

Gel Spinning of Polyacrylonitrile. In the late 1970s work started on the development of gel spinning of polyethylene. This technique allows production of high-tenacity fibers of high molecular weight from flexible molecular chains [38].

The synthetic fibers commonly produced have tenacities of only 4–12 g/den and an elastic modulus of about 40–120 g/den. However, these values are only 10–25% of what is theoretically possible. Much research has been focused on improving this situation.

The gel-spinning process is characterized by the use of

Polymers with ultrahigh molecular weights
Low-concentration spinning solutions

The tenacity of the fibers increases with their molecular weight. The low concentration of the spinning solution still allows high drawing ratios and this helps improve the technical properties of the fiber. Filaments produced by this method have a very high degree of crystallization and an almost perfect chain structure with an extremely high degree of orientation.

Various high-performance polyethylene fibers with tenacity values of around 30 g/den produced by gel spinning have been on the market for many years (e.g., Dyneema, Spectra). This led to the idea of using this technique for polyacrylonitrile fibers. This process is still at the development stage, but high-tenacity, high-modulus poylacrylonitrile fibers would be particularly interesting for the production of carbon fibers because the technological properties of carbon fibers are directly dependent on the quality of the precursor fibers. A number of patent applications have already been filed [39, 40].

According to Kwon [39] polyacrylonitrile homo- or copolymers with a mean molecular weight $\bar{M}_w$ of 500,000 (preferably between 1.5 and 2.5 million) are used to produce a 2–15% solution (depending on the molecular weight of the polymers). Dimethylformamide or dimethylsulfoxide is normally used as the solvent. The temperature is 140–180°C.

As Figure 20 shows, the solution is produced in two stages. Following thorough mixing in a conical double screw, the solution is pressed through the spinning pump with the aid of an extruder. The spinning pump conveys the solution through the spinneret where it is formed into filaments. The spinneret apertures should be 0.25–5 mm. The L/D ratio (length of capillary:diameter of apertures) should be at least 10 and preferably 20. The spun fiber is fed into a cooling bath at a predrawing ratio of 1:10 as in air-gap spinning. The bath temperature should be below the gel point of the polyacrylonitrile (0–50°C). It should be cooled at about 50°C/min.

The milky clear gel filaments are still extremely soft and sticky at this stage. In order to reduce the tendency to stick together, they are passed through a second bath in which solvent exchange takes place. They are then dried in mild conditions. After reduction of the primary solvent content (e.g., DMSO) in the second bath the filaments are in a fibrillar pseudoxerogel state, roughly equivalent to a solvent-free gel. These filaments can be wound onto a bobbin or drawn

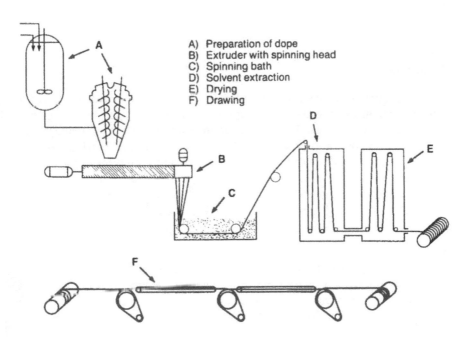

A) Preparation of dope
B) Extruder with spinning head
C) Spinning bath
D) Solvent extraction
E) Drying
F) Drawing

Figure 20 Gel spinning of polyacrylonitrile filaments (after [39]).

immediately. One- or two-stage drawing is carried out at 120–160°C at a drawing ratio of up to 1:30. Under optimal conditions, filaments with tenacities of 7–9 g/den and an elasticity modulus of 100–125 g/den can be produced [39]. Using the slightly modified process described by Sprenger [41], a tenacity of between 8 and about 20 g/den and an elastic modulus of 180–300 g/den can be achieved at 7% elongation.

C. Aftertreatment

The properties of the fiber in the spin cans at the end of the spinning process are unsuitable for their end use: the tenacity and modulus of elasticity are too low and the plastic elongation is too high. The fiber lacks the crimp needed to provide cohesion in processing and bulk in yarn. The fibers also contain residual amounts of solvent that have to be washed out. The fiber therefore must be aftertreated to obtain the required processing and performance properties.

Aftertreatment is a complex process made up of various steps. During this process the spinning material passes along a production line where it undergoes

Drawing (orientation and solidification of fiber)
Washing (removal of residual solvent)
Finish application (to protect surface and improve processing performance)
Drying
Crimping
Cutting
Heat setting (relaxation)
Packaging

The production line thus gives the fiber the properties required, for instance, tenacity, modulus of elasticity, elongation, shrinkage at the boil, degree of crimp, crimp stability, and bulk. However, in order to produce top-quality fibers it is just as important to ensure that all individual filaments passing along the line are treated as uniformly as possible across the entire working width.

Figure 21 is a schematic representation of a fiber production line [42]. At present, production lines for dry-spun acrylic fibers can process tows of up to 770 ktex (770 g/m) at speeds of up to 120 m/min. The output of such units is 110–120 metric tonnes a day [43].

Drawing

Before drawing, the chain molecules of polyacrylonitrile filaments spun at normal speeds have a low degree of orientation, so the filaments have high deformability. Drawing in thus one of the most important aftertreatment processes. It imparts the required strength and at the same time reduces breaking elongation and titer.

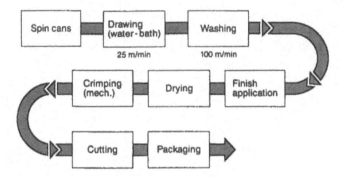

Figure 21 Conventional production process for polyacrylonitrile fibers (after [42]).

If the spun fiber is stretched between two clamps which move in opposite directions and the stress is recorded—i.e., if a stress-strain diagram is plotted—we can trace typical stress-strain paths. These vary according to the temperature and drawing conditions.

If the temperature is well below the glass transition point (104°C for polyacrylonitrile), the stress rises and the material has low breaking elongation (Figure 22a). However, if the temperature in not too far below the glass transition point (max. 40–50°C lower) increased elongation initially generates an elastic tension, until a draw point or bottleneck (Figure 23) is formed at a

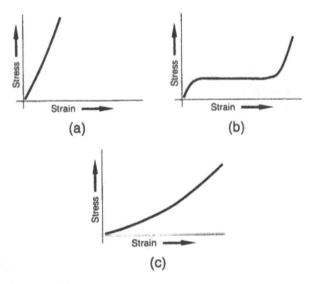

Figure 22 Stress-strain diagrams for spun fibers.

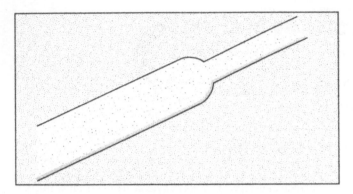

Figure 23 Draw point of a fiber (bottleneck).

random point. This bottleneck moves along the entire fiber. Meanwhile, the stress remains constant. It only starts to rise again when the bottleneck has moved the entire length of the fiber (Figure 22b).

The last case shown here (Figure 22c) occurs at drawing temperatures well above the glass transition point or if the spinning material is plasticized by the residual solvent content. The spun fiber stretches considerably with the stress increasing uniformly and far more slowly. This is what happens during drawing of polyacrylonitrile filaments with a residual solvent content of 5–25%. The drawing process is homogeneous and no draw point is formed.

At the molecular level, the drawing of spun fibers involves chain segments and aggregates (lamellae, fibrils) gliding past each other. This process is initiated by overcoming the initial resistance with the molecules "giving way" under tension. Once the "natural" draw ratio for heterogeneous drawing of the spun fiber has been exceeded, further deformation produces an oriented structure with relatively stable mechanical aggregates with a strong degree of cooperation separated by disordered intermediate areas. The oriented segments are joined by a varying number of tie molecules.

The drawing process pulls the molecules of the spun fiber so that they slide past one another until the molecular chains are aligned parallel to one another. The parallel arrangement of the molecular segments increases the number of lateral physical crosslinks due to dipole-dipole interaction and form a two-dimensional rod structure [44]. The polyacrylonitrile spun fiber is thus converted into a drawn fiber (amorphous-amorphous) with a larger number of physical crosslinks, but without undergoing crystallization. Structures of this type are particularly prone to cold flowing.

Drawing Technology. The spun fiber used in the production of staple fibers or tow is pulled upward out of a number of adjacent spinning cans over a can

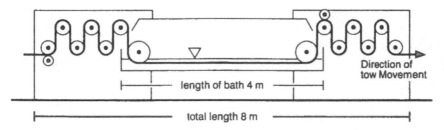

Figure 24 Wet-drawing unit.

creel and combined to form slivers with a spinning titer of 40–1200 g/m (ktex). Several slivers are conveyed side by side through the drawing unit.

Drawing is carried out as a one- or two-step process in drawing tanks containing boiling water (approx. 98°C) between two drawing frames. The drawing frames are fitted with pressure rollers that prevent the tow slipping (Figure 24). Each filament is surrounded by boiling water, ensuring that the material is heated uniformly. This also ensures dissipation of the heat generated during drawing. In addition to achieving the required drawing, the washing stage is initiated in the draw tanks. Circulation of the wash liquor ensures a uniform temperature throughout the tank.

The draw ratio is generally between 1:2 and 1:10 with a speed of 100 m/min leaving the draw unit. After stretching, the filaments have a boiling-water shrinkage of between 15% and 45% depending on the draw ratio, the copolymer composition and the draw temperature. By processing so as to retain this orientation, shrinkable fibers may be produced.

From a technical point of view, the conditions required for optimal drawing are

Smooth feed of the tow from the cans and formation of a homogeneous tow across the entire working width
Introduction of the tow into the drawing frame under very low, uniform tension
Maintaining the temperature in the drawing tanks as even as possible

These conditions are essential to ensure uniform drawing, thus preventing filament breakage, fluctuations in titer, elongation, and tenacity, and the occurrence of long fibers. After drawing, the sliver still contains residual solvent, which is removed at the next stage.

2. Washing

The washing stage is often carried out before drawing because the tow is conveyed relatively slowly at this stage, allowing a longer treatment time in the water, which in turn ensures more thorough washing [42].

On conventional units washing is carried out under tension. Using this method the solvent diffuses out of the fibers into the spaces between them. It is then transferred convectively to the washing liquor. The efficiency of this method is fairly low (around 20%), depending on the type of washing unit and the technology used.

In order to achieve a residual solvent content of <1% at the end of the washing stage on conventional aftertreatment lines, the fiber thus has to contain just <25% dimethylformamide at the start of the washing process. It has to pass through 20 washing zones at 90°C and the total treatment time should be 300 sec [42]. These theoretical figures illustrate the importance of the procedure and the design of the equipment in ensuring an effective washing process and realistic treatment times.

The washing unit shown in Figure 25 allows more effective removal of the solvent [43]. This unit comprises a series of baths or zones, each with a perforated drum or guide roller. The required number of baths can be selected by combining the necessary number on a modular principle.

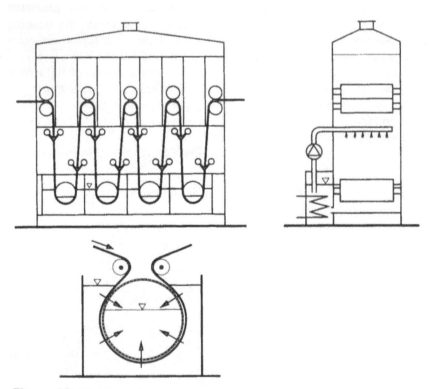

Figure 25 Washing unit.

The washing liquor, which flows countercurrent to the fiber material, flows into a container at the side of the washing unit where it is filtered, heated, and then pumped into spray units which spray it onto the oscillating tow. In this way the tow is kept at the right temperature even when it is not immersed in the liquor, the liquor is regularly drained, and there is no fluctuation in the concentration of the liquor. Before the tow is fed into the last zone, which contains clean water, the liquor is squeezed out by a pair of rollers. The washing units are closed systems, so any vapor released can be recovered and recycled.

3. Finish Application

Application of the finish (lubricant) is another important aftertreatment stage. A finish is applied to ensure smooth processing of the material on textile processing equipment. It should therefore smooth the surface of the fibers, but also ensure a certain degree of adhesion of the tow and prevent antistatic charges which impair the processing of water-repellent fabrics. Moreover, the finish should give the goods an attractive hand.

The amount of finish applied is as important as the type of finish. Depending on the intended processing route and end use, the solids content should be between 0.2% and 0.5%. A uniform amount of finish should remain on the fiber throughout subsequent processing stages.

The finish can be applied by roll coating, immersion, spraying or in a perforated drum. The roll-coating method is a two-bath system. Finishes applied by the immersion method are generally incorporated in the last bath on the aftertreatment line. A constant liquor level should be maintained by pumping finish solution continuously from the feed tanks into the bath and by using an overflow device. Application in a perforated drum allows the finish to penetrate the tow directly. It is thus a very efficient method of application.

4. Drying

After washing and application of finish, the tow still contains water and therefore has to be dried. Drying should reduce the moisture content to 1–2%.

The Drying Process. If drawn polyacrylonitrile tow is heated sufficiently for drying without tension it shrinks. This causes disorientation of the molecular chains in the amorphous regions and of the rodlike aggregates.

The structure of these regions also limits the degree of shrinkage. Shrinkage thus increases as the tow is heated above the glass transition point. The degree of shrinkage depends on the drying temperature and the density of the lateral physical crosslinking points formed during drawing.

Drying Technology. Drying of drawn tow can be regulated to allow free or controlled shrinkage or to maintain the original dimensions. Depending on the construction of the drying unit, the material can be dried by radiation, by contact heat, or by a gaseous medium. The type of contact heating common today involves running the tow over several large heated drums or calenders, changing

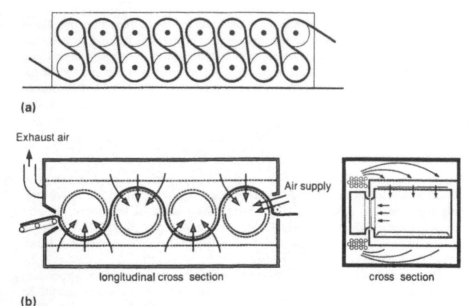

(a)

Exhaust air

longitudinal cross section   cross section

(b)

Figure 26 (a) Calender drier; (b) screen drum drier.

the side that comes into contact with the source of heat (Figure 26a). The speed at which these drums revolve can be regulated to control the degree of shrinkage.

Another commonly used drying unit is the screen drum drier (Figure 26b). This comprises a large rotating drum covered with a screen or sieve. Air is extracted from inside, heated and blown over the fiber material. Alternate sides of the material pass across half or more of each drum, depending on how they are arranged. The other half of the drier is covered. Normally up to 20–26 screen drums are combined to form a large drier with separate regulation of the heat and speed in each zone.

Polyacrylonitrile tow is usually dried at 120–170°C with throughput speeds of 100–150 m/min.

Since the tow contains water from the preceding washing process or finish application, the drying process also initiates the setting process. The degree of setting depends primarily on the drying conditions, i.e., temperature and time. Since a high degree of setting makes crimping more difficult, the drying temperature should not be too high.

5. Crimping

The crimp of staple fibers affects their running properties in spinning mills. In terms of importance, this stage is therefore comparable with the finish applica-

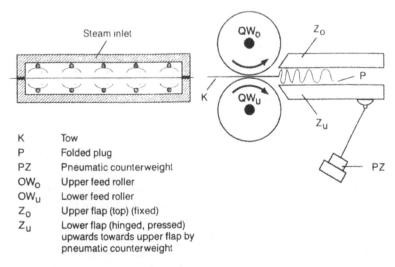

Steam inlet

K Tow
P Folded plug
PZ Pneumatic counterweight
OW_o Upper feed roller
OW_u Lower feed roller
Z_o Upper flap (top) (fixed)
Z_u Lower flap (hinged, pressed) upwards towards upper flap by pneumatic counterweight

Figure 27 Stuffer-box principle.

tion. The friction properties imparted by the finish and the crimp thus interact. The mechanism of interaction is somewhat complex.

Crimping. Crimping is generally carried out in a stuffer box crimper, where complex processes take place. The crimper basically comprises a pair of feed rollers followed by a stuffer box (Figure 27). The sides and top of the stuffer box are fixed, but the bottom flap is hinged.

The feed rollers draw the tow off the preceding units and push it—as a folded plug—against the pressure in the stuffer box. As the tow passes through the nip on the stuffer box side of the feed rollers it buckles (Figure 28). This forms the sawtooth crimp (microcrimp) needed for further processing. The "macrocrimp,"

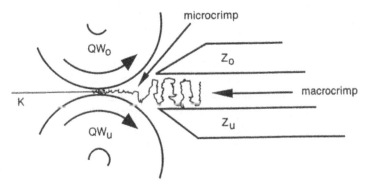

Figure 28 Crimp formation (microcrimp) and folding (macrocrimp).

i.e., the original fold, has no significance for processing technology. The stability of the crimp is as important as its appearance since the fibers are often exposed to strong forces during subsequent processing operations.

It has not yet been possible to produce a detailed description of what happens in the stuffer box. This is easy to understand if we look at the basic parameters:

Tow is fed into stuffer box at high speed (up to 150 m/min).

Tows have to be reduced from a width of approximately 1 m to around 100–300 mm; i.e., they have to divided and placed in layers.

The tow buckles constantly as it tries to avoid the pressure in the stuffer box.

The frequency recorded if it buckles at normal intervals is around 5 kHz.

Therefore, it is the nip between the feed rolls that is important as it is this that-imparts the crimp. The stuffer box itself only serves to build up the required pressure.

Crimping Technology. The width of the stuffer box varies between 40 and 360 mm. It can often be altered by changing the feed rollers and box.

Large units can crimp tows of up to 3×10^6 dtex. Tows with such high titers are usually wider than the feed rollers on the crimper. They can be narrowed by passing them over concave reels or, as described above, by separating them and layering as they are fed into the stuffer box.

Good-quality crimp can only be produced if the tow is homogeneous across the entire working width. It is also necessary to ensure that the temperature remains constant before and during crimping.

The tow must therefore be heated before it is fed into the stuffer box.

The crimping process basically follows the Euler buckling model according to which crimp capacity can be regulated by modifying the modulus of elasticity, i.e., by altering the temperature of the tow. For this purpose, a steaming zone can be incorporated before the stuffer box. Saturated steam is sprayed onto the tow in this zone to heat it to 60–70°C before it is fed into the stuffer box for crimping. The steaming zone must ensure uniform heating of the individual filaments so that the moisture content remains uniform. This is important for the slip/stick properties of the fibers. A constant temperature must be maintained during crimping.

During crimping, the fiber/fiber and fiber/metal friction generates heat, which must not be allowed to cause a localized increase in temperature. This would cause deformation of the filaments and fluctuations in the degree of crimp. The surfaces of the feed rollers must therefore be kept at a constant temperature by incorporating a closed heating/cooling circuit. This should ensure crimp uniformity even at working widths of 350 mm and tow titers of 3×10^6 dtex. A uniform moisture content and a uniform finish application simplify crimping by making it easier to avoid localized temperature increases.

Lastly comes preliminary setting of the crimp. This is done by spraying steam into the unit at some distance from the godet nip. Final setting of the fiber (described below), which gives the crimp its final stability, is carried out by steaming after cutting the tow into staple.

6. Cutting

Like other synthetic fiber yarns, polyacrylonitrile fiber yarns are normally produced by the conventional methods used to manufacture yarns from natural fibers. The thick tow, which may comprise several million filaments, must first be converted into staple fibers. This is generally done by cutting. Polyacrylonitrile staple fibers can then be processed into yarn either on their own or in blends with natural fibers. The staple lengths required for the production of staple yarns and flock range from 0.1 mm for flock to 180 mm for carpet yarns. For cotton system processing, the staple length should be as uniform as possible. For woolen and worsted processing, variable-cut staple is commonly sold.

Cutting Technology. A variety of different types of cutters are in use. For a long time, the tow was introduced into slotted wheels where it was cut by rotating blades. Now it is usually cut by a ring with outward-pointing blades arranged around the circumference. A pressure roll presses the material against the blades and after cutting the fibers fall inward through the center of the ring. The principle of this machine is illustrated in Figure 29. Machines of this type can process 3 metric tonnes/hr at titers of 1×10^6–3×10^6 ktex and throughput speeds of 120 m/min.

Production of Stretch Broken Tow. It is somewhat ironic that the parallel filaments in the tow are cut and supplied to processors as a disordered mass which has to be realigned before it can be turned into yarn. Quite early, the

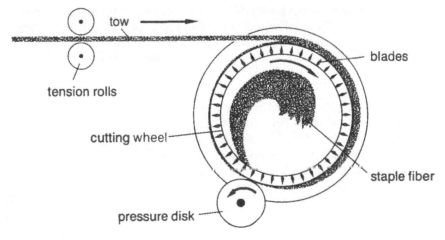

Figure 29 Principle of cutter.

industry therefore began supplying polyacrylonitrile fiber as uncut crimped tow. Processors then divide the tow into fibers without altering the parallel alignment. This is done either by passing the tow through screw-shaped blades under tension or by tearing it apart between rollers running at increasingly high speeds (stretch breaking). Chapter 10 provides a detailed description of the stretch breaking process. While staple polyacrylonitrile fibers can be blended with natural fibers before the first processing stage, when processing tow, blending can only be carried out after production of the sliver.

7. Heat Setting

The setting of the tow during drying is not sufficient to relieve all the internal stresses caused by drawing. The filaments only achieve their final state of thermodynamic equilibrium after tensionless heat setting.

The Heat-Setting Process. During heat setting, the bonds between the chain molecules, which were frozen under tension, are given greater mobility through heating. As a result, some chain segments change position. The chains are thus converted to a low-energy and thus more relaxed state. This process is carried out with steam and can reduce the boiling shrinkage of polyacrylonitrile to 0.5%–1.5% depending on the heat-setting conditions.

Tensionless heat setting thus alters the length and titer of the fiber. As demonstrated below, there is also a pronounced flattening of the stress-strain curve, tenacity decreases, and breaking elongation increases. Both of these effects are intensified by increasing the heat-setting temperature. Heat setting without tension thus improves the thermal stability of the fiber at the expense of mechanical stability.

Stress-strain diagrams or, to be more precise, their first derivatives are a useful source of information on the structural changes induced by heat setting. We have already seen how drawing affects the mechanical properties of polyacrylonitrile fibers. These influences are manifested, for instance, by a steep stress-strain curve. Even more information can be obtained from the first derivatives, which we know correspond to the effective modulus of elasticity at a given elongation.

Figure 30 shows the stress-strain diagrams and their first derivations for drawn specimens and for specimens after relaxation in saturated steam at 130°C. Fibers made from pure PAN (specimen A) have the steepest stress-strain curve, followed by the curve for the copolymer with a low comonomer content (specimen B) and lastly the copolymer with the high comonomer content (specimen C). Heat setting only causes a slight change in the curve for the fibers made of PAN, while the curves for the steamed fibers made of the two copolymers flatten out markedly.

In the derivative stress-strain diagram, the elastic modulus first falls but then rises to a maximum. This is more pronounced in some cases than in others. It is highest for A (PAN), followed by B and C (copolymers). During heat setting in

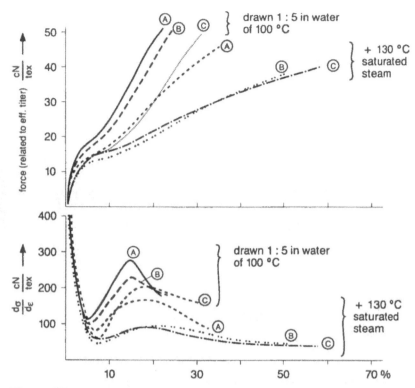

Figure 30 Stress-strain diagrams, together with their first derivatives, of untreated and heat-treated fibers.

saturated steam the maximum of the copolymers B and C flattens out far more than the maximum of A.

The first drop in the elastic modulus can be explained by the disentangling of the molecule segments within the elongation time. The subsequent rise is due mainly to an increase in the tension in the tie molecules between the regions where the chains are more densely packed. These tie molecules are stressed on further elongation, so the modulus must increase by the same amount until the tie molecules tear or are pulled out of the densely packed regions.

On heat setting with saturated steam, the fiber structural differences are accentuated: the more densely packed areas can be distinguished more clearly from the others. The number of tie molecules decreases during relaxation so the elastic modulus maximum based on their tension flattens out and shifts toward higher temperatures. This effect is much less marked with the PAN than with the copolymers. This can be explained by the fact that, in the case of the PAN, a more strongly cooperating molecular network is formed during the drawing

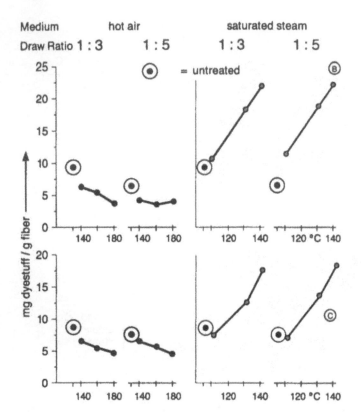

Figure 31 Effect of heat setting on dyeing rate.

process because the molecular structure is not disturbed to the same extent. This reduces the scope for mobile segments to change their position [45]. The structural change that takes place during heat setting can also be identified by x-ray technology [21].

An examination of dyestuff diffusion shows that the amorphous regions are loosened by heat treatment. Figure 31 shows how heat setting increases the dyeing rate of fibers as a function of the dyeing temperature for copolymers B and C [46].

Heat-Setting Technology. The drying units shown in Figure 26 are basically suitable for heat setting too. However, one of the most widely used is the perforated conveyor steamer (Figure 32). The staple fiber is laid evenly on a perforated conveyor belt which transports it through the various zones in the steamer. Steam is fed separately into each zone so that it comes into contact with the fiber uniformly at right angles.

PAN tow

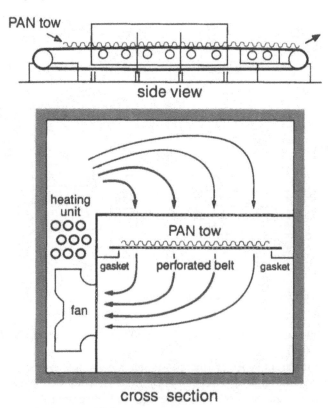

side view

cross section

Figure 32 Perforated belt steamer.

8. Packaging

As with all other industrial products, the last stage in the production process is packaging. Polyacrylonitrile staple fibers are pressed into cubic or rectangular bales in a bale press. The bale press (Figure 33) comprises a prepress and a final press (baler). The material is first conveyed into the filler pneumatically down a chute, then pressed into the baling box of the prepress. Prepressing compresses the fiber material until it is compact enough to form a bale. The press box then swings round and deposits the material under the final press where it is stamped into its final shape. Meanwhile, the empty prepress box is refilled in the prepress (Figure 33). The bales are wrapped and strapped before leaving the bale press. A fully automatic bale press has an output of 120–140 metric tonnes a day and can produce bales weighing up to 500 kg.

Tow is coiled sideways by a special machine and stamped into shape. These days tow bales generally weigh 400 kg. Bale weights as high as 1,000 kg are

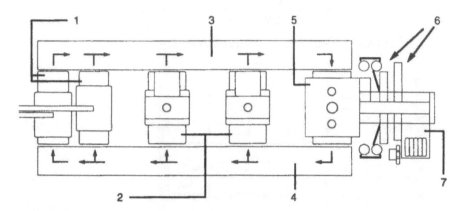

1) Tow plaiter
2) Full can conveying system
3) Staple fiber prebaler
4) Conveyor for empty cans

5) Central mainbaler
6) Wrapping and strapping devices
7) Bale discharge

Figure 33 Combined system for staple fibers and tow baling.

now being produced as the larger bales save packaging and allow an increase in the stretch breaking time [47].

The larger number of different types of fibers available means that synthetic fiber producers market a wide range of products. They therefore need extensive storage capacity. However, since both warehousing and dispatching are extremely labor intensive, more and more manufacturers are investing in automated high-bay warehouses. Computerization of these warehouses ensures that the goods are stored correctly and that orders are dispatched on time. Moreover, continuous monitoring of inventories also allows more cost-effective utilization of production capacities.

Having looked at the technology used in the various aftertreatment stages, let us now look at some of the process parameters, e.g., during the drawing, drying and crimping processes, which can be varied to produce a range of commercially interesting fibers.

Shrinkable Fibers. On account of their rodlike paracrystalline structure, acrylic fibers can easily be used to manufacture shrinkable fibers. Shrinkable fibers are blended with nonshrink fibers to produce yarns which obtain high bulk once their shrink capacity has been activated by heat treatment. There are two methods of producing shrinkable fibers:

As staple fiber by selection of appropriate drawing and aftertreatment conditions

As tow by processing on a stretch-breaking machine

In the production of shrinkable staple fiber, the desired degree of shrinkage can be obtained by regulating the drawing ratio and temperature and by subsequent drying in mild conditions. Two different levels of shrinkage are normally available: fibers with about 20% shrinkage and high-shrinkage fibers with 35–40% shrinkage.

There are a number of patents containing detailed descriptions of the production process [48–50]. It has proved advantageous, for example, to carry out drawing in two stages using different temperatures and draw ratios. The fiber should be crimped in the wet state. Drying temperatures should not exceed 70°C. Two important stages in the production of shrinkable fibers are the densification of the filament and the removal of the solvent by mild drying and treatment in hot water or steam, partly before drawing and partly between the two drawing stages. The preconditions for manufacturing fibers without macrovoids are negative birefringence and a density of over 1.180 g/cm^3 [37]. Shrinkable fibers produced using the process described in [50] have good textile properties and have a tenacity of at least 2 g/dtex at 35% to 40% shrinkage.

Shrinkable fiber can also be produced as follows: The tow made after the common spinning and aftertreatment process can be afterstretched on a calander.

A considerable proportion of shrinkable acrylic fibers are produced by drawing tow on stretch-breaking machines. The stretch-broken tow gives a shrinkable sliver, which is processed together with nonshrink slivers to produce high bulk yarns. Generally speaking, the sliver shrinkage obtained on the stretch breaker lies between 15% and 24%. By taking special measures during production of the tow and during stretch breaking, it is possible to produce slivers with shrinkage values of up to 24% and 40%.

Colored shrinkable slivers can also be produced from dope-dyed tow after the stretch-breaking process.

D. Continuous Production

In an effort to gain an advantage in the fierce competition on the market for synthetic fibers, over the past few decades R & D efforts have mainly been concentrated on reducing production costs and increasing the quality of fibers. There are basically two ways of achieving this:

1. Improve the productivity of existing technology, but the returns on this approach diminish with each incremental improvement.
2. Reduce production costs through a fundamental change in production technology.

In order to assess the advantages and disadvantages of new technology, we must first look at the characteristics of both the old and the new production

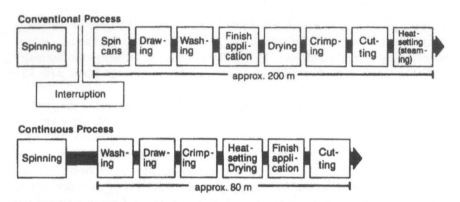

Figure 34 Conventional (discontinuous) and continuous process for the production of polyacrylonitrile fibers (after [42]).

processes. Figure 34 outlines both processes. The main characteristics of the conventional (discontinuous) process are

Interruption between spinning and aftertreatment
Linear or drawn product guidance under tension
Low processing speeds (up to 150 m/min)
High tow weight

Some processes also feature a second interruption between the wash-draw and drying stages.

Because the product is transported under tension, productivity is restricted by the treatment times required to remove the solvent and for drying and by the length of the plant.

In order to reduce these negative factors, the working width and feed density of aftertreatment lines were increased. However, the possibilities were soon exhausted. Attention thus turned to the second option, leading to the development of a new continuous dry-spinning process [42, 51] reported by Bayer.

In the continuous process the situation is completely different:

No interruption in production
Tensionless product guidance
High processing speeds
Low tow weight
Combining of several steps

As Figure 34 shows, this type of plant is far smaller than one utilizing conventional machines (lower capital expenditure). Continuous plants also allow cost-effective production of specialty fibers. Figure 35 shows the operating costs of

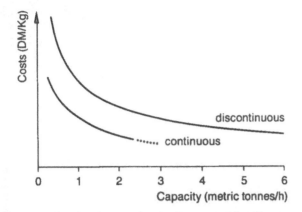

Figure 35 Influence of aftertreatment capacity and comparison of process alternatives in relation to production costs.

continuous and discontinuous processes in relation to throughput. The comparison shows that the continuous process is more economical, even for low production rates.

Completely new machinery had to be developed for this process. For example, a new washing unit had to be developed to improve the tensionless removal of the solvent. By means of a mathematical model and extensive tests, the conditions needed for a new type of washing unit for the continuous process were established. The unit illustrated in Figure 36 is able to process the product directly from a machine spinning at a rate of 300 m/min and at the same time the treatment times can be varied from 4–8 min as required.

The top diagram shows a longitudinal cross section of the machine. A special device loads the tow directly into an oscillating container which acts as a conveyor. At various stages during its passage through the machine the material is sprayed with wash water which is circulated through the machine countercurrent to the product. After washing, the tow is removed, any residual liquor is squeezed out and it then proceeds to the next stage, drawing. The washing efficiency of this apparatus is 50–60%.

Similarly, new machines had to be produced for the drawing, crimping, and drying processes. Crimping of high tow weights at high speeds required the development of a modified aerodynamic crimping nozzle. This makes it possible to reduce the high drawing speed of 1000–1500 m/min to the necessary level (Figure 37) [52].

By laying the crimped fiber sideways in a drier with a larger working width, transportation speed was reduced even further to 0.5–5 m/min. As in washing, this allows variable treatment times of 5–15 minutes.

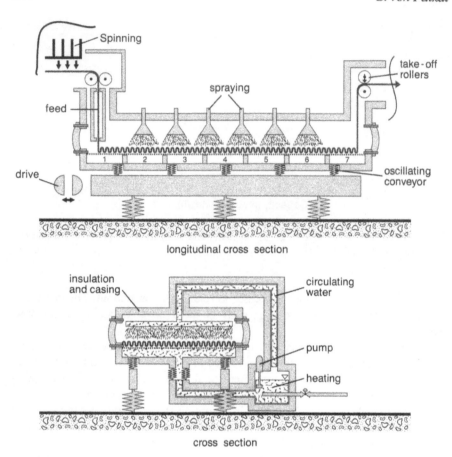

longitudinal cross section

cross section

Figure 36 Apparatus for low-tension transportation of product during washing process (after [42]).

Positive experience of the use of oscillating conveyors in the washing and drying steps of fibers, thereby obviating the need for moving components to come into contact with the fiber, and ensuring good densification of the fiber at the washing stage, led to the development of a new drier based on this principle.

It was this development that brought the breakthrough to more realistic treatment times. The entire drier operates in a partial vacuum so that

No process air can escape into the production facility.
Minor leakages hardly dilute the process air at all.

Figure 38 shows a cross section of the drier. The illustration shows that the ventilator casing, heat exchanger, and oscillating conveyor/ventilation channel

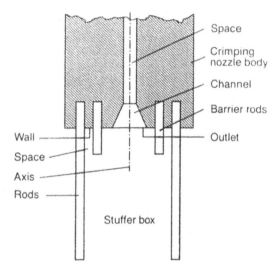

Figure 37 Diagrammatic representation of an aerodynamic crimping nozzle for high crimping speeds (after [52]).

for each zone form a separate, completely closed system within the machine. The extraction side of the ventilation system is open inside the insulated outer machine casing and is used to superpose the air which is transported in counter-current to the product.

Figure 39 compares the costs of the continuous and discontinuous processes [42]. One advantage of the discontinuous process is that it has high capacity at the aftertreatment stage. It also makes lower demands on spinning technology because the interruption of the process makes it possible to check the quality of the spun fiber and to ensure a satisfactory level of consistency. This last advantage, however, also produces certain disadvantages:

Higher personnel requirements
Lower product yield as a result of container weight nonuniformity
Lower on-stream time
Less consistent quality due to the nonsteady product treatment as it enters and leaves the line

The continuous process eliminates these limitations of dry spinning which have existed since its inception.

The continuous dry-spinning process thus combines the advantages of the wet-spinning process, i.e., lower personnel requirements and higher yields, with the advantages of dry spinning, i.e., smoother touch and feeling and also agree-

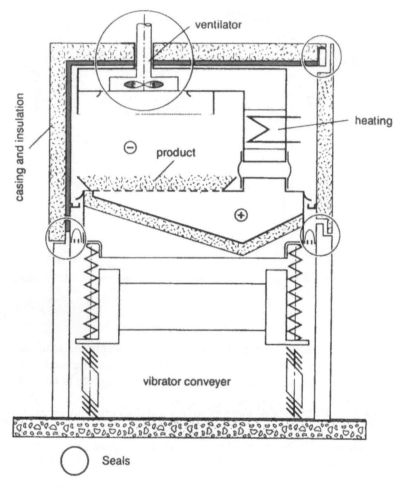

Figure 38 Cross section of a new drying system (after [42]).

able appearance products and considerably lower energy requirements, as illustrated in Figure 39.

A further step in this direction would be removal of the solvent from the fiber
by a dry process instead of using water [53].

E. SOLVENT RECOVERY

Dry-spinning processes require two to four times as much solvent as PAN, while
wet-spinning processes require three to six times as much solvent. Since sol-

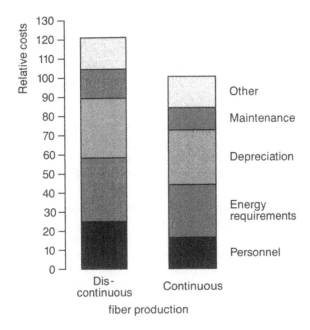

Figure 39 Comparison of costs (after [42]).

vents are comparatively expensive and a great deal of energy is needed to recycle them, the cost efficiency of the complex production process depends on the efficiency of the solvent recovery process. Since dimethylformamide is most commonly used for the dry-spinning process, we will take this solvent as our example in an overview of the solvent recovery processes currently available.

The solutions used in the dry-spinning process contain about 30% PAN by weight; i.e., the solutions contain 2.33 kg DMF/kg polymer of which 97–99% can be recovered. The remaining 1–3% is lost through emissions from the dissolving, spinning, and washing units, in the effluent, as part of the product (fiber) or in production waste.

Two different solvent concentrations are released from dry-spinning units. Eighty five to 90% of the total solvent content is evaporated in the spinning tube, where it condenses. This solvent is relatively pure but contains some water.

The other 10–15% is recovered from the subtows and in exhaust air. At the washing stage, which takes considerably longer for the fibers produced by dry spinning than for the more porous wet-spun types, low-molecular polymers and salts are washed out of the fiber, as well as dimethylformamide. The wash water contains 5–20% dimethylformamide and about 0.5% suspended solids.

DMF tends to hydrolyse in aqueous solutions at elevated temperatures to dimethylamine and formic acid. This process is catalyzed by acids or alkalis. The mixture should therefore be neutralized before distillation [54]. The normal rectification process is carried out in a column under 0.3 bar pressure to reduce the thermal exposure. Despite this mild treatment, about 0.04% of the DMF decomposes.

Separate rectification of the different concentrations is not economically viable. Mixing the two solvent-water mixtures together gives a solution with a dimethylformamide content of 50–55% by weight. This mixture can then be rectified. Alternatively, the strong and weak liquors are fed into the distillation column at different levels [55].

Some manufacturers evaporate dilute dimethylformamide solutions (to a concentration of about 60% DMF) before rectificatien to reduce the amount of water in the recovery units. A multistep vacuum evaporation unit is needed for this preliminary treatment [56]. Alternative methods of concentrating the solution could be considered, for example distillation with a vapor compressor or a reverse osmosis process, analogous to the desalination of seawater [55].

The preconcentration unit comprises at least two distillation columns in series. Solutions with solvent contents of up to 60% by weight can be treated economically in these columns. The concentrated mixture is then mixed with the condensate from the spinning tube, producing a solution with a DMF concentration of 85–90%. This is fed into a separate distillation column where the water is drawn off as the head product in the form of steam. Dimethylformamide is removed as a side product from the lower third of the unit. The less volatile components are found in the sludge in liquid or solid form. This is removed from the system separately. It is also in the interests of the manufacturers that the water from distillation plants is as pure as possible so it can be reused in the production process. The heat energy produced during distillation is recycled; condensing the overhead water may be done using the column feed.

Figure 40 shows a solvent recovery plant comprising a preconcentration unit and a distillation column. At the end of the distillation process the quality of the dimethylformamide and the water is checked. The water content of the dimethylformamide should not exceed 0.02% by weight, while the acid content (calculated as formic acid) should not exceed 50 mg/L. If the water content of the dimethylformamide is too high, the viscosity of polyacrylonitrile spinning solutions is reduced, causing problems during production. The head water should not contain more than 0.1% by weight DMF.

These days, multivariable adaptive control is used to set the process parameters and minimize energy consumption in the distillation column.

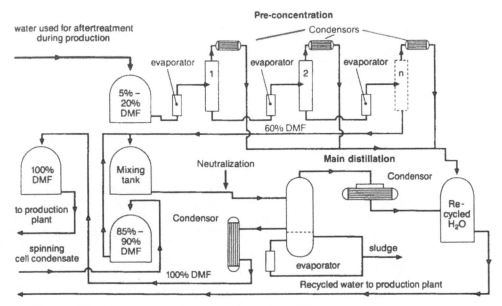

Figure 40 Diagram of a DMF recovery unit (after [56]).

Air, steam, and water mixtures with a very low solvent content are generated during the production of polyacrylonitrile fibers. Economic considerations determine whether these go through the normal solvent recovery process or whether a special recovery process is used, e.g., absorption. Alternatively, liquid waste can be treated in a biological wastewater treatment plant.

F. INFORMATION SYSTEMS AND QUALITY ASSURANCE IN PRODUCTION

The success of a product on the market depends largely on its quality. If it does not meet today's exacting demands it has little chance of surviving in our competitive world. Therefore all large fiber manufacturers, and most acrylic fiber manufacturers are large because of the relative complexity of the production process, have set up systems to help them monitor and assess the quality of their products. Two of the main instruments used are:

Information systems
Quality assurance systems

These days, marketing tends to follow this pattern: an enquiry about fiber types, quantities, specifications, delivery times, etc., is received electronically and should automatically initiate a response giving details of the above, together with the price. The price is determined by production costs and the competitive market price. The customer then decides whether to accept the offer. If accepted, the manufacturer supplies the quantity and quality ordered on time. These days, all these stages should be automated and require little manual input, thereby reducing the likelihood of errors. This can, however, only be ensured if the fiber manufacturer has an extensive information and quality assurance system.

1. Information Systems

Anyone who knows how complicated it is to produce acrylic fibers will realize just how difficult it is to manage all the data from the administrative, technical, and marketing sides. Producers therefore structure their data in logical levels:

All processes requiring an extensive interactive data exchange are grouped in levels.
All tasks carried out with the same equipment and programs are grouped within these levels.

Figure 41 shows the various levels and the flow of information [57]. When comparing production and marketing, it should be noted that in the production plant processes are largely automated, relying on computerized controls, sensors, and activators while marketing still relies far more heavily on personnel-intensive processing of information.

Figure 41 Flow of information model showing various levels (after [57]).

This description of how an information system functions is based on the production of dry-spun acrylic. The following factors were taken into account when developing the system:

User friendliness
Reliability
Possibility of continuous recording
Rapid response
Possibility of expansion
Compatibility

The description of this system can be broken down into the following subsections:

The components of the system in the production plant
Hardware
Interaction with other information systems

The Components of the System in the Production Plant. A production plant for acrylic fiber is normally divided into various sections:

Polymerization
Dissolving
Spinning
Aftertreatment
Distillation
Laboratory
Warehouse

The objective is to provide each level in the hierarchy, from the control room to the plant manager, with the necessary information.

The information system used in the production plant is part of the range of information systems available in the company. In practice its functions are broken down to serve the various sections within the plant.

Hardware. The various programs are stored at various levels in the computer system. Figure 42 shows how the functions are assigned to the various hardware levels. All data are stored in the central computer, which also houses the evaluation system and sets up links with other systems. The control room is equipped with computers which are suitable for the operating conditions. The process control systems run off their own computers. These regulate and monitor the plant on the basis of predefined values and recipes, measure tolerances, and indicate when the process deviates from these.

Links with Other Information Systems. In principle, any information system can be linked to any other system. This allows information to be relayed to and from the marketing, production, and quality assurance systems.

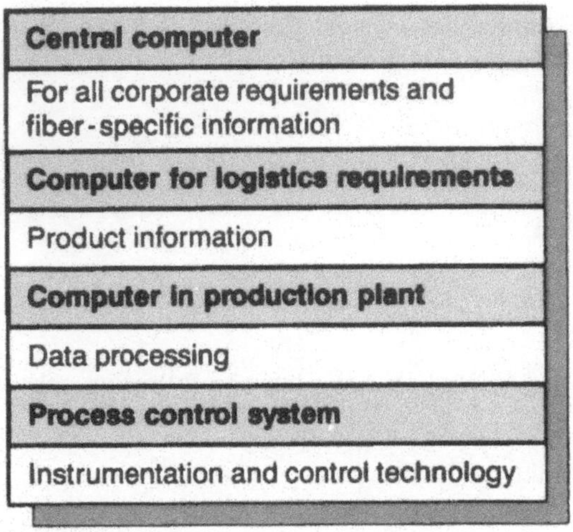

Central computer
For all corporate requirements and fiber-specific information
Computer for logistics requirements
Product information
Computer in production plant
Data processing
Process control system
Instrumentation and control technology

Figure 42 Breakdown of functional levels.

The production system regularly provides information for quality assurance and to the test laboratories and pilot plants. Similarly, data from quality assurance is transmitted to the production system. For example, inspection of a fiber lot requires a large number of tests in the laboratories. These tests are carried out on samples taken at the polymerization, dissolving, spinnning, aftertreatment, and distillation stages. Sometimes tests are carried out in the pilot plant, too. The results are transmitted to various parts of the company. Rapid and reliable collation of this extensive body of data, transmission of data to the production plants, and release of the batch for sale can all be optimized by using a computerized system.

Experience. Computerized data processing speeds up the flow of information to the plant management and allows a more rapid response to errors. Production data can also be monitored 24 hr a day without any manual intervention. This reduces fluctuations in the production process, thus reducing errors and improving quality and yield.

2. Quality Assurance

Quality assurance has become an increasingly important aspect of industrial production. Fiber manufacturers have therefore had to set up a proper quality assurance system. Quality assurance helps

Reduce the risk of supplying faulty goods
Companies counter complaints and claims for damages
Optimize costs
Manufacturers remain competitive by improving the quality of their goods

Quality assurance helps manufacturers ensure that they supply products that meet specifications.

The large amount of data collected for quality assurance purposes in the production, marketing, customer service and R&D sections makes a computerized data processing system virtually essential in large factories of the sort used to produce acrylic fibers.

III. SPECIAL PROPERTIES OF DRY-SPUN POLYACRYLONITRILE FIBERS

Chapter 4 describes the relationship between the chemical and physical structure of wet-spun polyacrylonitrile fibers and their textile properties. Let us now look at the properties that are characteristic of dry-spun fibers. The spinning method used, i.e., wet or dry spinning, does not necessarily provide any clue as to the properties of a fiber. What is important is the overall production process (type of comonomer, polymerization, spinning, drawing and aftertreatment conditions), since in the final analysis this is what determines the properties of the product. This makes it difficult to compare the properties of wet- and dry-spun fibers since in both methods it is possible to vary the textile properties of the product. However, here is an overview of the main properties of the fibers produced by the two methods.

- Good lateral tenacity is easier to achieve in dry-spun fibers because the void structure is finer which implies more lateral bonds. This allows more freedom of choice of aftertreatment lines and process parameters and the order in which the various steps are performed. Any differences in properties can, of course, be evened out by heat setting (relaxation) at the end of the process.
- The dog-bone-shaped cross section of dry-spun fibers means that they generally have better covering power, better bulk, better elastic recovery, and a fuller head. Wet-spun fibers tend to be softer and fluffier, especially at fine titers of between 1.2 and 1.3 dtex.
- Dry-spun fibers have a smooth to slightly structured surface (see Figure 14), whereas wet-spun types are rougher and more clearly structured. One direct result of this difference is that the amount of finish needed to ensure good processing properties varies between the two types. Dry-spun fibers need add-ons of 0.2–0.4%, while wet-spun types need far more, 0.4–0.7%. The surface also affects the luster of the fibers. Smooth, dry-spun fibers are generally brighter than their rougher wet-spun counterparts.
- There are also slight differences in stretch-broken tow produced by the two spinning processes. Wet-spun tow has a slight advantage in that it allows more reliable processing, but it does not give the same volume and bulk as dry-spun material.

These differences mainly affect the appearance of the fibers since they manifest themselves in the volume, bulk, luster, and hand of the finished goods.

ACKNOWLEDGMENTS

I would like to express my gratitude to Dr. U. Reinehr and Dr. C. Pieper, who were of great help to me in writing this chapter. I would also like to thank Dr. G. Spilgies of Bayer AG, Dormagen, Germany, who provided the electron scanning microscope images.

REFERENCES

General References

Ullmann's Encyclopedia of Industrial Chemistry, 5th ed., Vol. A10, *Fibres*, 3. *General Production Technology*, 4. *Synthetic Organic*, VCH Verlagsgesellschaft mbH. D-6940 Weinheim, 1987.

Ziabicki, A., Physical fundamentals of the fiber-spinning processes, in *Man-made Fibers, Science & Technology*, 3rd ed. (H. Mark et al., eds.), Interscience, New York, 1968.

Cernia, E., Acrylic fibers, in *Man-made Fibers, Science & Technology*, 3rd ed. (H. Mark et al., eds.), Interscience, New York, 1968.

1. Lathan, R. G., and Houtz, R. C., Du Pont de Nemours & Comp., USP 2,404.714, 2,404,717, 2,404,720, June 17, 1942.
2. Klare, H., *Geschichte der Chemiefaserforschung*, Berlin, 1985, pp. 180–186.
3. Rein, H., I. G. Farbenindustrie AG. DPA 72.024 IVC/39b, April 13, 1942.
4. Fourné, F., *Chemisfasern Textilind*, *21*, 369–379 (1971).
5. Reinehr, U. et al., Bayer AG., USP 4,497,868 (1985).
6. Reinehr, U. et al., Bayer AG., USP 4,239,722 (1980).
7. Reinehr, U. et al., Bayer AG., USP 4,810,448 (1989).
8. Reinehr, U. et al., Bayer AG., USP 4,432,923 (1984).
9. Runge J., *Acta Polym.*, *33*, 708–713 (1982); *34*, 631–636 (1983).
10. Engelhard, A. et al., Bayer AG., German Federal Republic patent 3,333,145 (1985).
11. König, J. et al., Bayer AG., German Federal Republic patent 3,333,146 (1985).
12. Aurich J. et al., VEB Friedrich Engels, German Democratic Republic patent 215,342 (1984).
13. David K. H. et al., Bayer AG., USP 4,286,076 (1980).
14. Höroldt, E. et al., Hoechst AG., German Federal Republic patent 3,048,059 (1982).
15. Jost, K., Reyon, Zellwolle u.a., *Chemiefasern*, *9/29*, 320–327 (1959).
16. Falkai, von B. (ed.), *Synthesefasern*, Verlag-Chemie, Weinheim, 1981, p. 38.
17. Labudzinska, A., and Ziabicki, A.:*Kolloid Z.*, *21*, 243 (1971).
18. Petersen, K.: *Dissertation der Universität Hamburg*, 1986, p. 16.
19. Falkai, von B. (ed.), *Synthesefasern*, Verlag-Chemie, Weinheim, 1981, p. 204.
20. Ohzawa, Y.. Nagano, Y., and Matsuo, T.: *J. Appl. Polym. Sci.*, *13*, 257–283 (1969).
21. Falkai, von B. (ed.) *Synthesefasern*, Verlag-Chemie, Weinheim, 1981, p. 208.
22. Reinehr, U., Häfner, G. and Nogaj, A.: *Chemiefasern, Textilind.*, *35/87*, 588–592 (1985).

23. Eastman Kodak Company, USP 2,811,409 (1957).
24. Reinehr, U. et al., Bayer AG., USP 5,015,428 (1991).
25. Du Pont de Nemours & Comp. USP 3,458,616 (1969).
26. Reinehr, U. et al., Bayer AG., USP 4,842,793 (1989).
27. Okamoto, M.: *Chemiefasern Textilind.*, *29/81*, 30–34 (1979).
28. Jap. Exlan., German Federal Republic patent 2,434,232 (1974). Japan Prior (1973).
29. Körner, W. et al., *Chemiefasern/Textilind.*, *29/81*, 453–462 (1979).
30. Hinrichsen, G., *Angew. Makromol. Chem.*, *20*, 121 (1971).
31. Mitsubishi, Japan patent 1,160,416, 27.12 (1984).
32. Sumitomo, German Federal Republic patent 1,221,802, 15.06 (1964).
33. Du Pont, USP 2,585,444 (1948).
34. Monsanto, USP 3,402,231 (1964).
35. Du Pont, OS 2,248,244 (1973).
36. American Cyanamid, *OS* 2,403,947 (1973).
37. American Cyanamid, USP 4,296,059, 20.07 (1979).
38. Pennings, A. J., *J. Polym. Sci. Polym. Symp.*, *58*, 395 (1977).
39. Kwon, Young D., Allied Corporation, US 557,984, 05.12 (1983).
40. Stamicarbon B. V., NL 8.304.263, 10.12 (1983).
41. Sprenger, S. et al., *Chemiefasern Textilind.*, *41/93*, 226 (1991).
42. Wagner, W., Bayer AG., *2nd Int. Conf. Man-Made Fibres.* Beijing, China, 1987.
43. Schmitt, F., *Melliand Textilber.*, *12*, 830–835 (1982).
44. Heinrici-Olive, G., and Olive, S., *Advances in Polymer Science*, Vol. 32, Springer-Verlag, Berlin, 1979, p. 127.
45. Schultze-Gebhardt, F., in *Synthesefaser* (B. von Falkai, ed.), Verlag-Chemie Weinheim, 1981, p. 73.
46. Falkai, von B., Bayer AG., *2nd Int. Conf. Man-Made Fibres*, Beijing, China, 1987.
47. Wittorf. E., *Chemiefasern/Textilind. Man-Made Fibre Year Book*, 1989, pp. 62–64.
48. Du Pont de Nemours & Comp., DE-AS 1,435,611 (1970).
49. Du Pont de Nemours & Comp., DE-AS 1,435,619 (1969).
50. Reinehr, U. et al., Bayer AG., USP 4,108,845 (1978).
51. Bueb, D. et al., Bayer AG., USP 4,622,195 (1984).
52. Paulini, D. et al., Bayer AG., USP 4.891,873 (1988).
53. Reinehr, U. et al., Bayer AG., European patent 0/098,477 (1985).
54. *A Review of Catalytic and Synthetic Applications for DMF*, Du Pont, 1961.
55. Brochure, Lurgi Gas- und Mineralöltechnik GmbH, Frankfurt/Main, 1962.
56. Fourné, F., *Synthetische Faser*, Wissenschaftliche Verlagsgesellschaft, Stuttgart, 1964.
57. Friedrich, K. F. et al., *Chem.-Ing.-Tech.*, *63*, 123–127 (1991).

6

Product Variants

James C. Masson

JCM Consulting
Mooresville, North Carolina

I. INTRODUCTION

Acrylic fibers (including modacrylic) encompass a broad range of products, more diverse in composition, and probably in end use, than any other synthetic fiber. The major reason for this is that acrylonitrile is readily copolymerized with a wide range of ethylenically unsaturated monomers to create polymers with different characteristics. Also acrylonitrile polymers of differing composition may be blended, usually without phase separation, and the blends do not reequilibrate when heated as is the case with nylon and polyester. These compositionally different polymers may then be spun to fibers with special properties. Examples of products produced by compositional variation and, in certain cases, blending, include flame retardant (both acrylic and modacrylic), bicomponent, acid dyeable, and carbon fiber precursor.

Additional product variants may be made using the same base polymer, by variation in fiber processing conditions and/or inclusion of additives in the spinning or drawing step. Examples include moisture absorbent, microdenier, pigmented for outdoor application, producer dyed and optically brightened fibers. Other product variants that are designed to be processed on specific textile equipment, such as high bulk, tow, and continuous filament, are discussed in the chapters on fiber spinning.

The product variants or specialty products constitute less than 25% of total acrylic fiber sales, but, because they incorporate unique technology, generally sell for substantially more than the commodity products and thus are very important to the profitability of the business. The technology underlying the more important of these product variants will be discussed in detail.

II. PRODUCTS

A. Modacrylic and Flame-Retardant Acrylic Fiber

Modacrylic fibers (defined in the United States by the Federal Trade commission as having 35–85% polymerized acrylonitrile) are produced for one purpose —to be fire retardant. The polymer compositions incorporate chlorine, bromine or both, in some cases using antimony oxide as a synergist, to pass the relevant flammability test. U.S. tests for flammability of textile materials include [1]

NFPA 702-1975	45° test for wearing apparel
DOC FF 5-74	Children's sleepwear
NFPA 701-1976	Drapery and awnings
FAA Reg. 28.853	Aircraft interiors
NFPA 253	Radiant panel for floor coverings
DOC FF 1-70	Pill test for residential carpets

The 45° test is intended to eliminate "superflammable" fabrics and may be met by unmodified acrylics. Depending on carpet construction, the floor coverings tests may be met by halogenated polymers within the acrylic category (vide infra), but the other three generally require a modacrylic product to "pass."

Modacrylics of commerce are copolymers of acrylonitrile with one or more of the following monomers:

vinyl chloride (VCl)	$CH_2=CHCl$	56.7%	Cl
vinylidene chloride (VCl$_2$)	$CH_2=CCl_2$	73.1%	Cl
vinyl bromide (VBr)	$CH_2=CHBr$	74.7%	Br

The halogen content of commercially available modacrylics is in the range of 25–34%, corresponding to 34–51% halogen monomer. In addition, many modacrylics incorporate a sulfonate monomer for dyeability enhancement and improved structure. A neutral monomer such as acrylamide is also sometimes present to tailor shrinkage and physical properties. Table 1 lists the most important modacrylic fibers and their compositions. Of these, the most significant commercially is Kanecaron SE, which is reported [2] to have a 40,000-ton annual rate, 64% of the estimated worldwide total of 62,500 tons.

Many other monomers have been reported in the patent literature as flame retardants for acrylonitrile polymers. Aromatic halogen monomers such as *p*-

Table 1 Modacrylic Fibers

Product	Producer	Halogen	(%)	Other monomer	Comments
Dolan 88	Hoechst	VCl$_2$	(16.7)	Methyl acrylate Sulfonate	
Dralon MA	Bayer	VCl$_2$	(36.1)	Sulfonate	
Dynel	Union Carbide	VCl	(60.0)		Stopped production
Kanekaron SE	Kanegafuchi	VCl	(50.8)		
Lufnen	Kanebo	VCl$_2$	(35.6)		
SEF	Monsanto	VCl$_2$	(23.2)	Sulfonate	
		VBr	(11.8)		
Teklan	Courtaulds	VCl$_2$	(45.6)	Sulfonate	
Velicren FRS	Enichem	VCl$_2$	(36.9)	Sulfonate	
Verel	Tennessee Eastman	VCl$_2$	(39.1)	Substituted acrylamide	Stopped production

chlorostyrene, aliphatics such as α-chloroacrylonitrile, and 2,3-dibromopropyl acrylate are halogen examples; bis(β-chloroethyl)vinyl phosphonate is a halogen plus phosphorous example. Phosphorus-containing polymers have been reported [3] to achieve oxygen indices as high as 28 when incorporated as polymer blends in acrylic fibers. None of the above examples are used in commercial acrylics or modacrylics.

Polymerization of modacrylics is generally done under pressure as all three halogen monomers have boiling points below ambient temperature. For a vinyl chloride copolymer process, reactor pressure of about 5.3 kg/cm^2 (75 psig) will be reached at 40°C and 9.8 kg/cm^2 (140 psig) at 60°C. For the other halogen monomers, operating pressures will be significantly lower. Vinyl chloride is extremely unreactive in copolymerization with acrylonitrile (r_1(AN) has been reported to be 2.55–3.65 and r_2 0.02–0.07] [4]. As a consequence, polymerization methods must maintain a large excess of vinyl chloride, for example 82% of the monomer to incorporate 60% in the polymer. A consequence of the reactivities and the low AN concentration is that radicals ending in an AN unit do not often encounter an AN monomer and they are unlikely to react with a VCl monomer. Thus the overall rate is controlled by this slow reaction shown below [5]:

Poly-VCl• + VCl → fast k_p
Poly-VCl• + AN → very fast k_p
Poly-AN• + VCl → slow k_p
Poly-AN• + AN → fast k_p

Commercial acrylonitrile-vinyl chloride processes are often run in batch fashion, but with incremental addition of acrylonitrile to maintain uniform composition through the long batch cycle. Additional difficulties in the process are molecular weight control and separation of the toxic unreacted monomer.

In contrast, copolymerization of acrylonitrile with vinylidene chloride presents few problems. Pressure is still needed if a polymerization temperature of 50°C or higher is employed, but pressure will be less than 1 kg/cm^2 (15 psig). Reactivity ratios are reported as r_1(AN) = 0.44–1.04 and r_2 = 0.28–1.8(3). In practice, the polymerization is nearly azeotropic, with vinylidene chloride incorporated in slight excess over the monomer concentration. Vinyl bromide has reactivity ratios close to that of vinyl chloride (r_1 = 2.25–2.79, r_2 = 0.06), but no fiber producers use solely vinyl bromide to flame retard a modacrylic composition. In low concentrations, such as for flame-retarding a carpet fiber, the low r_2 is not a problem. All three of the monomers, in common with other halogenated organics, exhibit significant chain transfer to monomer and polymer. To produce polymer of useful molecular weight for fibers (number average molecular weight >35,000) low initiator concentration may be required, limiting reaction rate.

Modacrylic fibers are wet spun; the exception was the former Dynel process which used dry spinning from acetone, a solvent not suitable for any but the acrylonitrile-vinyl chloride composition. Any acrylic fiber solvent will also be applicable to the modacrylic compositions. Most modacrylics have antimony oxide either in the trioxide, Sb$_2$O$_3$, or the pentoxide, Sb$_2$O$_5$, form added to enhance the flame retardant action. These compounds are added to the spinning solution in concentrations of 3% or more based on polymer. The trioxide tends to deluster the fiber, whereas the pentoxide is a colloid and does not scatter light. In addition, light stabilizers and other delusterants such as titanium dioxide may be added. Color stabilizers [6] may also be incorporated in the solution. Pigments may be added to modacrylic solutions to make light stable outdoor fabrics; a more detailed discussion of pigmented acrylics is discussed later in this chapter.

Wet spinning of modacrylics is similar to that of acrylics (Chapter 4). Because the halogens make the fiber more heat sensitive, additional measures to minimize thermal history are important both in the solution and fiber stages. Even so, tiny quantities of HCl and/or HBr are evolved during hot processing steps. The presence of this acid on the fiber tends to make the fiber corrosive in downstream processing. Addition of a weak base or an epoxide after the heating steps neutralizes the acidity. If done before the heating steps, halogen acids will be evolved at an increased rate. Customer processes that involve heating may evolve additional acid. A weak base overspray or use of corrosion-resistant materials may be required.

In addition to the pigmented modacrylic variant, other modacrylic product variants include special cross section such as ribbon to produce hand effects similar to animal hair, uncrimped very short staple (floc) for use as lead battery plate reinforcement (Chapter 11), and producer-dyed modacrylic.

Flame-retardant acrylic fibers are those with 85% or more acrylonitrile. Because of the limited amount of flame retardance which can be conferred with 15% or less substitution of a flame-retardant species, only the less demanding tests such as DOC FF1-70 (pill test for carpets) can be met with this category of fibers.

With the decline in the acrylic carpet market beginning in the mid 1970s, most producers have discontinued production of flame-retarded acrylics, but recently Monsanto has reintroduced their 90-type product line. Several compositions which have been commercially available in the past used a blend of a non-flame-retardant acrylic polymer with 12% poly(vinyl chloride) [7]; a copolymer with vinyl bromide; and a copolymer with vinylidene chloride. Product variants in flame retardant carpet fiber include bicomponent, high bulk, pigmented, and producer dyed.

B. Bicomponent Fiber

Bulk is achieved in fibers through crimp. When crimped and then subjected to a mixing step so that the crimp is no longer in register, individual filaments cannot lie in closest proximity to one another. The air spaces thus introduced provide the bulk. In staple fibers, crimp performs another crucial function—it gives cohesion to the fiber bundle forming the yarn and thus imparts strength to the yarn. In addition, crimp affects the tactile aesthetics—the hand—of the fiber. There are several mechanical crimping devices, such as the "stuffing box" crimper and the gear crimper, that impart crimp and are routinely used on almost all commercial acrylic fibers. But the mechanical crimp is a planar crimp. All of the crimp is in an X-Y plane; there is no component in the Z direction.

In contrast, wool has a spiral crimp as a result of the bicomponent character of the cortex of the fiber [8]. The *ortho* and *para* cortex are in a bilateral arrangement with the *ortho* cortex present on the inside of the crimp. The first synthetic fiber to be produced as a bicomponent was viscose rayon [9].

In 1958, DuPont introduced the first acrylic bicomponent fiber now known as Type 21. The technology, revealed in a series of patents [10], involved the dry spinning of two polymers: polyacrylonitrile (PAN) and a copolymer of AN and a sulfonate monomer through a spinnerette (Figure 1) that ensures each filament will possess the two polymer components. Although the spinnerete holes are cirular, the forces acting on each polymer in gellation produce mushroom-shaped filaments comprising a "cap" of PAN and a "stem" of the copolymer (Figure 2).

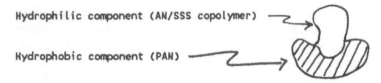

Hydrophilic component (AN/SSS copolymer)

Hydrophobic component (PAN)

Figure 1 Cross section of Type 21 Orlon.

The fiber, as sold to the textile mill, has mainly mechanical crimp, but in the mill processes of dyeing and drying, the spiral crimp is developed. The sulfonate copolymer by virtue of its many ionic sulfonate groups is hydrophilic and shrinks as the water is removed, whereas the PAN portion is stable. This difference in length creates the spiral crimp with the copolymer portion being on the inside of the spiral. The mechanical forces creating the crimp are of low magnitude, however, so that the yarn or garment must be relatively free of restraint for full crimp development. This type fiber has the unique property of being able to repeat the cycle. When a garment containing Type 21 is laundered—the hydrophilic portion elongates, straightening out the crimp; then on tumble drying, the crimp redevelops. This acts as a renewal process to bring the garment back to shape. Again, the crimping force is low and will not develop hanging on a wash line or in blocking.

The so-called water-reversible crimp bicomponent products were originally introduced and sold for hand knitting and machine-knit sweaters. At this time, virtually all sales are to the hand-knit market. DuPont Orlon Type 21 was sold in the United States and Canada under the certification marks Sayelle and Wintuk. In the United States, the Sayelle name denotes yarn of 100% Type 21, whereas Wintuk denotes at least 60% Type 21 with the remainder being Type 42 monocomponent fiber. Type 21 is also used in lower proportions with monocomponent fiber. Blend levels are chosen to obtain the desired hand, performance, and economics in the final yarn. In 1991, as part of their exit from the acrylic fiber business, DuPont sold technology and certification marks for their bicomponent acrylic fiber to Monsanto.

Because garments from water-reversible crimp acrylics will only recover their shape when tumble-dried, the market, to date, has been largely the United States. An early effort by DuPont to sell the product in Europe was unsuccessful. Mitsubishi Rayon sells a small volume of a similar product, Finel 513, in Japan. Monsanto introduced wet spun water reversible crimp products made by laminar flow technology under the tradenames Remember™ and Paqel™ in 1980, but withdrew them in 1988.

In addition to the water-reversible crimp bicomponents, there are a wide variety of nonreversible bicomponents. In these products the two polymers differ in comonomer composition, but neither polymer has a substantial quantity of a hydrophilic unit. For example [11], the two components may be "Component A is . . . a terpolymer of 93.8% acrylonitrile, 6% methyl acrylate and 0.2% sodium styrene sulfonate; Component B is . . . a mixture of 90% polyacrylonitrile and 10% of the terpolymer of Component A." The fiber was spun using the type of dry spinning bicomponent spinnerette shown in Figure 2 and processed as usual for a dry-spun product. Spiral crimp develops during drying. This crimp is not lost on exposure to water and thus fabrics made from such a fiber do not recover their shape on tumble drying. The crimp thus developed gives yarns and fabrics a crisper (harsher) hand than does the crimp from a water-reversible crimp bicomponent. DuPont's Type 24 Orlon (discontinued in 1990) was an example of a dry spun nonwater-reversible crimp bicomponent.

Bicomponent fibers can also be prepared by wet spinning; however the much higher number of holes per spinnerette makes design of a wet-spinning jet a formidable task. Recently, a patent has issued [12] to Monsanto claiming such a jet. Other means exist for making a bicomponent product using wet-spinning technology. Fitzgerald and Knudsen [13] reported use of a device which creates a multiplicity of laminar dope streams from two feed streams. The feed pattern within the spinnerette is stable. The individual filaments vary in cross section from 100% of component A to equal amounts of A and B to 100% of component B. A small proportion of "sandwich" (A-B-A or B-A-B) cross section may also be produced. These so-called random bicomponent fibers have roughly 50% of the bicomponent character of a "true" bicomponent fiber, but for many end uses this is satisfactory.

Wet-spun non-water-reversible crimp fibers may use the same polymer types as their dry-spun counterparts. Depending on the fiber manufacturing process, the bicomponent crimp may be reserved or latent when delivered to the textile mill. Wet processing will activate the crimp, however, and produce a bulky product. Most non-water-reversible crimp textile fibers are used in blends with monocomponent fibers to achieve the best compromise between bulk and hand. Bicomponent carpet fiber however is frequently used in 100% form. Examples of wet spun non-water-reversible crimp fibers for apparel applications are American Cyanamid Creslan Type 68 and Mitsubishi Rayon Vonnel Types 57, 59, and H-525. No carpet bicomponents are currently marketed.

C. Acid-Dyeable Fiber

Acid dyes are commonly used in wool and nylon dyeing, and in the beginning of the acrylic fiber industry were intended to be a major coloration route. Chemstrand's (now Monsanto) first product, Acrilan 36, was acid dyeable. Acid dyeing of acrylics had significant drawbacks however. The dyesites were based on weak bases such as 2-vinyl pyridine, which required a strong acid in the dyebath to be protonated and thus dyeable. The presence of the amine group reduced the heat stability of the fiber; original color of acid-dyeable fibers

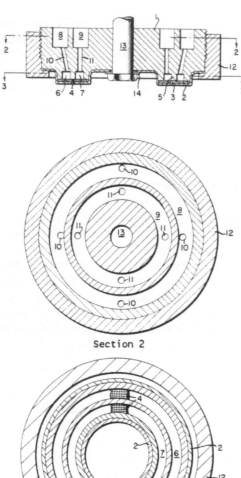

Section 2

Section 3

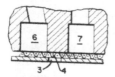

Detail of cross section

Figure 2 Bicomponent spinnerette for dry spinning [10].

tended to be "cream" and the fastness of acid dyes was not outstanding. From the producer's viewpoint the cost of the specialty monomer, especially in the high percentage required, was a drawback.

In the 1950s DuPont developed and introduced the Sevron cationic (basic) dyes and Bayer introduced the Astrazon dyes. These dyes had affinity for the sulfonic and sulfatic dyesites introduced via the polymerization initiator (see Chapter 3) and greatly expanded the range of cationic dyes available. Fibers with only cationic dyeability did not require the expensive basic monomer and hence were less expensive to make, were whiter and could be dyed at a more neutral pH to brilliant colors of better fastness. Producers began offering cationic dyeable acrylics as the standard product and thus the acid dyeable product became a "specialty."

As a specialty product, acid dyeable acrylic yarns can be knit or woven in patterns with cationic dyeable fiber and then piece dyed to give stripes or other patterns, or fiber blends of the acid and basic dyeable products can produce heather effects. It is this niche in which the business still exists.

Three types of basic monomers have been employed commercially to make acid-dyeable acrylic polymers. They are pyridines, tertiary amines and quaternary ammonium salts. Examples are shown in Figure 3. Many other variants have appeared in the patent literature. The main difference among these bases is the dyeing pH which can be employed. As previously mentioned, pyridines require a very acid pH—1 to 2; tertiary amines can be dyed at pH 3, and quaternary ammonium substituted units up to pH 6. The more neutral pH is desireable, particularly in cross-dyeing since fiber affinity for cationic dyes is reduced at low pH.

Several problems are encountered in producing a polymer for an acid dyeable acrylic:

1. The monomers are either nonvolatile or only difficultly so, making recovery by distillation impractical. This makes a batch process which can achieve higher monomer-to-polymer conversion more attractive than a continuous one.
2. The usual persulfate-bisulfite initiator system provides acidic end-groups which must be "neutralized" by some of the basic monomer. Even so, in cross-dyeing conditions some staining of the acid dyeable product by basic dye may occur at any but the lowest pH, except in the quaternary ammonium monomer case. Other initiator systems such as hydrogen peroxide-1-thio-glycerol [14] which do not produce acid end groups, lead to lower monomer-to-polymer conversion.

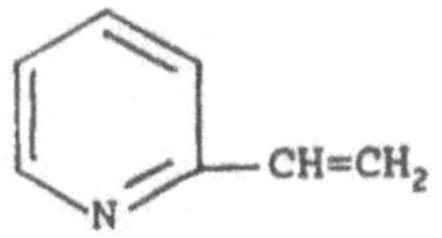

CH$_2$=CH—⟨pyridine⟩—CH=CH$_2$

2-vinyl pyridine

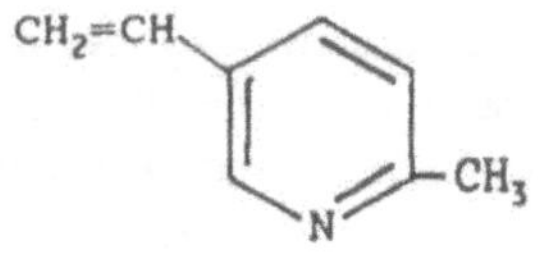

CH$_2$=CH—⟨pyridine⟩—CH$_3$

2-methyl-5-vinyl pyridine

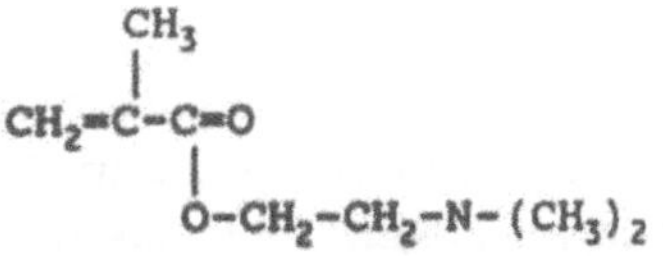

$$
\begin{array}{c}
CH_3 \\
| \\
CH_2\text{=}C\text{-}C\text{=}O \\
| \\
O\text{-}CH_2\text{-}CH_2\text{-}N\text{-}(CH_3)_2
\end{array}
$$

N,N-dimethylaminoethyl methacrylate

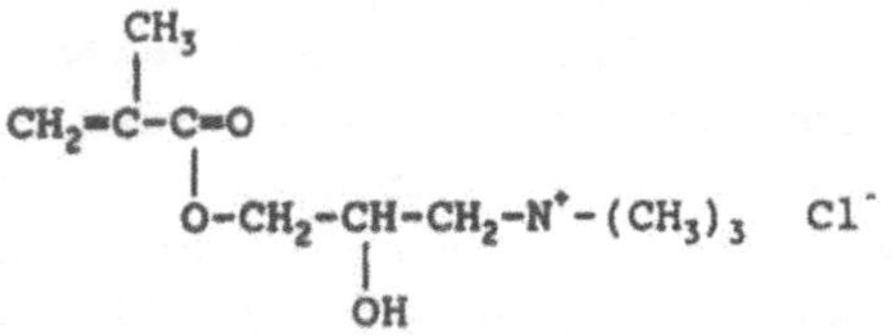

$$
\begin{array}{c}
CH_3 \\
| \\
CH_2\text{=}C\text{-}C\text{=}O \\
| \\
O\text{-}CH_2\text{-}CH\text{-}CH_2\text{-}N^+\text{-}(CH_3)_3 \quad Cl^- \\
| \\
OH
\end{array}
$$

2- hydroxy-3-methacryloylpropyl trimethylammonium chloride

Figure 3 Basic comonomers.

3. To minimize problems in spinning, copolymers of acrylonitrile and the basic
   monomer are often made incorporating a high proportion of the basic
   monomer. These compositions can then be blended with other acrylonitrile
   copolymers to produce the final product. This polymer, which may have
   from 20% to 50% of the basic monomer is often produced in the form of a
   partial emulsion. Several techniques have been reported to be effective in
   isolating the product. For example, a copolymer of 50 parts acrylonitrile and
   50 parts 2-hydroxy-3-methacryloyloxypropyltrimethylammonium chloride
   was precipitated from its polymerization mixture by addition of an aqueous
   solution containing 20 parts sodium thiocyanate [15]. Similarly, caustic
   neutralization was found to be useful for isolating polymers containing
   tertiary amines [16]. In each case the polymer is converted into a less water
   soluble form—the thiocyanate salt or the free amine.

   Spinning of acid-dyeable acrylics is not fundamentally different from that of
the basic-dyeable ones, but acid dyeable fibers may be expected to have more
spinning faults. In addition, fiber whiteness is difficult to maintain owing to the
presence of the basic groups. Acidic compounds which neutralize the amine
groups are commonly reported as stabilizers [17], but other actions such as

exclusion of oxygen and minimization of heat exposure in polymer drying, solution preparation and fiber processing, can also be effective.

In the past, two bicomponent, acid dyeable acrylic carpet fibers were sold: Creslan T-84 and Acrilan B-93. These products also contained halogen for fire retardency. Polymeric halogens will react with tertiary amine or pyridine groups to form a quaternary. In doing so, a crosslink is formed which is detrimental to fiber properties. One way to limit the extent of this reaction is to have the halogen and amine functions on different polymers, and blend them immediately prior to spinning. Fiber drying and relaxation may result in minor crosslinking, but at that stage the effect is small.

Worldwide only a few acid dyeable acrylic fibers remain on the market. Cashmilon AD from Asahi and Dralon types A100 and A800 are the most prominent.

D. Producer-Dyed Fibers

Although on a worldwide basis, producer-dyed fibers are clearly "specialty" products, for one producer, they are the dominant product amounting to 75–85% of Courtaulds total production [19]. Three methods of producer dyeing have been practiced: gel state dyeing, solution (dope) dyeing, and tow dyeing. The latter, using a device such as a Serricant Tow-Fix-R is not unique to fiber producers. Serricant dyeing is discussed in Chapter 9.

Wet-spun fibers prior to drying have a porous structure which may be 50% or more void volume. This "spongelike" structure is ideal for dyeing. Penetration of dyestuff is rapid and complete—perhaps as little as 1 sec is required. This means the process can be operated without a reduction of the normal wet processing speeds using a dye applicator [20] or a dip bath. In contrast, dry-spun fibers do not have a sufficiently porous structure to readily dye in the gel state. DuPont introduced a producer-dyed fiber named Accucolor [21], which was based on addition of dyestuff to the polymer solution prior to extrusion, but it was not commercially important at the time they withdrew from the acrylic market.

Dye can be applied to a wet-spun fiber at almost any point during the processing: in the coagulation bath, during washing and stretching, and after stretching [22]. Jenny [23] showed (Figure 4) that the dyeing rate is related to the surface area of the coagulated gel, and that it changes through the processing steps.

Until the fiber is dry, however, the rate is sufficient to support a short exposure time process. The preferred option is in the washing-stretching operation. The dye applicator is positioned so there is sufficient washing after the dye application to remove unfixed dye and other impurities such as solubilizing agents introduced with the dye. Uptake of dye is even more rapid if it is applied

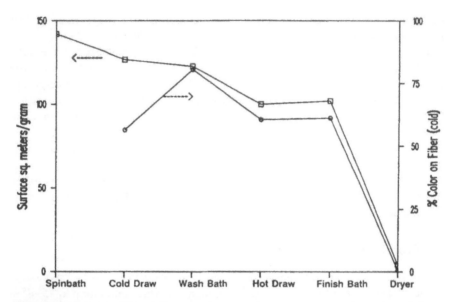

Figure 4 Relation of dyeability to surface area during processing of the fiber [23].

before the fiber is oriented. Mullinax [24] revealed Monsanto's system in block form (Figure 5).

The preferred dyestuffs for producer application are the "liquid dyes"—concentrated aqueous solutions of cationic dyestuffs. These can be fed to the applicator either as individual streams of one to four dyes or as a mixed solution of the desired components. The dye concentration, in either case, may be adjusted from that of the received dye by addition of water. Color control is achieved in the first case by adjusting the individual rates, and in the second by adjustment of the ratio of the components in the mixed solution to adjust hue plus changes in feed rate to adjust chroma. Feeding of individual dye streams has the possibility of more precise color control if the product color can be determined rapidly in order to feed back flow corrections. The Monsanto scheme utilizing a noncontact spectrophotometer [24] to rapidly determine fiber color is shown in Figure 6. An output from the computer controlling the product color is shown in Figure 7

Many fiber producers are able to guarantee color matching from one production run to the next, even on the most critical applications.

Producer-dyed colors exist for one reason only—lower systems cost to the mill compared to alternative dyeing methods. Because they require a colored fiber and yarn inventory, the products are most economical on large volume colors, and mills often use traditional dyeing methods to produce specialty colors. Even so the number of individual colored products (considering color,

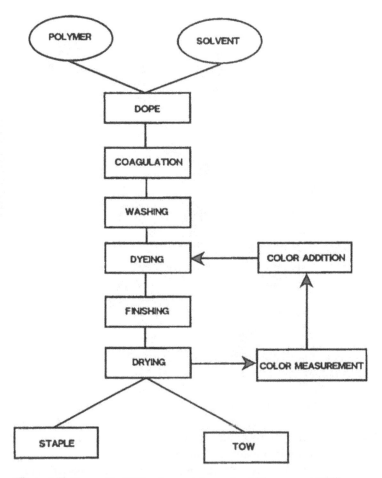

Figure 5 Dyestuff addition in a producer-dyed fiber process [24].

luster, and denier) produced is large. In 1987, Monsanto was reported [25] to offer over 250 colored products. As early as 1982, Courtaulds reported [19] running 100 shades per week with 10,000 in their computer.

E. Pigmented Fiber

Acrylic fibers have long been reported to have the best lightfastness of all organic fibers. A comparison is shown graphically in Figure 8 [26]. This advantage is capitalized on by producing pigmented acrylic fibers. Through pigmentation, the fabric color as well as the strength may be maintained. This is especially important for applications such as awnings where appearance is

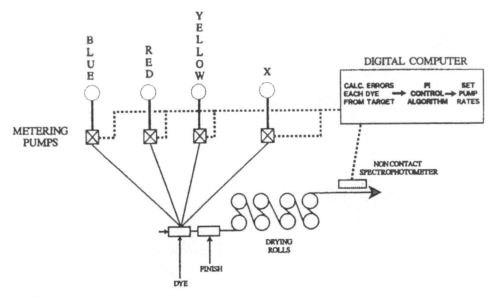

Figure 6 Color control in a producer-dyed fiber process [24].

paramount. Modacrylic fibers also have good lightfastness although it is inferior to that of acrylics as the halogen groups can lose HX (X = Cl or Br) in exposure to heat plus light. The HX can initiate nitrile polymerization which both embrittles and colors the fabric. In addition, the high comonomer content increases oxygen permeability of the fiber. Oxygen reacts at weak bonds on the polymer backbone to form hydroperoxides which in turn decompose to cleave the chain. These mechanisms are treated in detail in Chapter 8.

The acrylic and modacrylic polymers developed to produce "indoor" products have been used for pigmented applications as well. Although in principle some polymer engineering might be possible to produce a superior pigmented product, the small market for outdoor acrylic fibers mitigates against it. In spinning, the pigments and any insoluble light stabilizers are added to the polymer solution. Pigment selection for weathering stability in the expected environment is important. Teige [26] recommended the following pigment list for awnings:

Carbon Black
Flavanthrohe Yellow   Pigment Yellow 24
PV Fast Orange GRL   Pigment Orange 43
PV Fast Brown HFR   Pigment Brown 25
PV Fast Red B   Pigment Red 149
PV Fast Maroon HFM 01   Pigment Red 171
Novoperm Red Violet MRS   Pigment Red 88

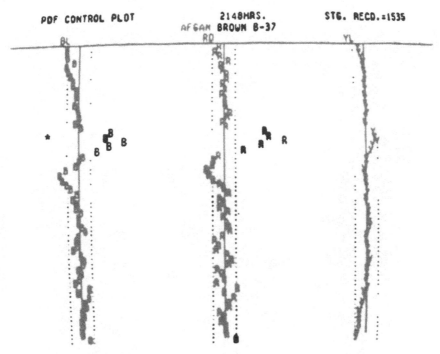

Figure 7 Color control in a producer-dyed fiber process showing the effect of a 10-min stoppage on all metering pumps [24].

PV Fast Violet RL spec.	Pigment Violet 23
PV Fast Blue A2R	Pigment Blue 15:1
PV Fast Blue B2G 01	Pigment Blue 15:3
PV Fast Green GG 01	Pigment Green 7
Hostaperm Green 8G	Pigment Green 36

The pigments are metered into the spinning solution as a dispersion in a dilute polymer solution. Usually the polymer employed is acrylic, but cellulose acetate dispersions are also used. As with the producer dyed colors discussed previously, either one suspension of all pigments, or individual pigment suspensions, is added. With addition of individual pigments, color control on the spinning machine using a noncontact spectrophotometer, as shown in Figure 6, is possible.

The spinning process for pigmented products does not differ substantially from the ecru (white) versions, but additional orientation may be introduced to provide greater strength for the expected applications. A comparison of fiber and yarn physical properties of SEF and SEF Plus (pigmented) are shown in the following table [27]:

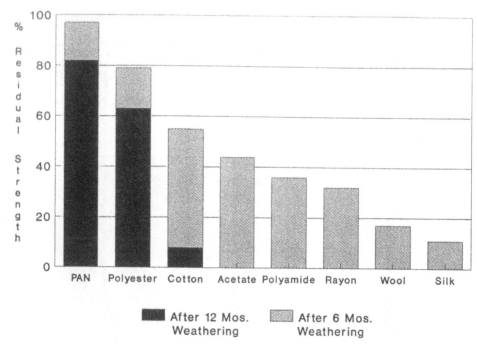

Figure 8 Comparison of strength retention on weathering for synthetic and natural fibers [26].

Property	SEF Plus	SEF
Tenacity, g/d	3.2–3.3	2.3–2.4
Elongation, %	45–50	50–55
Skein break		
Factor (18/1 cc)	2600–2800	1600–1800

Properties reported [28] for 1.6 d/tex, 40-mm Dralon Dorcolor X 270 pigmented acrylic are

Density	1. 18 g/cm^3
Tenacity	30–35 cn/tex
Elongation	35–40%
Loop elongation	25–30%
Moisture regain	
(20°C, 65% r.h.)	1–2%
Stability temperature	ca. 100–120°C.
Softening point	ca. 230°C.

In addition to providing superior light stability, pigmented acrylic and modacrylic fibers will retain color and strength after exposure to a variety of solvents, acids and bases. Gustafson [27] reported over 90% strength retention and no color change to SEF Plus after 20 hr exposure at ambient temperature to the following chemicals:

Sulfuric acid	1% and 70%
Hydrochloric acid	1% and 37%
Nitric acid	1% and 40%
Phosphoric acid	1% and 85%
Sodium hydroxide	1% and 25%
Sodium hypochlorite	0.5% and 5%
Sodium chlorite	0.5% and 5%
Hydrogen peroxide	0.3% and 30%
Acetone	100%
Methyl ethyl ketone	100%
Perchloroethylene	100%
Gasoline	100%

The color stability to a variety of chemicals suggests the possibility of pigmented acrylic and modacrylic applications in uniform and work clothes fabrics as well as outdoor applications.

F. Reinforcing (Asbestos Replacement) Fiber

The finding that workers exposed over long periods to fibrous-asbestos-developed mesothelioma (lung cancer) launched a search for substitute materials in the many applications of asbestos. One end use is as a reinforcing material in cement and concrete products. In this application, properties that are important are the tenacity and modulus of the fiber, the retention of modulus in warm-moist conditions, the resistance of the fiber to alkaline degradation, and the ability of the fiber to bond to the cement. Cost is also a factor! Acrylic fibers can be engineered to perform well in this application, but considerable changes are required compared to the usual textile product.

To meet the physical property requirements, the polymer should be poly-acrylonitrile. Even small amounts of comonomer significantly reduce the modulus and tenacity under warm-moist conditions. Polyacrylonitrile used for asbestos replacement fiber may have a higher molecular weight than is usual for a textile fiber, in the range of 100,000 to 500,000 [29]. These higher molecular weights coupled with higher draw ratios in spinning lead to increased strength and modulus compared to textile-grade acrylic fibers. Physical properties of several commercial acrylic fibers used for reinforcing applications were

Table 2 Properties of Industrial PAN Fibers Produced in Western Europe

Fiber	Type	Linear density (dtx)	Tenacity (cN//tex)	Initial modulus (cN/tex)	Elongation at break (%)
Sekril (Courtaulds)	100	3.2	50–60	800–1200	—
Dolanit (Hoechst)	12	2.2	54–58	—	13–16
		2.8	54–58	—	13–16
		8.2	43–47	—	14–17
	10	3	70–80	1400–1600	9–11
	VF 11	1.5–25	70–80	1400–1600	9–11
Dralon (Bayer)	Staple	—	27–31	—	30–35
	Filament	—	37–42	—	17–22

compiled by Maslowski and Urbanska [30] and are shown in Table 2. The effect of a hot, aqueous alkaline environment on the tenacity of Dolanit 10, an acrylic reinforcing fiber made by Hoechst, is shown in Figure 9 [31].

Spinning solutions of these higher molecular weight homopolymers have reduced solids content and require higher temperature and longer heating time than a conventional acrylic polymer solution. Spinning technology is conventional, except that several stretch zones may be employed, including stretching of the dried, collapsed filament to achieve the desired physical properties. Total stretch may be as high as 20×. Fiber finish may be markedly different than for a conventional fiber, since its function is to promote dispersion in the cement matrix. A "size" may be added to the fiber which retains the staple product in clumps to reduce handling problems. This size is dissolved by the cement, freeing the individual filaments. Reinforcing fibers are not crimped or steam relaxed since the properties imparted by such steps are not important to this end use and they reduce tenacity and modulus. Fiber is cut to lengths of 3–7 mm using conventional reel cutters.

Gel spinning has been reported to yield fibers of exceptional tenacity and modulus. Although no commercial acrylic reinforcing fiber is yet made by this technology, it appears to have great merit for extending the range of physical properties beyond that attainable with conventional polymers and spinning technology. This process uses a polyacrylonitrile of greater than 1 million molecular weight. The relationship between molecular weight and tenacity achieved is shown in the table [29].

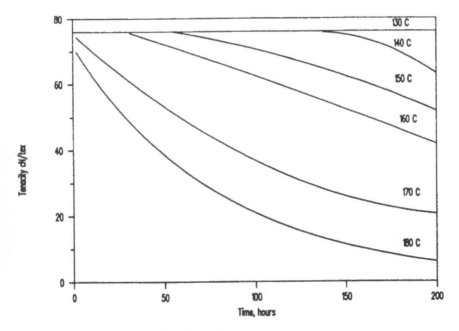

Figure 9 Tensile strength of Dolanit 10 fibers after a 24-hr treatment with an aqueous slurry of cement [31]

Molecular weight	Tenacity
M_w	(cN/tex)
2,280,000	222.0
1,350,000	166.0
530,000	151.9
320,000	126.3
120,000	75.9
70,000	33.0

A dilute solution of polymer (3–8%) is made using one of the PAN solvents. The solution is spun through an air gap into a spinbath, a process termed dry-jet wet spinning. The cooling and solvent evaporation in the air gap cause the solution to rapidly gel. The spin bath may be an water/spin solvent mixture, or another nonsolvent. Drawing may be accomplished before or after solvent removal; hot nonaqueous baths may be employed. Table 3 compiled by Maslowski and Urbanska [32] shows the range of conditions and properties reported for gel spinning in the patent literature. Compared to the tenacity of 75 cN/tex reported for Dolanit 10, the claimed processes offer increases up to 3×.

Table 3 Mechanical Properties of PAN Fibers obtained by Various Gel Processes [32]

Company	M_w or limit viscosity	Solvent for polymer	Polymer conc. (%)	Cooling bath	Gelation temp.	Draw conditions (ratio/temp.)	Total draw ratio	Tenacity (cN/tex)	Initial modulus (cN/tex)	Elong. at bk. (%)
Allied	1.63×10⁶	DMSO	6	DMSO/H₂O (50/50)	ambient	4.07 at 135–50 2.11 at 146–60	8.6	70	2200	—
Stamicarbon	1.3×10⁶	DMF/ZnCl₂	5	CH₂Cl₂	ambient	dry, 180	28	153	1860	10
Toyobo	1.4×10⁶	DMF	7	Alcohol/ dry ice	−40	135&150 on hot plate	>12	very high	very high	—
Japan Exlan	1.35×10⁶	NaSCN (50% aq.)	5	15% NaSCN	5	2 at 20 in H₂O 2 at 85 in H₂O 2.5 at 100 in H₂O 1.8 at 130 in EGᵃ 1.6 at 160 in EG	28.8	164	2700	—
Japan Exlan	2.28×10⁶	NaSCN (50% aq.)	5	15% NaSCN	5	2 at 20 in H₂O 2 at 85 in H₂O 2.5 at 100 in H₂O 1.8 at 130 in EGᵃ 1.6 at 160 in EG	28.8	222	—	—
Toray	[n]=3.2–5.5	DMSO	5–20	55–60% DMSO	15–20	hot H₂O superheated steam dry, 190–240	12–15	100–130	1600–2000	—
Mitsubishi	[n]=4	DMF	—	—	—	—	16	180	—	—
Kuraray	1.65×10⁶	NaSCN (50% aq.)	2.9	15% NaSCN + 5% CH₃ OH	−10	2.5 at 20 in 15% NaSCN 9 at 200	22.5	196	2490	8.2

ᵃEthylene glycol.

G. Carbon Fiber Precursor

Acrylic fiber is the major precursor for carbon fiber production, with rayon and pitch also being used. In some respects the desired fiber properties are similar to those required for a reinforcing fiber (vide infra), but there are also specific requirements for a good precursor fiber. Chief among these is ease of producing the "stabilized" product. This "stabilization" is achieved by formation of a ladder polymer through nitrile polymerization; see Figure 10. This can be a radical process as shown or an ionic process. This ionic process is initiated by nucleophilic agents which can be incorporated either as monomers or additives.

Preferred polymers for precursor fiber are copolymers of AN with small amounts of monomers such as carboxylic acids or vinyl bromide which cause ionic initiation and thus lower the temperature for the onset of the nitrile polymerization and moderate the exotherm. Figure 11 shows the exotherm of three compositions [33]. A smaller exotherm is desireable because it eliminates localized heat buildup and charring in large yarn bundles. Monomers reported to be effective include acrylic acid, methacrylic acid, itaconic acid [34], and vinyl bromide [33]. Comonomers commonly incorporated in textile acrylic fibers such as methyl acrylate and vinyl acetate do not appreciably catalyze the exotherm, but serve to interrupt the nitrile polymerization [35]. Tsai and Lin [36] have shown that incorporation of 2-ethylhexyl acrylate increases the temperature of onset of nitrile polymerization and the peak temperature, but lowers the heat released in the exotherm, presumably by limiting the chain length of the nitrile reaction. Courtelle, which contains both itaconic acid and methyl acrylate, is used commercially as a precursor for both carbon fiber and PANOX (the stabilized intermediate).

There is a positive relationship between precursor fiber strength and modulus and carbon fiber strength and modulus [30]. Thus it is advantageous to have precursor polymer of higher molecular weight than is usual for textile applications. Molecular weights of greater than 1 million have reportedly been used [37]. Both suspension and solution polymerization are employed. Because of the chain transfer effect of organic PAN solvents, high molecular weights are more difficult to attain in solution polymerization.

Commercial polymer processes for precursor production generally utilize free radical initiators, usually azonitriles, but Asahi has reported [38] use of

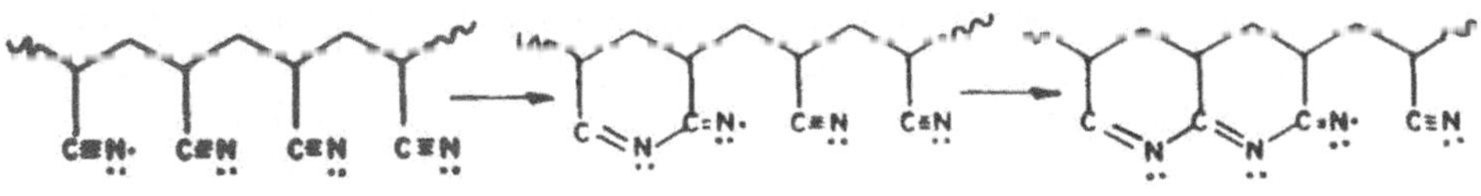

Figure 10 Ladder polymerization of PAN nitrile groups by free-radical mechanism [37].

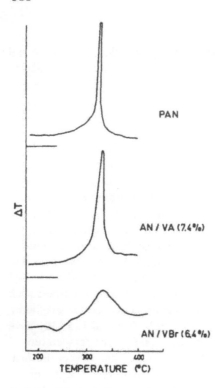

Figure 11 Differential thermal analysis of PAN and copolymers [33].

organometal compounds in an organic solvent such as benzene or toluene. This process produces a significant fraction of isotactic triads which are reported to enhance the properties of the resultant carbon fiber.

All conventional acrylic fiber-spinning technologies, dry, wet, and dry-jet wet spinning are used to make acrylic precursor fibers. The usual trade-offs among velocity and productivity per jet, discussed in the chapters on dry and wet spinning are in effect here, but there is an additional parameter: for some end uses, a carbon fiber bundle of relatively low total denier is required. This eliminates one of wet spinning's chief advantages: high productivity per spinnerette. One process, originally developed by American Cyanamid and later sold to BASF, used a form of melt spinning. In this process, water (33% or more) is added to an acrylic polymer of low molecular weight ($M_n = 6000-15750$) and the mixture heated under pressure to form a plasticized melt; for PAN, a temperature of >185°C. is required [39]. The fiber is extruded into a steam pressurized solidification zone and stretched up to 25 times while still in the solidification zone to develop high tenacity and modulus [40].

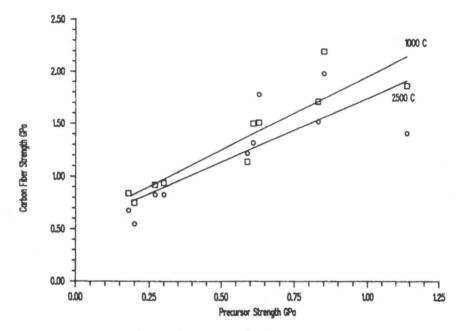

Figure 12 Strength of carbon fibers as a function of precursor strength [41].

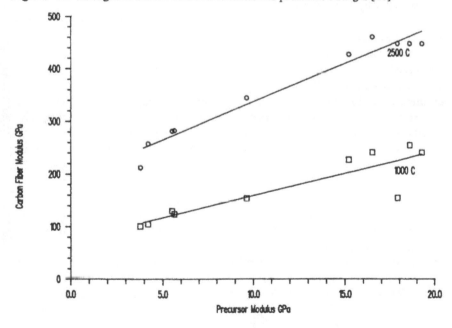

Figure 13 Modulus of carbon fibers as a function of precursor modulus [41].

In addition to using a PAN of high molecular weight, precursor fiber processes utilize high orientation stretches to attain maximum tenacity and modulus as these properties can be passed on to the resultant carbon fiber. Maslowski and Urbanska [41] gave the data depicted in Figures 12 and 13 for the relationship.

Precursor fiber of large total denier is packaged in bales or boxes as conventional tow. Smaller total denier product may be packaged on cones or beams. Since the linear speeds of acrylic fiber spinning are much higher than those of fiber stabilization for carbon fiber production, no integrated processes are known.

H. Moisture Absorbent Fiber

In many end uses, the comfort of a fabric relates to the ability of the fiber to either absorb moisture without feeling wet or to transport moisture to the outer surface where it can be evaporated. Dunova, a dry-spun product produced by Bayer is in the first category. The fiber has an interior consisting of thousands of pores and channels which can store water [42]; the interior is connected to the surface by fine capillaries. The surface layer is 1 to 2 μM [43] thick. Figure 14 shows an overall view of the Dunova filament; Figure 15 shows the detail of the skin-core structure. Capillary action draws in the perspiration from the skin; it is released only by evaporation.

Dunova is produced by addition of a nonsolvent of low volatility such as glycerine to a DMF solution of the acrylic copolymer; the polymer is similar to that used in standard Dralon production except for additional sulfonate dyesites presumably added to achieve deep shades in dyeing despite the presence of the light-scattering pores. The nonsolvent remains in the filament through the dry-spinning operation, but is removed during fiber washing, leaving behind a void or pore. Care must be taken in processing to avoid collapsing the voids. Specific gravity of Dunova as measured by flotation was found to be 1.167 g/cm^3, in the normal range for an acrylic [44], indicating that the liquid had penetrated the voids. When the measurement was made with mercury, a specific gravity of 0.90 g/cm^3 was found; a pore volume of 23% is indicated. A water-holding capacity of 30–40% has been claimed, without fiber swelling [42]. Bayer has also reported the production of fiber with similar water holding/transport characteristics using wet-spinning technology [45].

Other companies have also produced moisture transport acrylics: Montefibre's product was called Leaglor and Kanebo's Aqualon or Lumiza. Both utilize the same principle of incorporating stable voids in the fiber structure to hold water. The technology used to incorporate the voids has not been disclosed, but in the case of Leaglor, appears to be addition of a high molecular weight polyethyleneoxide to the spinning solution as some residue is found in the product [43]. A high-moisture-regain fiber based on a solution blend of an acrylic polymer and up to 20% of an unspecified water soluble polymer has been reported [46] by Courtaulds. The change in tensile and moisture absorbing properties is shown below:

Level of hydrophilic polymer(%)	0	5	10	20
Tenacity (cN/tex)	30	28	26	21
Extension (%)	40	38	37	32
Moisture regain (37°C/90% RH)	3.2	4.6	5.6	7.2
Water imbibition (%)	5	12	20	38

No commercial product based on this technology is known.

I. Pill-Resistant Fiber

Pilling is the accumulation of balls of fiber at the fabric surface, and is an unsightly reminder that the garment is wearing. In carpet, the term "fuzzing" is usually employed. In both cases, the cause is a loss of fiber, due to disentanglement of short staple lengths or a breaking of longer filaments, followed by a binding of the loose fiber to the fabric surface by other, unbroken filaments. Pilling occurs to some extent with all staple derived fabrics and is affected by fabric construction, yarn construction, fiber denier, and staple length, in addition to intrinsic fiber physical properties. Yarn and fabric treatments are marketed to reduce pilling.

Since acrylic fiber is used extensively in sweaters and jersey knits where pilling is a well-known problem, it is not surprising that a criticism of acrylic performance relates to pilling propensity. The tendency toward finer deniers also increases the tendency to pill, so the perception of the problem has increased. Since it is inevitable that, in a staple product, some short filaments will move out of the yarn, the cure for pilling must be for the fabric surface to release the loose fiber before the "pill" is noticeable. A kinetic model of pilling has been proposed by Doria and Trevisan [47] as shown in Figure 16. These authors assumed that $k_4 = 0$ (no fuzz was lost without forming a pill). Based on this assumption plus experimental data, they concluded:

The initial slope of the pill curve is proportional to the original yarn hairiness and the entanglement tendency.

Fuzz formation rate (k_1) becomes more important for open fabric construction where $A_t \neq 0$.

For tight fabrics, the overall performance is related to pill wearoff (k_3).

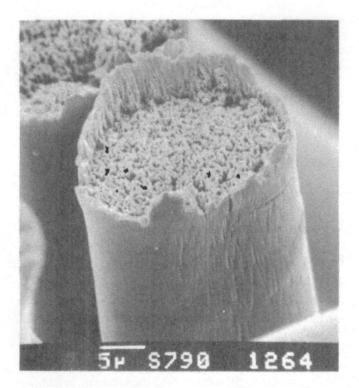

Figure 14 Cross section of Dunova moisture-absorbent fiber (courtesy Bayer AG).

The authors found that the best pilling performance was achieved with a fiber that had a corrugated surface, probably due to the association of this surface type with structural defects which contribute to easy pill breakoff. Lower orientation stretch also favored good pilling performance. Pilling performance was found to be inversely related to fiber tenacity, loop tenacity and elongation, flex life, and yarn strength. Fibers which yielded fabrics with good pill performance showed an increase in "fly" in yarn processing.

Several manufacturers market low pill versions of their acrylic fiber: among them are Montefibre's Leacril NP, Mitsubishi Rayon's H-613 and Monsanto's Pil-Trol™ (now called HP).

III. SUMMARY

Compared to other synthetic fibers, acrylics stand far ahead in the number of product variants available, where more than the mere fiber diameter, length, and color have been altered. New variants, not considered here, will undoubtedly be commercial by the time this volume is available.

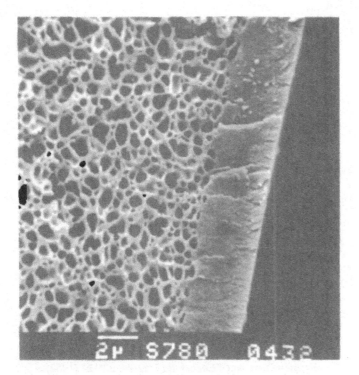

Figure 15 Skin-core structure of Dunova (courtesy Bayer AG).

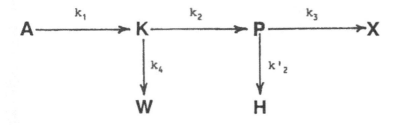

A = Available fiber for fuzz formation
F = Fuzz fiber
H = Hairiness
P = Pills
X = Pills removed
W = Fuzz removed

Figure 16 Pilling kinetic model [47].

REFERENCES

1. Drake, G. L., Flame retardants for textiles, in *Encyclopedia of Chemical Technology*, vol. 10, Wiley, New York, pp. 426–428.
2. *Textile Month*, April 1991.
3. Herlinger, H., Hardtmann, G., Hermanutz, F., Schneider, R., and Einsele, U., *Melliand Textilberichte, 72*, 353 (1991).
4. Greenley, R. Z., in *Polymer Handbook* (J. Brandrup and E. H. Immergut, eds)., Wiley, New York, 1989, pp. II, 169–170.
5. Knorr, R. S., Monsanto Co., unpublished data.
6. For example, see G. Palethorpe, U.S. Patent 3,642,628 (2/15/72) to Monsanto.
7. Leonard, R. L., Veitch, J. D., and Veasey, T. M., U.S. Patent 3,193,602 (7/6/65) to Monsanto.
8. *Encyclopedia Britannica Macropaedia 7*, Encyclopedia Britannica Inc., 1981, p. 285.
9. Moncrieff, R. W., *Man-Made Fibers*, Wiley, New York, 1975, p. 193.
10. Kovarik, F. J., U.S. Patent 3,038,240 (6/12/60).
11. Huffman Jr., H. A., U.S. Patent 3,547,763 (12/15/70) to DuPont.
12. Carter, D. E., U.S. Patent 5,017,116, (5/21/91) to Monsanto.
13. Fitzgerald, W. E., and Knudsen, J. P., *Textile Res. J., 37*, 447 (1967).
14. Wilkinson, W. K., U.S. Patent 4,585,844 (4/29/86) to DuPont.
15. Coleman, D., South African Patent Application 686,986 (8/30/68) to American Cyanamid.
16. Masson, J. C., and Self, L. A., U.S. Patent 3,736,304 (5/29/73) to Monsanto.
17. Masson, J. C., and Hall, A. M., U.S. Patent 3,784,512 (1/8/74) to Monsanto.
19. Titheridge, D. E., *Chemiefasern/Textilind. (English Ed.), 32/84*, E26-27 (1982).
20. Bittle, D. F., and McPeters, A. L., Canadian Patent 954,256 to Monsanto.
21. Anon., DuPont introduces Accucolor producer colored fiber, *Knitting Times, 57*, 39 (1983).
22. Sharma, N. D., Sharma, P., and Mehta, R., *Textile Asia, 20*, 60 (1989).
23. Jenny, R., *Tinctoria*, No. 9, 77 (1990).
24. Mullinax, G. B., New developments in colored acrylic fibers, Textile Research Institute presentation, Charlotte NC, April 13, 1988.
25. Anon., Monsanto producer-dyed fiber expansion widens color, denier range, *Inside Textiles, 8*, 10 (1987).
26. Teige, W., *Chemiefasern/Textilind., 33/85*, 636–642 (1983).
27. Gustafson, J. H., Presentation to Industrial Fabrics Association, Textile Technology Forum, Las Vegas, NV, 1987.
28. Teichmueller, G., *Chemiefasern/Textilind., 32/84*, 43 (1982).
29. Maslowski, E., and Urbanska, A., *Am. Textiles Int. 18*, FW6 (1989).
30. Maslowski, E., and Urbanska, A., *Am. Textiles Int. 18*, FW2 (1989).
31. Hahne, H., *Chemiefasern/Textilind., 12*, 839 (1983).
32. Maslowski, E., and Urbanska, A., *Am. Textiles Int. 18*, FW2 (1989).
33. Henrici-Olivé, G., and Olivé, S., *Adv. Polym. Sci., 51*, 36 (1983).
34. Gupta, A. K., Paliwal, D. K., and Bajaj, P., *Rev. Macromol. Chem. Phys., C31* (1), 7 (1991).
35. Grassie, N., and McGuchan, R., *Eur. Polym. J. 8*, 865 (1972).
36. Tsai, Jin-Shy, and Lin, Chung-Hua, *J. Appl. Polym. Sci., 43*, 679 (1991).

37. Gupta, A. K., Paliwal, D. K., and Bajaj, P., *Rev. Macromol. Chem. Phys.*, *C31* (1), 13 (1991).
38. Asahi Chemical Ind. KK, Japan Patents 3,068,608 and 3,076,823 (3/25/91 and 4/2/91).
39. Frushour, B. G., *Polym. Bull.*, *7*, 1 (1982).
40. Porosoff, H., U.S. Patent 4,163,770 (8/7/79) to American Cyanamid.
41. Maslowski, E., and Urbanska, A., *Am. Textiles Int.*, *18*, FW2 (1989).
42. Wilschinsky, H., *Pakistan Textile J.*, 80 (April 1990).
43. Korner, W., et al., *Chemiefasern/Textilind.*, *29/81*, E62 (1979).
44. Preston, J., and Hofferbert, W. L., unpublished data.
45. Gröbe, V., et al., *Faserforschung und Textiltechnik, 9/28*, 482 (1977).
46. James, J. R., Akers, P. J., and Picker, J., *Textile Asia*, 70 (1988).
47. Doria, G., and Trevisan, E., *Textile Asia*, 64, (1989).

7

Acrylic Polymer Characterization in the Solid State and in Solution

Bruce G. Frushour

Monsanto Company
St. Louis, Missouri

I. POLYMER SOLID STATE STRUCTURE

A. Introduction

The majority of textile fibers have a morphology that can be described by the classical two-phase model for semicrystalline polymers [1]. In this model discrete crystalline domains on the order of several hundred angstroms (Å) are mixed with amorphous domains having similar size. The individual polymer chains have end-to-end distances on the order of 1000 to 2000 Å, and a single chain can span two or more adjacent crystallites. Such a chain is called a "tie molecule," and their assembly in the intercrystalline regions forms the amorphous phase. In a sense, then, the fiber morphology can be viewed as a micro-composite structure of the crystalline and amorphous domains, and the tie molecules provide the connectivity that supports tensile loads on the fiber.

The two-phase model is often used as the framework for understanding physical properties of fibers, and the correlations and explanations of the fiber properties in terms of the model are referred to as structure-property relationships [2]. This endeavor has been successful because the descriptive parameters of the model can be precisely defined and measured. Physical characterization techniques have been developed for determining the volume fractions of the crystalline and amorphous domains, the size and perfection of the crystalline domains, and the degree of orientation of the polymer chains with respect to the fiber axis. These parameters are very sensitive to the fiber spinning and

197

processing conditions used in the manufacture of most synthetic fibers. It is well known that the tensile properties and dyeing behavior of synthetic fibers, such as nylon and polyester, can be manipulated by controlling the degree of drawing and relaxation after initial fiber formation, and one of the triumphs of modern fiber physics has been the rationalization of the fiber properties in terms of the constitutive parameters of the two-phase model.

Developing useful structure-property relationships for acrylic fibers has proven difficult because the polymer is not semicrystalline in the conventional sense, and discrete crystalline and amorphous phases are simply not found. The classical two-phase model is not entirely adequate to describe the morphology of polyacrylonitrile (PAN) and the fiber-forming PAN copolymers. This PAN morphology has been described as "amorphous with a high degree of lateral bonding," or as a "two-dimensional liquid crystalline-like" structure with many defects. Fortunately, the recent application of newer polymer characterization techniques has revealed a surprising degree of structural detail, and should provide the basis for a better understanding of acrylic fiber properties.

The current views on the of the acrylic morphology are summarized in this chapter. We begin with the chemical structure of the isolated PAN chain and the influence of the highly polar nitrile side chain on the chain conformation. The importance of the chain conformation becomes apparent in subsequent discussions of x-ray and electron diffraction studies of polymer, fibers, and solution-grown single crystals. Various models of the acrylic morphology are described at this point utilizing additional information gleaned by structure sensitive techniques such as solvent swelling and wide-line NMR measurements. The next part of the chapter deals with the thermal properties of PAN and how they are influenced by addition of comonomers. Included under thermal properties are the melting point, glass transition, and molecular relaxations. These studies are also relevant to the understanding of the morphology since any explanation of the influence of copolymerization on properties must be done in reference to a model of the morphology. A section on molecular weight studies of PAN concludes the chapter.

B. Stereoregularity and Chain Conformation

1. Stereoregularity

It is well known that the stereoregularity of a vinyl polymer influences its crystallizability [3, 4]. A good example of this is polypropylene. Isotactic polypropylene is highly crystalline, whereas the atactic polymer is completely amorphous. In order to crystallize, the backbone chain of the polymer must first be able to assume some regular chain conformation such as a planar zigzag or helical structure, and then the individual chains will pack together to form a crystalline unit cell. The chain conformation allows maximization of various intermolecular and intramolecular interactions, e.g., van der Waals forces, hydrogen bonding and dipolar bonding in the crystal. The length of a single polymer chain is far longer than the longest axis of the unit cell,* and the polymer chains contained within a unit cell actually enter through one face and leave through the opposite face [1]. For the following reasons this crystalline packing is not possible if the chain is atactic. First, it may not be possible for the chain to form the same regular structure as the corresponding isotactic or syndiotactic chain because steric hindrance between substituent groups on adjacent monomer units may occur when the backbone attempts to assume this structure. Secondly, even if a regular conformation is found, then there still will be hindrance between groups on adjacent chains when the chains pack to form the crystalline unit cell. Most commercial vinyl polymers such as polystyrene and polymethyl methacrylate are atactic and form amorphous glasses that have no long-range crystalline order. Fibers can be melt-spun from these glassy polymers, but they cannot be used for textile applications because of the lack of crystallinity. As soon as these fibers are heated near the glass transition temperature, which is approximately 100°C for the latter two polymers, they begin to shrink. When a crystalline phase is present the fiber shrinkage is minimal until it is heated near the melting point. There are, however, exceptions. Some atactic polymers are not truly random but possess a surprisingly high degree of order or pseudocrystallinity, and PAN falls into this interesting class of polymers. Common among these polymers is a substituent group capable of forming strong interactions through secondary bonding. Other examples of commercial importance are polyvinyl alcohol and polyvinyl chloride; commercial fibers have been spun from both.

All commercial PAN and attendant fiber-forming copolymers are manufactured using free-radical polymerization processes (see Chapter 3). One expects free-radical polymerization to produce polymers with little or no stereoregularity, and nuclear magnetic resonance (NMR) analysis of these polymers confirms this expectation. Schaefer [5] has shown that the ^{13}C NMR spectrum of the nitrile carbon can be interpreted in terms of steric triads and pentads. Three NMR lines are clearly resolved and are attributed to the three possible steric triad configurations, i.e., the hetero, syndio, and isotactic triad configurations. A completely atactic polymer should have concentrations of the hetero, syndio, and isotactic triads in the ratio of 2:1:1 [6]. Schaefer observed a ratio of 5:2:3, which is not markedly different from expectations for the atactic case and therefore he concluded that stereoregularity was low.

Over the years attempts have been made to develop polymerization methods that will increase stereoregularity. Presumably increasing the stereoregularity would yield a more crystalline polymer with improved fiber tensile properties especially under hot-wet conditions. Most of these efforts have either given only

---

*A polymer chain with a degree of polymerization of 1000 can have an extended length on the order of 1000 Å, whereas the unit cell dimensions are on typically less than 50 Å.

marginal improvements, or involved approaches that could not easily be commercialized for fiber manufacturing. Stereoregular polymers such as polypropylene are made with Ziegler-Natta catalysts. This type of polymerization has not been successfully applied to acrylonitrile, vinyl chloride, and other monomers having polar substituents [7].

Chiang [8] claimed to have produced a PAN with enhanced stereoregularity by using a proprietary organometallic catalyst. The basis for the claim of higher stereoregularity was an increase in the temperature required to dissolve the polymer in propylene carbonate. The solubility temperature was increased from 135°C for the conventional free-radical polymer to 175°C, though no corresponding improvement in the crystalline perfection as measured by x-ray diffraction was reported. The theoretical basis for using dissolution temperature as a means for characterizing the degree of stereoregularity and crystalline perfection was also described by Chiang [9].

2. Chain Conformation

The fact that functional synthetic fibers can be made from acrylic polymers should suggest that some degree of order must be present. All textile fibers, whether they are synthetic or natural, are crystalline, and this crystallinity imparts good tensile properties and prevents the fiber from shrinking when it is heated during dyeing, washing, and ironing. We will now consider how PAN, essentially atactic, could be crystalline.

We begin by comparing PAN with atactic polypropylene.* Both are monosubstituted vinyl polymers with carbon-carbon backbones as shown in Figure 1. In polypropylene the α-carbon is bonded to three hydrogen atoms to form the methyl group and in PAN it is bonded to a nitrogen atom to form the nitrile group. The molar volumes can be estimated (from group contributions to the molar volume of organic liquids) and the values are 32.3 cm^3/mole for the methyl group and 27.2 cm^3/mole for the nitrile; i.e., there is not a great deal of difference [10]. The distinguishing feature of the nitrile group is the large dipole moment with a magnitude of 3.9 Debyes, thus ranking it among the most polar organic functional groups. The methyl group has little or no dipole moment, so when we compare these two polymers were are seeing directly the influence of the large dipole moment, and the way an assembly of them can interact. The glass transitions of the atactic polypropylene and PAN are approximately −20°C and 95°C, respectively [11], and the former is a waxy material readily soluble in many solvents while PAN is soluble only in very polar solvents. Billmeyer [11] contends that the differences in T_g's can be expected given that the solubility parameters of polypropylene and PAN are 16.6 and 31.5 $(J/cm^3)^{1/2}$, respectively. The magnitude of the solubility parameter is a measure of the

---

*Atactic polypropylene is the small soluble fraction recovered from the Ziegler-Natta polymerization of isotactic polypropylene, which is mainly crystalline and has a melting point of 165°C.

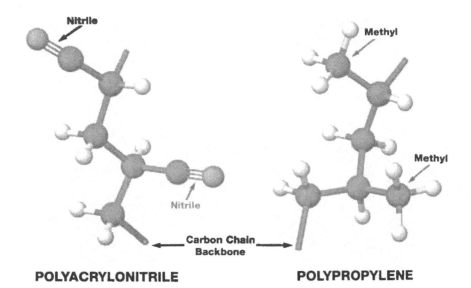

Figure 1 Random dimer segments of polyacrylonitrile and polypropylene polymer chains showing the relative sizes of the nitrile group in polyacrylonitrile versus the methyl group in polypropylene. The small light-colored atoms are hydrogen.

intermolecular attractions and can be calculated from molar attraction constants that are available for various functional groups that comprise the polymer structure. The molar attraction constants for the methyl and nitrile groups are 303 and 725 $(J/cm^3)^{1/2}$/mole, respectively [11]. Clearly if the dipole moment of the nitrile group were much lower, then the properties of PAN would be different, and it would probably resemble the atactic polypropylene.

The interaction energy between two nitrile groups has been described by Olivé and Olivé [12]. The interaction energy can be either attractive or repulsive, depending upon the spatial orientation of the nitriles, while the magnitude of the interaction depends upon the distance of separation. Possible modes of interaction between two nitrile groups are shown in Figure 2. We have to consider both intrachain and interchain dipolar interactions in order to understand the solid-state structure of PAN. In the isolated chain, the preferred conformation will be one that minimizes the chain potential energy [13]. If the backbone chain were placed in a planar zigzag arrangement then the adjacent nitrile groups would fall into a parallel alignment thus giving a net repulsion. The chain potential energy can be lowered by placing the adjacent nitrile groups as far apart as possible; this will require that the backbone chain become helical. In a helical conformation all of the nitrile groups will be pointing away from the axis of the helix, and if several chains are grouped together, then some of the

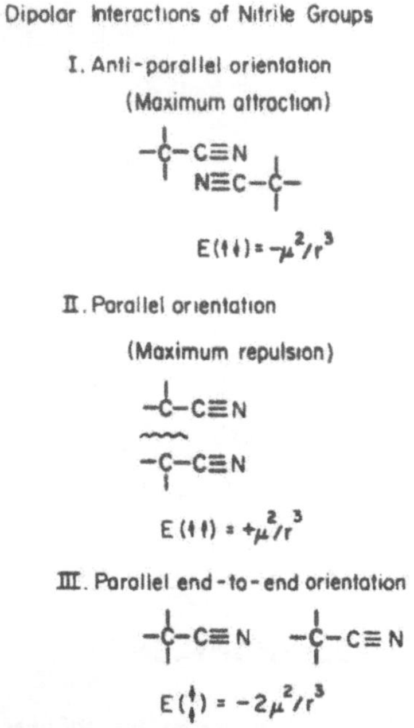

Dipolar Interactions of Nitrile Groups

I. Anti-parallel orientation

   (Maximum attraction)

$$-\overset{|}{\underset{|}{C}}-C\equiv N$$
$$\qquad N\equiv C-\overset{|}{\underset{|}{C}}-$$

$$E(\uparrow\downarrow) = -\mu^2/r^3$$

II. Parallel orientation

   (Maximum repulsion)

$$-\overset{|}{\underset{|}{C}}-C\equiv N$$

$$-\overset{|}{\underset{|}{C}}-C\equiv N$$

$$E(\uparrow\uparrow) = +\mu^2/r^3$$

III. Parallel end-to-end orientation

$$-\overset{|}{\underset{|}{C}}-C\equiv N \qquad -\overset{|}{\underset{|}{C}}-C\equiv N$$

$$E(\substack{\uparrow\\\uparrow}) = -2\mu^2/r^3$$

Figure 2 Types of dipolar interactions between nitrile groups: (E) dipolar interaction; (μ) dipole moment; and (r) vector between dipoles [12].

nitriles on adjacent chains will be positioned in the antiparallel orientation to produce an attractive interaction. Thus we can begin to see how chain conformation leads to a solid-state structure where the repulsive interactions are minimized and attractive interactions are maximized.

Krigbaum and Tokita [14] performed potential energy calculations on both the syndiotactic and isotactic isolated PAN chain to determine the net dipolar interactions among the nitrile groups, and in both cases the net interaction was always repulsive. They also cited the melting behavior of PAN gels and solution properties of PAN in concluding that the chain conformation must be extended and highly irregular. The above mentioned work supports the concept of strong dipolar interactions controlling the individual chain conformation and solid-state structure. In the next section we discuss the x-ray and electron diffraction studies of PAN fibers to understand how crystalline order could emerge from the packing of these highly polar atactic chains.

C. Diffraction Studies of the Crystalline Structure

1. Analysis by X-ray and Electron Diffraction

The crystalline structure of a fiber is determined by analysis of the x-ray diffraction pattern. In textile fibers the polymer chains are always oriented parallel to the fiber axis so that the tensile forces lie along the covalent bonds of the polymer chains.* The diffraction pattern is generated by mounting the fiber vertically and impinging the x-ray beam at 90° to the fiber, as shown in Figure 3a. Scattered x-rays are detected by the film or x-ray detector mounted behind the fiber (Figure 3b). If crystallites are present then discrete reflections off of the Bragg mirror planes will generate a pattern of spots on the film. Proper analysis of the film pattern gives the unit cell structure and conformation of the individual chains.

The relationship between the spatial orientation of the crystalline phase and the diffraction pattern can best be seen by reference to Figures 3a and 3b. We let axis X be collinear with the incident x-ray beam (horizontal), axis Z be collinear with the fiber (vertical), and finally axis Y be mutually normal to X and Z, and hence normal to the fiber.

Now consider how various reflection planes within the fiber might reflect the x-rays. The incident x-ray beam is in the X-Y plane. Any Bragg planes that are parallel to Z, the fiber axis, would-reflect the x-rays such that they remain in this plane. These are called the equatorial reflections. In order for the xrays to be reflected above or below the X-Y plane, the corresponding Bragg planes must be tilted with respect to the fiber axis, and these reflections out of the X-Y plane are termed off-equatorial. There is a major distinction between equatorial and off-equatorial reflections. The Bragg planes responsible for the equatorial reflections can be uniquely defined by a two-dimensional unit cell arising from a high degree of packing in the direction normal to the fiber axis [1]. In other words, a regular packing in the X-Y plane is sufficient to define Bragg planes that will produce the equatorial reflections. The requirement for off-equatorial reflections is stricter, for now the Bragg planes will be defined by intersections with all three unit cell axes, including now the axis parallel to the fiber. The existence of a three-dimensional unit cell requires a high degree of order along the fiber axis, which means that all of the polymer chains must assume the identical regular conformation, whether it happens to be planar zigzag or some helical conformational. If these requirements are not quite met, then the off-equatorial reflections will begin to broaden and eventually vanish.

At this point it is instructive to compare the fiber diffraction patterns of polyethylene terephthalate and acrylic fibers, which are shown in Figure 4. The fiber axis is vertical and the equatorial direction is horizontal. The pattern for polyester shows both the equatorial and off-equatorial reflections indicative of three-dimensional order, whereas only equatorial reflections are found for the acrylic fiber. The unit cell of the polyester is triclinic [15]. The polymer chain is planar zigzag and is parallel with the c-axis of the unit cell. Bohn, Schaefgen,

---

*This applies to both synthetic and natural fibers. In synthetics the orientation is imparted by the spinning and drawing operations and in natural fibers the orientation is developed during the fiber formation by the living organism.

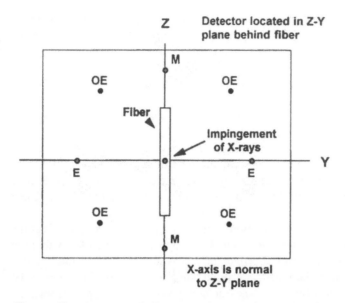

Figure 3a Geometry of fiber scattering experiment showing Z-Y plane. The fiber is mounted along the Z-axis and the incident x-ray beam arrives along the X-axis and is normal to the Z-Y plane. The polymer chains are oriented in the direction parallel to the long axis of the fiber, hence they are also oriented along the Z-axis. The x-ray detector, either film or electronic, is mounted in the Z-Y plane behind the fiber. The three possible types of Bragg reflections for a fiber are shown: E, equatorial; OE, off-equatorial, and M, meridional.

and Statton [16] published the first detailed model of the PAN fiber morphology, using combined analysis of the diffraction pattern with infrared dichroic and volumetric expansion measurements. They observed several sharp equatorial reflections, but none that were off-equatorial. The equatorial reflections could be indexed in a two-dimensional hexagonal lattice such that the intense 5.2-Å reflection (the most intense equatorial reflection seen in the acrylic fiber pattern) would correspond to an interchain distance of 6 Å. The hexagonal lattice is thought to arise from the packing of adjacent chains. It is assumed that a single chain will, on the average, occupy a cylinder with a diameter of 6 Å, and the equatorial reflections then arise from the packing of these cylinders somewhat like the packing of sticks of chalk in a box. The conformation of the chain within the cylinder would be extended in a highly irregular fashion, and therefore true crystallographic order in the direction of the fiber axes would be highly unlikely. The absence of off-equatorial reflections would be expected with this fiber given that the polymer is atactic.

This model explains the origin of the intermolecular bonding among neighboring chains. The proposed extended, kinked chain will not fit into a 6-Å

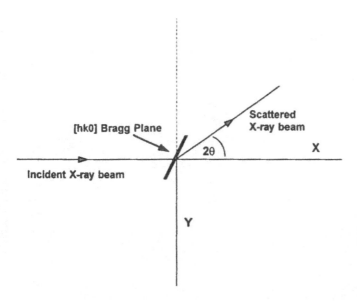

Figure 3b Geometry of the scattering experiment showing the *X-Y* plane. The [*hk*0] Bragg planes (1) are parallel to the fiber direction, which is mounted along the *Z*-axis. Therefore all scattering falls within the *X-Y* plane, and results in equatorial reflections as shown in Figure 3a. Similarly, for the meridional reflections in Figure 3a, the Bragg planes must be normal to the *Z*-axis and hence be of the [001] type, and for the off-equatorial reflections in Figure 3a the planes must be inclined with respect to the *Z*-axis and hence be of the [*hkl*] type. The angle 2Θ is half of the Bragg angle Θ.

cylinder unless some of the nitrile groups extend beyond the confines of the cylinder. This is shown in Figure 5. These outlying groups are potentially available for intermolecular bonding because they will be oriented in the antiparallel orientation that produces a net attraction (see Figure 2). In support of this model one can cite the work of Saum [17] who has shown that dipolar interactions of the antiparallel type in simple nitriles can lead to the formation of dimers. While relatively few of the nitriles in the PAN chain will be able to participate in perfect antiparallel bonding pairs, still many attractions will be present. The term "laterally bonded" has been used to describe this two-dimensional crystalline morphology. It has also been compared to the liquid-crystalline state where stiff rodlike molecules will pack together, giving two-dimensional order [18].

Many efforts have been made to coax more information from the acrylic fiber pattern by analyzing samples with very high polymer chain alignment. This approach generally improves the resolution of the fiber diffraction patterns so that mathematical relationships between the diffraction pattern and the chain conformation can be applied. One method of obtaining this alignment is to grow

(a)   (b)

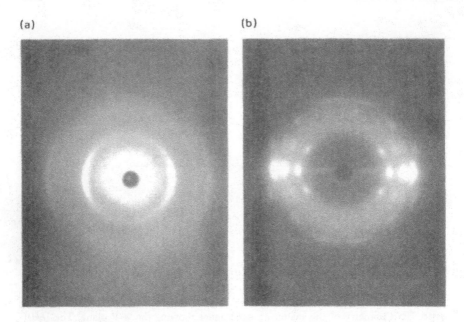

Figure 4 X-ray diffraction patterns of (a) commercial acrylic and (b) polyethylene terephthalate fibers. The acrylic fiber was a commercial sample containing 7% vinyl acetate comonomer. In both cases, the fiber axis is vertical.

individual crystals of polymers isothermally from solution [19]. The highest levels of crystallinity and crystalline perfection in synthetic polymers have been obtained in this manner because the polymer concentration is kept very low (less than 1%) during the crystallization, thus virtually eliminating the chain entanglements that limit crystalline perfection during crystallization from the melt or more concentrated solutions. These polymer single crystals generally have a lamellae shape, and are typically 10–20 microns wide and several hundred angstroms thick. Electron diffraction analysis shows that the individual polymer chains are oriented parallel to the thickness direction. Since individual polymer chains are many times longer than the crystal thickness, it must be concluded that the chains are folded back and forth in a regular manner. Holland et al. [20] were able to grow elliptically shaped, lamellar, single crystals of PAN from propylene carbonate. The crystals were approximately 100 Å thick and the electron diffraction patterns indicated that the polymer chain axis was normal to the lamellae surface.

2. Analysis of PAN Single Crystals and Fibers

One might expect some evidence of order in the chain axis direction in these PAN single crystals, since this is generally the way to produce polymers with

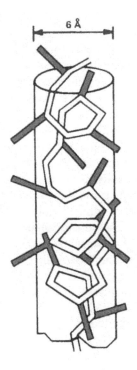

6 Å

Figure 5 Model of assumed rigid, irregularly helical conformational of the polyacrylonitrile chain as it would exist in the solid-state polymer [12].

the highest degree of crystalline perfection. Klement and Geil [21] carefully analyzed the x-ray and electron diffraction patterns of PAN single crystals. Up to nine equatorial reflections were reported, which means that the lateral packing of the chains in these single crystals is much improved over the packing in the fibers discussed earlier, but still no off-equatorial reflections were observed. The equatorial reflections were indexed as a two-dimensional orthorhombic unit cell instead of a hexagonal cell. This was required since the intense 5.2-Å reflection observed in the fiber patterns (see Figure 3) appeared as a doublet. The unit cell dimensions were as follows: $a = 21.18 \pm 0.04$ Å and $b = 11.60 \pm 0.02$ Å.

Other methods for forming PAN single crystals have been reported. Patel et al. [22–24] prepared single crystals from propylene carbonate using a film formation method. Isothermal crystallization was carried out on glass slides by evaporating solvent from a 0.5–1% solution in an atmosphere of the same solvent. The electron diffraction spots could be indexed on the basis of the orthorhombic unit cell proposed by Klement and Geil [21]. Kumamaru et al. [25] prepared single crystals during the process of solution polymerization. The acrylonitrile monomer was dissolved in PAN solvents (dimethylformamide (DMF), dimethylacetamide (DMAC), and dimethylsulfoxide) and heated without polymerization catalyst at elevated temperatures. They were able to obtain a few weak and diffuse off-equatorial reflections, in addition to the expected equitorial reflections observed by Klement and Geil [21]. Using the diffuse off-equatorial reflections they proposed an orthorhombic unit cell of $a = 21$ Å, $b = 1.19$ Å, and $c = 0.504$ Å, where the c-axis is parallel to the fiber and polymer chain axes. Electron diffraction analysis indicated that the polymer chains were oriented normal to the lamellae surfaces, and fibrous materials spanning cracks in the crystal strongly suggested that the chains within the crystal were folded. The authors commented on the tendency of "atactic" PAN to form chain-folded lamellar crystals and attributed this to the formation of strong dipolar interactions among adjacent chains, with the lack of tacticity reducing the overall crystallinity. An analogy was made with atactic polyvinyl alcohol, which has fairly high crystallinity on account of strong intermolecular hydrogen bonding.

Rather striking evidence of true three-dimensional crystallinity in PAN was reported by Colvin and Storr [26] in very highly oriented PAN fibers. These fibers were spun from an aqueous sodium thiocyanate solution, drawn 10× in steam, and then after drying they were given an additional 1.5× stretch by drawing over a hot surface. They observed four sharp off-equatorial reflections in addition to the equatorial reflections discussed previously. All of the reflections could be indexed in an orthorhombic unit cell having the following dimensions: $a = 21.48 \pm 0.02$ Å, $b = 11.55 \pm 0.03$ Å, and $c = 7.096 \pm 0.03$ Å. Note that the a and b dimensions are identical to those reported by Klement and Geil [21]. The new off-equatorial reflections indicate an increase in perfection along the chain axis, apparently a consequence of the extremely high orientation in the fibers. This unit cell will accommodate 24 monomer units with 6 located on the base plane formed by the a and b axes and then 4 monomer units in the c direction. The c-axis repeat distance of 7.096 Å and the appearance of a strong [004] reflection indicates a monomer-to-monomer translation distance of 1.774 Å, which gives a structure reminiscent of syndiotactic polypropylene [27]. This structure locates the nitriles in a manner that will minimize the repulsions among adjacent groups. The calculated density is 1.199 ± 0.002 g/cm^3. This agrees well with experimental values of 1.197–1.20 g/cm^3 reported for fibers in the absence of microvoids [28].

The apparent success of Colvin and Storr in achieving three-dimensional crystallinity in PAN is somewhat surprising given the experience of previous investigators, particularly those using single crystals. However their method of fiber preparation is one that would be expected to produce very high levels of chain orientation, which may have resulted in the segregation of chain segments rich in syndiotactic sequences, even though the polymer itself is thought to have a low degree of stereoregularity. It is apparently the syndiotactic isomer, by virture of the location of the nitriles, that allows the chains to assume a regular conformation without excessive intramolecular nitrile repulsions.

3. Evidence for Two-Phase Morphology and Determination of Crystallinity

Single-Phase Morphology Having established the presence of a rather well-defined laterally bonded structure, the next issue is the characterization of the overall polymer and fiber morphology. Many investigations have focused on the existence of multiple levels of order somewhat akin to the classical two-phase structure. The classical work of Bohn et al. [16] indicated that the laterally bonded crystalline structure existed throughout the fiber, and that a typical amorphous phase was absent. They noted the absence of the characteristic amorphous scattering halo in the fiber patterns, but instead they found diffuse scattering throughout the pattern. This suggested that the laterally bonded structure was essentially a single phase that contained many defects. One can see in Figure 4 that the amorphous scattering halo is quite pronounced in the polyester fiber diffraction pattern, but is largely absent in the corresponding pattern of the acrylic fiber. Instead, one observes diffuse scattering located in a narrow band at a Bragg spacing of approximately 3.4 Å. This diffuse scattering is sometimes concentrated in the quadrants of the diffraction pattern, and although the origin of the scattering is not completely understood, it is generally agreed that it is not the typical scattering of an amorphous polymer phase. Lindenmeyer and Hoseman [29] applied the theory of paracrystallinity to the acrylic fiber diffraction pattern and concluded that the diffuse off-equatorial scattering could arise as the net result of having different conformations distributed along the chain, with no single conformation persisting over a long distance. The same authors point out that attempts to alter the density of PAN by drawing and annealing have been unsuccessful. Such treatments normally will affect the degree of crystallinity and hence the density of a semicrystalline polymer.

If separate amorphous and crystalline phases did exist for PAN, then one ought to be able to observe the glass transition of the amorphous phase. The measured glass-transition temperature for PAN is generally considered to be in the range of 80–90°C when low-frequency techniques are used (glass transition studies will be reviewed in Section IE1). Bohn et al. [16] compared the polymer volumetric expansion coefficient with the temperature dependence of the 5.2-Å Bragg spacing and observed an increase in both parameters at 85°C. This comparison, shown in Figure 6, is inconsistent with the classical two-phase model, since the glass transition of the amorphous phase should have little effect on the chain packing within the crystalline phase. These observations are taken in

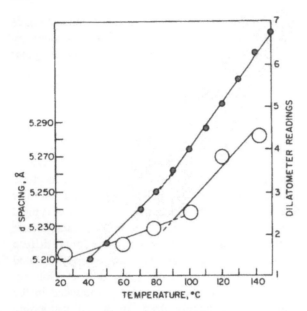

Figure 6 Thermal expansion of polyacrylonitrile. Open circles refer to diffraction *d* spacings, closed circles to dilatometer readings [16].

support of a hybrid single-phase morphology that has both crystalline and amorphous polymer properties.

Determination of Crystallinity Attempts have been made to determine a degree of crystallinity for acrylic polymers and fibers [30–33] via the classical crystallinity analysis of the x-ray diffraction patterns, where the total scattering intensity is divided among the crystalline and amorphous scattering contributions [1]. The percent crystallinity by this method is defined as the ratio of the intensity under the crystalline peaks divided by the total intensity; for the two-phase model the latter must be partitioned among the crystalline and amorphous scattering. In order for this approach to give absolute values it is necessary to have standards for the polymer in the completely amorphous state.* The method is useful even if good standards are not available because it yields relative differences in crystallinity among samples, e.g., a relative crystallinity index can be defined.

Gupta and Singhal [32] have reviewed methods for defining a crystallinity index for acrylic polymers. These methods are compared in Figure 7. The meth-

---

*This method works well for polymers such as polyethylene terephthalate where completely amorphous standards can be prepared by rapid quenching of the molten polymer. The x-ray diffraction of a polymer in the molten state can also serve as the amorphous standard.

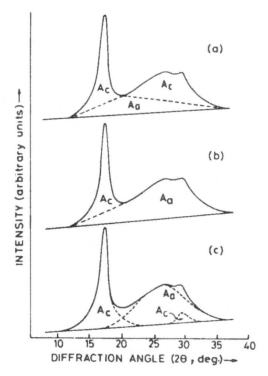

Figure 7 Schematic representation of extrapolation of crystalline and amorphous components in the x-ray diffraction pattern of PAN: (a) Hinrichsen's method [31], (b) Bell and Dumbleton's method [30], and (c) method of Gupta and Singhal [32].

ods of Hinrichsen [31], coefficient of crystallinity, and Bell and Dumbleton [30], crystalline index are shown along with the authors' preferred method, degree of order. The differences among the methods concern the proper identification of the scattering components from the crystalline and amorphous regions. Hinrichsen's coefficient of crystallinity gives higher values on a given sample because a large fraction of the broad hump centered near $2\Theta = 27°$ is apportioned to the crystalline component. In the other two methods most of this scattering near $2\Theta = 27°$ is considered to be amorphous. The degree of disorder method of Gupta and Singhal can be considered to be a refinement of the crystalline index method of Bell and Dumbleton since the intensity of the small crystalline peak at $2\Theta = 30°$ is now included in the crystalline scattering intensity.

Matta et al. [33] have utilized an amorphous standard in determining the crystallinity of acrylic fibers used as precursors in the manufacturing of carbon fibers. The amorphous standard was prepared by dissolving PAN in dimethylformamide, casting a film of the solution, and immediately pouring ice cold water on the film. The resulting precipitate was taken to be essentially amorphous because the major crystalline peak appearing at $2\Theta = 17°$ was very weak. It was assumed that the precipitate was essentially free of solvent. The diffraction pattern of this material was then superimposed on the patterns of the actual samples for location of the amorphous scattering contribution. By this method crystallinities from 64% to 66% were reported for two commercial PAN fibers commonly used as precursors to make carbon fibers. Modification of one of the precursors by stretching an additional 60% in nitrogen increases the crystallinity to 70%, by this method, and the authors also report improvements in physical, mechanical, and thermal properties [34]. Amorphous PAN has also been prepared by Imai et al. [35] and Joh [36] using the catalyst bis(pentamethyleneimino)magnesium in heptane at 70°C. This material was used in the analysis of the dynamic-mechanical properties of PAN, which will be discussed later. A good comparison of the x-ray diffraction curves of this material and conventional "crystalline" PAN materials is shown in Figure 8. The amorphous material did not become more crystalline upon heating, which does occur to a limited extent with normal PAN [31, 32] polymer and fiber, and thus supports the claim that it is amorphous. The main crystalline peak at $2\Theta = 17°$ is very

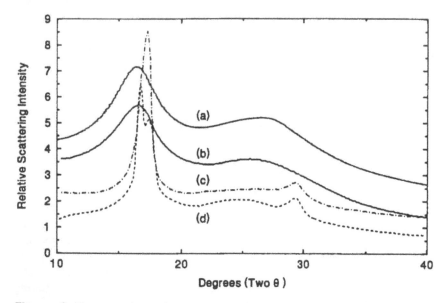

Figure 8 The x-ray diffraction pattern for the amorphous polyacrylonitrile as polymerized (a); for this polymer heat-treated at 180°C for 3 min in an air oven (b); hexagonal crystalline form obtained from the conventional PAN (c); and orthorhombic form from the single crystal lamellae mats of conventional PAN (d) [36].

broad in the amorphous material and the crystalline peak at $2\Theta = 29°$ is absent. The cause of the amorphous nature was not clear to the investigators; a high degree of chain branching was given as the most likely cause.

Annealing Studies A number of investigators have performed drawing and annealing experiments on acrylic polymers, copolymers, and fibers that provide support for a limited two-phase model of the structure [31, 37–41]. Hinrichsen and Orth [37], for example, have observed periodicities on the order of 120–105 Å in the small-angle x-ray pattern of drawn films of the PAN homopolymer. Electron micrographs of the surface of films etched with activated oxygen reveal striations approximately 105 Å thick that are oriented in a normal position to the draw direction. The small angle period is qualitatively similar to the long period observed in semicrystalline fibers such as nylon and polyester where it is associated with the thickness of chain-folded crystalline blocks [42]. Recall that the thickness of the PAN single crystals is also approximately 100 Å, and the electron diffraction pattern confirmed the existence of chain folding. These observations strongly suggest the presence of a two-phase structure, perhaps with lamellar crystalline units being separated by less ordered material.

The increase in the long period of semicrystalline polymers and fibers upon annealing is well known and is a consequence of the corresponding increase in the thickness of the chain-folded lamellae [43]. Hinrichsen [31] has observed a similar behavior in drawn PAN fibers. The small angle patterns corresponding to different annealing times appear in Figure 9 and in Table 1. The long period is given for different annealing temperatures.

A two-phase model for the structure of acrylic fibers has been proposed by Warner et al. [40], who studied the oxidative stabilization step of these fibers

Table 1 Long Period and Maximum Intensity of Small-Angle Reflections of Drawn Polyacrylonitrile Fibers

Annealing conditions	Long period (Å)	Intensity (arbitrary units)
Original (drawn 1:10 at 110°C)	105	43
60 min at 140°C	110	40
10 min at 180°C	106	30
60 min at 180°C	108	50
300 min at 180°C	106	80
10 min at 220°C	127	30
60 min at 220°C	134	145
300 min at 220°C	149	170

Source: Data by Hinrichsen [31].

(a)   (b)

(c)   (d)

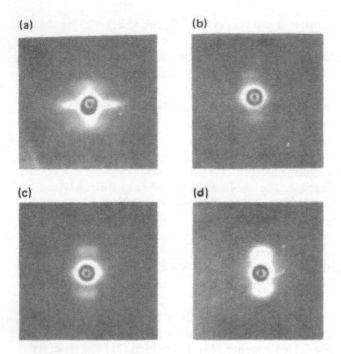

Figure 9 X-ray small-angle diagrams of polyacrylonitrile films: (a) unannealed; (b) annealed for 1 hr at 140°C; (c) annealed for 1 hr at 180°C; and (d) annealed for 1 hr at 220°C [31].

that is used during the preparation of graphite fibers. Acrylic fibers are commonly used as precursors for graphite fibers, and the oxidative stabilization is a chemical change that is part of the conversion process. They investigated the development of periodic density fluctuations along the fiber axis under conditions of oxidative stabilization, with the expectation that the rate of chemical change would be influenced by the degree of order, i.e., in analogy to chain scission reactions in semicrystalline polymers where the reaction proceeds more rapidly in the amorphous regions.

In the model of Warner et al., the fiber is composed of fibrillar subunits that contain distinct regions of amorphous and partially ordered material. The model utilizes the laterally bonded concept of Bohn et al. [16] for the partially ordered phase, though the comparison to liquid-crystalline order is made. A drawing of this model is shown in Figure 10. The ordered or laterally bonded crystalline phase is thought to have a lamellar texture oriented perpendicular to the fiber axis. Evidence for the lamellar texture comes from small-angle x-ray diffraction patterns of the fiber and transmission electron micrographs of fibrils showing striations with periodicities ranging from 80 to 140 Å. The fibrils were prepared

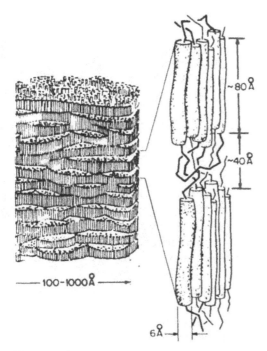

Figure 10 Model of molecular structure of highly oriented acrylic fibers [40].

by subjecting the fibers to ultrasonic treatment in aqueous dimethylformamide solutions. The agreement with similar periodicities found by other investigators is quite good.

Additional small-angle x-ray diffraction patterns by Warner [41] suggest that rows of lamellae can be formed that are oriented at an angle with respect to the fiber axis. A four-point pattern was observed in a heat-treated Orlon 42 type acrylic fiber. Although four point patterns are often observed in oriented semicrystalline polymers, this particular pattern was unusual in several regards. The angular inclination of the lamellae normals (the direction normal to the chain-folded lamellae surface) from the fiber axis was estimated to be approximately 15 Å, which is much less than is commonly observed in polymers such as polyethylene, and the largest lamellae dimension appeared to be the thickness rather than the width, which is usually the case [19].

Dielectric Analysis Additional evidence for a two-phase morphology can be taken from the work of Gupta and co-workers who systematically investigated the influence of thermal treatment and incorporation of comonomers on the structure [44–52, 32]. Their analysis primarily utilized dielectric relaxation measurements, x-ray diffraction and infrared spectroscopy. The dielectric constant and dielectric loss were typically measured from 40 to 160°C at frequencies from 10^2 to 10^5 on disks of compressed polymer powder. The single peak in the tan δ loss curve shifted from 110 to 150°C with increasing frequency in a manner consistent with the WLF theory of viscoelastic behavior [53, 54] (this will discussed in more detail in Section IE). A large increase in the height of the tan δ loss maximum was observed with increasing temperature; this is not normally the case, and the following explanation is proposed. Below the T_g the molecular chains are not only immobile but tightly bound at some points due to dipole-dipole interactions of the nitrile groups. As the temperature increases, more and more nitrile groups are released from their bound state and at the same time polymer chains begin to develop segmental mobility as the T_g is approached. Above the T_g not only does the segmental mobility increase with temperature but the net dipole moment of the segments also increases on account of the dipole moment of the free nitrile groups attached to the segments. As the temperature further increases, more and more bound nitrile groups are released, presumably due to higher segmental mobility, and thus contribute to further increase in the net dipole moment of the segments.

The tan δ loss peak characteristics (temperature, height, width, and dielectric strength) were found to vary systematically with temperature and duration of heat treatments [48]. Measurements were made of PAN disks that were unheated and heated at 120 and 160°C for 24 hr in air. At each frequency the heat treatment caused the height of the tan δ loss peak to decrease in comparison to the unheated control, and unexpectantly, the magnitude of the decrease was greater for the 120°C treatment (tan δ maximum at 1 kHz: unheated control, 0.27; 120°C, 0.12; and 160°C, 0.21). The other dielectric parameters showed a similar reversal in magnitude. The possibility that chemical degradation could be an interfering factor was ruled out by conducting the heat treatment under vacuum, and the same phenomenon was observed; in fact the magnitude of the changes were slightly greater [50]. The heating also caused the crystallinity and crystalline perfection to increase. The crystallinity index increased from 28.5% for the unheated polymer to 32.0 and 34.5% for the polymers heated in air and vacuum, respectively, at 120°C for 16 hr, and the diffraction peaks were sharper, which is evidence of larger and more ordered crystals. The intensity of the 2240 cm^{-1} nitrile infrared stretching vibration also decreased after the heat treatment; chemical changes were ruled out since the polymer did not become discolored. It was proposed that the heat treatments caused a slight crystallization of the samples. This crystallization increased the degree of nitrile dipolar bonding, hence the reduction in the height of the tan δ loss peak. The reason given for the effect being stronger at 120°C versus 160°C was that the crystallization occurs more slowly at the higher temperature since the degree of molecular mobility exceeds that which is required for maximum crystallization rate. The maximum rate will be observed at a temperature that is sufficient to provide the molecular mobility required for the chain segments to orient so that the degree of nitrile dipolar bonding can become maximized. At higher temperatures this process will be less efficient because the increased segmental mobility will be disrupting some dipolar bonds at the same time that others are being formed.

This effect of heating time on the dielectric parameters was also examined [52]. The tan δ loss maximum decreased continuously up to 16 hr and then the trend reversed; a similar pattern was observed in the other dielectric parameters. The T_g, derived from WLF analysis of the loss curves, increased continuously up to 16 hr and then decreased. The origin of the reversal is not clear, but analogous effects are observed in dynamic-mechanical spectra of conventional semicrystalline polymers [55]. The T_g of amorphous PET increases upon crystallization, reaches a maximum, and then decreases with further crystallization. This is explained in terms of the amorphous volume per crystallite. The initial crystals are very small and as the crystallinity increases, the average volume of the amorphous domains decreases. The T_g arises from these amorphous domains and their small size puts a constraint on the chain and hence the T_g increases. This continues until rather high levels of crystallinity are reached at which point the average crystallite size begins to increase (the crystals are larger and fewer in number). Therefore the average size of the amorphous domains also increases. This reduces the constraints on the chains and the T_g now begins to decrease.

Gupta and Maiti [49] studied the effects of heat treatment on the structure and properties of acrylic fibers. Drawn PAN fibers were heated at 110 and 150°C for 24 hr. Again, the order increased with heat treatment and the trend reversed slightly at the higher temperature. Measurement of the tensile properties showed that the fibers broke at higher loads with heat treatment and the breaking elongation decreased slightly as shown below:

Annealing temperature	Stress at break	Elongation at break
Unheated	33.5 (N/tex × 10^2)	38%
110°C	40 (N/tex × 10^2)	35%
150°C	35 (N/tex × 10^2)	33%

Effects of Comonomers on Morphology The effect of comonomers on the morphology has been the subject of several investigations. Gupta and coworkers investigated 2-hydroxyethyl methacrylate [44, 45], methacrylonitrile [51, 32], and methyl methacrylate [32], Kulshreshtha *et. al.* investigated methyl acrylate [56], and Frushour did a similar study with vinyl acetate [57]. In all cases addition of comonomer reduces the crystallinity and crystalline perfection. The changes in the structure caused by addition of the comonomer are not uniform

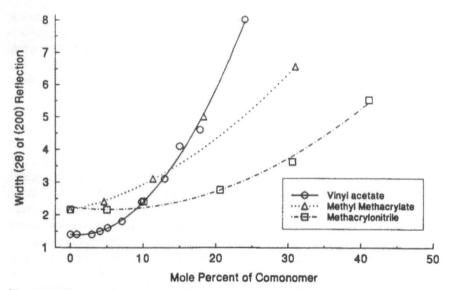

Figure 11 Response of the major equatorial [200] reflection to incorporation of the following comonomers: vinyl acetate, methyl methacrylate, and methacrylonitrile.

over the entire range of composition, but instead a critical range of comonomer is reached at which the characteristic PAN morphology becomes highly disordered. This can be seen in Figure 11 where the line width of the most intense reflection is plotted against the mole fraction of comonomer for three copolymers. This reflection originates from the [200] Bragg plane and corresponds to an average chain separation of 6 Å. An increase in the half-width indicates a loss in crystallite size and perfection. The increase in peak width becomes significant at approximately 8–10 mole % for the vinyl acetate and methyl methacrylate copolymers, but significantly more methacrylonitrile, e.g., approximately 20 mole % is required to achieve approximately the same level of disorder. Vinyl acetate and methyl methacrylate are similar in size, and both larger than the methacrylonitrile group, hence would be more effective in disrupting the crystalline morphology. Also the methacrylonitrile unit retains the nitrile dipole, so replacement of acrylonitrile groups with the latter would not reduce the level of intermolecular dipolar bonding to the extent that would be expected with the other comonomers. This effect of comonomers on the morphology will be discussed again later in Section ID along with the melting behavior of copolymers. The crystallinity of the copolymers can be increased by heat treatment, though even after heating the copolymers are less crystalline than the heated homopolymer [51, 32].

Swelling Studies Additional evidence for a multiphase morphology is the complex swelling behavior observed by Andrews et al. [58–60], and Grobelny and co-workers [61–65]. The equilibrium weight gain in aqueous solutions of iodine-potassium iodide was measured as a function of concentration. Multiple steps in the absorption curve were observed. These were interpreted as the penetration of the solution into domains of increasingly higher order within the polymer.

In summary, one can conclude that a strong case can me made for the existence of at least a limited two-phase morphology for acrylic fibers. The differences in order between the two phases may be much less when compared to conventional melt-spun fibers. Furthermore, the two phases may be highly coupled since the chains in the less ordered region may be rather stiff and extended due to the presence of the intermolecular dipolar bonds. We now move on to a discussion of thermal properties and molecular relaxations. A number of these studies provide additional support for the concept of a two-phase morphology in PAN homopolymers and copolymers.

D. Thermal Properties: Melting, Gelation, and Crystallization

1. Evidence for Melting of Acrylic Polymers

Semicrystalline fibers such as polyethylene terephthalate (PET), nylon 6,6, and isotactic polypropylene will melt upon heating and then recrystallize upon cooling. Only the crystalline domains participate in these phase changes. The amorphous domains, however, can undergo a glassy to rubbery transition in a manner similar to completely amorphous polymers such as atactic polystyrene. The nature of the glass transition will be discussed in more detail in Section IE. A distinction should be made between polymers that are completely amorphous by nature of their chemical structure and the amorphous domains in semicrystalline polymers. Glassy polymers such as polystyrene are amorphous from their lack of stereoregularity. The chains simply cannot be packed into a regular lattice. However, in PET and other semicrystalline fibers, the chemical structure of the polymer chains in the crystalline and amorphous regions are identical and the amorphous domains exist only because of the steric restraints such as entanglements that develop upon crystallization from the melt. These amorphous domains do undergo a typical glass transition, and in the case of PET and nylon 6,6 the T_g's are approximately 80°C and 50°C, respectively. In fact, when a slow crystallizing polymer like PET is quenched from the melt it can be completely amorphous, and crystallinity will develop upon annealing at temperatures above the glass transition.

The melting of polymers is conveniently studied using differential scanning calorimetry (DSC) and from these DSC thermograms one can extract much information about the crystalline morphology.* From the area under the melting endotherm one can obtain the experimental heat of fusion, $\Delta H_{f(exp)}$, which is a relative indicator of crystallinity. Processes such as drawing and annealing that increase crystallinity will lead to a corresponding increase in the $\Delta H_{f(exp)}$. The glass transition of the amorphous domains can also be followed using DSC. The T_g appears as an inflection in the baseline of the DSC thermogram.

The existence of DSC melting endotherms and crystallization exotherms could be taken as good evidence that some type of crystalline order exists in PAN. However, obtaining good melting data for acrylics is difficult. PAN homopolymer and the fiber-forming copolymers belong to the class of polymers that degrade prior to melting.† PAN undergoes a degradation reaction at elevated temperatures in which the adjacent nitrile groups on the polymer chain cyclize to form six-membered rings [66]. This reaction obviously precludes normal melt processing for acrylics, and happens to be the basis for production of graphite fibers from PAN. Chapters 8 and 11 further discuss the degradation mechanism and its applications in carbon fiber production.

When PAN is heated in the DSC at normal heating rates of 20°C/min. the cyclization reaction produces a large exotherm. However, if the heating rate is sufficiently high, then some melting will occur before the polymer degrades, and this melting endotherm can detected in the thermogram. This can be clearly seen in Figure 12 where differential scanning calorimetric (DSC) thermograms of PAN are compared at heating rates of 20, 40, and 160°C/min. At the lowest heating rate the reaction exotherm appears at 315°C and no trace of a melting endotherm is seen. Increasing the heating rate to 40°C/min. shifts the exotherm to 330°C and now the beginning of an endotherm near 319°C can be detected. The melting endotherm becomes quite evident at a heating rate of 160°C/min., because the exotherm has shifted to 390°C. Obviously no crystallization

---

*In DSC the sample and an inert reference material are heated at a constant rate of rise and the differential heat flow between them is continuously adjusted to keep the difference in temperature equal to zero and the average temperature equal to the programmed rate. When the sample undergoes an endothermic transition, such as melting, the direction of net heat flow is into the sample and out of the reference. When the sample is completely melted the heat flow reverses until the state of zero temperature difference is regained. The instrument produces a plot of differential heat flow verus programmed temperature. An endothermic transition produces a peak and the area under the peak is equal to the amount of energy absobed by the sample. Dividing by the sample weight gives the $\Delta H_{(exp)}$, the experimental heat of fusion. The breadth of the transition reflects the melting range. For polymers, the melting point is usually taken as the peak maximum. Exothermic transitions give rise to peaks in the opposite direction, with respect to the heat flow axis. DSC has largely supplanted an older thermal technique known as differential thermal analysis (DTA), in which the difference in sample and reference temperature are measured as they both are heated at the programmed rate. Both DSC and DTA give accurate transition temperatures, but heats of transition are generally only obtained using the DSC.

†In this regard acrylics are similar to the natural fibers, cotton and wool, and synthetics made from cellulose, such as rayon and cellulose triacetate.

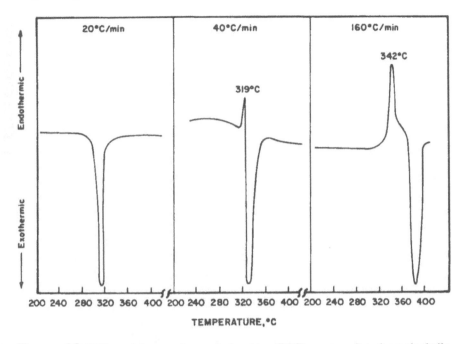

Figure 12 Differential scanning calorimetric (DSC) scans of polyacrylonitrile homopolymer in nitrogen atmosphere [83]. Exothermic peak arises from the cyclic degradation reaction [66].

exotherm will be observed upon cooling because the polymer will have completely degraded after completion of the cyclization exotherm.

The use of very high heating rates in DSC measurements of melting are generally undesirable because thermal lag and possible superheating of the sample will tend to increase the apparent melting point and broaden the endotherm. However, these problems are unavoidable in this case; the experiment should not be interpreted quantitatively, but rather as evidence that a phase transition resembling melting can be observed.

Ultrafast heating experiments by Hinrichsen [38], provide corroborating evidence that the DSC endotherm can be attributed to melting. PAN fibers were immersed in a bath of molten metal for several seconds, quenched in an ice bath, and then examined by scanning electron microscopy and x-ray diffraction for evidence of melting. The heating rate was estimated to be near 1000°C/min. Clear evidence of melting was detected for a metal bath temperature of 320°C.

We conclude that PAN undergoes a melting-like transition in the range of 320–330°C. This represents a relatively high melting point for polymers, and is characteristic of stiff chains. Both nylon 6,6 and PET fibers melt near 265°C, and polyolefins such as polyethylene and polypropylene melt around 135°C and 165°C respectively [67].

2. Effects of Diluents and Comonomers on Melting Behavior

One goal in studying the melting behavior of a polymer is to derive values of basic thermodynamic parameters for the pure crystalline polymer. These include the melting point, and the heat and entropy of fusion associated with melting. The latter are essential if the thermal behavior of the polymer is to be understood in the context of the chain structure and morphology. We also desire to understand what happens during the melting of the unique PAN morphology. Fortunately, it is not necessary to work exclusively with the pure homopolymer to obtain this information. We will now discuss some investigations into the melting and crystallization behavior of the PAN homopolymer and copolymers in the presence of diluents.

When PAN solutions of sufficient concentration are allowed to stand at room temperature the viscosity begins to increase, and eventually a rubbery gel is formed. This gelation is thermoreversible, i.e., the material will return to the original viscosity when heated. The gel is formed because some of the polymer chains phase separate and form very small ordered regions, or microcrystallites, that serve as reversible crosslinks. The melting of the gel upon heating corresponds to the melting or redissolution of these crystallites.*

Krigbaum and Tokita [14] studied the melting behavior and gelation of PAN gels prepared from concentrated solutions of the polymer in dimethylformamide and γ-butyrolactone. A recording dilatometer was used to follow the volume change during gelation and subsequent gel melting. The PAN solution will contract upon gelation and expand upon melting in accordance with the crystallization and melting behavior of semicrystalline polymers. Polymer solutions were prepared over a wide concentration range and then allowed to gel by cooling to a temperature where the gelation rate was maximized. The gelled solution was then transferred to the dilatometer and heated. In most cases, two transitions in the thermal expansion coefficient could be detected. The lower temperature transition was the smaller of the two and attributed to the polymer glass transition, and the larger and higher temperature transition was the melting of the gel. In Figure 13 the transition temperatures are plotted as a function of polymer volume fraction for the two solvent systems. The gel melting points increase rapidly with polymer concentration, and extrapolation to pure polymer gives a PAN melting point of 317°C, which is in good agreement with the directly measured values discussed previously.

---

*Dissolution and melting of a crystalline polymer are very similar. In order for a crystalline polymer to dissolve when put into a solvent the crystalline phase must melt. Solvents will depress the melting point, and if the polymer dissolves at a specific temperature, then the melting point will have been depressed below that temperature. If not, the amorphous domains will swell with the solvent but the crystallites will remain intact.

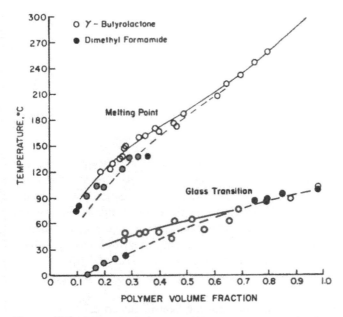

Figure 13 Melting and glass transition temperatures of polyacrylonitrile gels plotted against the volume fraction of polymer [14].

The gel melting data were analyzed using the Flory theory for the melting point depression of a polymer by a diluent [68]. This theory predicts the following dependence of melting point on the volume fraction of the diluent, which in this case is the solvent:

$$\frac{1}{T_m} - \frac{1}{T_m{}^\circ} = \frac{RV_u}{\Delta H_u V_1}\, (v_f - X v_f{}^2) \tag{7.1}$$

where T_m and $T_m{}^\circ$ are the melting points in Kelvin of the polymer with diluent present and the pure polymer, respectively, R is the gas constant, ΔH_u is the heat of fusion per mole of crystalline repeat unit, V_u and V_1 are the molar volumes of the repeat unit and diluent, v_f is the volume fraction of diluent, and X is the polymer-diluent interaction parameter.

The theoretical heat of fusion of pure crystalline PAN polymer, ΔH_u, is derived from the gel-melting data by plotting the reciprocal melting point against the diluent volume fraction and substituting the initial slope of that line into (7.1).

When a polymer reaches the melting point, the change in free energy upon melting will be equal to zero, and we can write the following equation [69]:

$$\Delta G_M = 0 = \Delta H_m - T_m \Delta S_m$$

or

$$T_m = \frac{\Delta H_m}{\Delta S_m} \qquad\qquad (7.2)$$

where T_m is the melting point, and ΔH_m and ΔS_m are the enthalpy (or heat) and entropy of fusion, respectively.

We see from this equation that the melting point is the ratio of the enthalpy and entropy changes. The enthalpy change reflects the strength of the inter-molecular bonds in the crystalline phase. These bonds resist the lattice vibra-tions that develop upon heating, but are broken at the melting point. The entropy change is primarily a measure of the increase in chain flexibility that occurs upon melting. All polymer chains are rigid in the crystalline state, but those that are inherently flexible in the melt undergo a much larger change in entropy than do those polymers that are relatively inflexible. Thus, a high polymer melting point can be obtained through either strong bonding or stiff chains. In Table 2 the PAN melting point and heat and entropy of fusion derived from the gel-melting experiments are given and compared with those of other semicrystalline polymers.

Krigbaum and Tokita [14] point out that the heat of fusion of PAN is about the same as that of polychlorotrifluroethylene and is considerably less than that of a polyethyiene or polyethylene oxide. Since it is difficult to reconcile strong attractive forces with the chemical structures of these three polymers, it appears unlikely that strong intramolecular interactions are responsible for the high melting point of PAN. For many polymers, the entropy of fusion per single chain bond lies in the range of 1.5–2 cal./deg. This quantity is related in at least a

Table 2 Thermodynamic Parameters Derived from Melting Point Measurements

Solvent	T_m° (K)	ΔH_u (kcal)	ΔS (cal/deg)	ΔS per bond
A. Polyacrylonitrile				
γ-Butyrolactone	590	1.25	2.1	1.0
Dimethylformaide	(590)	1.16	2.0	1.0
B. Other polymers				
Polyethylene oxide	339	1.98	5.85	1.95
Polyethylene	410	1.57	3.80	1.90
Polychlorotrifuoro-				
ethylene	483	1.20	2.50	1.25
Cellulose 2.44-				
nitrate	890	1.35	1.51	0.76
Cellulose trinitrate	970	0.9–1.5	1.5	0.75

Source: Data by Krigbaum and Tokita [14].

general way to the flexibility of the polymer chain in the melt or in solution. For example, cellulose trinitrate has a value of only 0.75 cal./deg.; this polymer is known to be highly extended in solution [70]. Similarly, the entropy per bond of the PAN chain lies below the normal range of values. Furthermore, the unperturbed dimensions estimated for this polymer from dilute solution measurements indicate a highly extended conformation [71], thus supporting the above arguments.

This analysis of the melting behavior is also consistent with the model of the structure based on the x-ray diffraction fiber patterns discussed in Section IC1. The rather sporadic nature of the dipolar bonding between the nitrile groups on adjacent chains would not be expected to yield large values for the heat of fusion.

Up to this point we have discussed the structure and melting behavior of the PAN homopolymer. However, only rarely is the homopolymer used for fiber spinning, and virtually all commercial acrylic fibers are spun from acrylonitrile polymers containing from 5–10% comonomer. The comonomers are introduced for several reasons. A fiber spun from the PAN homopolymer is very difficult to dye because the laterally bonded structure is not easily penetrated by the relatively large dye molecules. Incorporation of comonomers such as methyl acrylate and vinyl acetate disrupts the laterally bonded structure in a manner that allows for more rapid diffusion of the dye molecules during dyeing. Several other benefits accrue from their use. The polymer becomes much more soluble, thus making the preparation and storage of the spinning dopes much easier, and the resulting fiber is also more extensible and therefore less prone to fibrillation, albeit the hot-wet strength and modulus is decreased in comparison with fiber spun from the homopolymer. A more extensive discussion of the benefits from incorporation of comonomers is in Chapter 3.

Incorporation of another monomer into an otherwise crystalline homopolymer will generally reduce the crystallinity and lower the melting point in a continuous manner. This occurs because the new comonomer will only rarely be isomorphous, i.e., will fit into the crystalline lattice of the homopolymer. Several theoretical models have been developed addressing the reduction of crystallinity and melting point. In discussing these models we will designate the major comonomer as the monomer in the base homopolymer, and any others will be minor comonomers. When we discuss fiber-forming acrylic copolymers the major comonomer will be acrylonitrile and other comonomers such as vinyl acetate are the minor ones.

In the classical model developed by Flory, the assumption is made that the copolymer crystallizes in a manner that places the minor comonomer in the amorphous domain [72, 73]. The crystalline phase is formed by parallel alignment of chain segments containing uninterrupted sequences of the major monomer. The presence of the minor comonomer in the chain reduces the average sequence length of the major monomer, which in turn reduces the number and average size of the crystallites. It simply becomes less probable that large sequences of the major comonomer will associate in the melt and crystallize. This reduction in size increases the surface to volume ratio of the crystallites. The chain segments on the surface of the crystallite have a larger free energy than those away from the surface, and hence increasing the surface to volume ratio also increases the free energy of the crystal; this leads to a lower melting point.

The following equation derived by Flory gives the theoretical dependence of the melting point on the amount of minor comonomer added to the crystalline homopolymer [72].

$$\frac{1}{T_m} - \frac{1}{T_m{}^\circ} = \frac{R}{\Delta H_u} X_B \qquad (7.3)$$

where T_m and $T_m{}^\circ$ are the melting points of the copolymer and homopolymer, respectively, R is the gas constant, ΔH_u is the heat of fusion per mole of crystalline repeat unit, and X_B is the mole fraction of minor comonomer.

The equation predicts that the melting point depression will be independent of the size, or any other structural feature of the minor comonomer. The only role played by the comonomer is purely statistical, and that is to restrict the average sequence length of the major crystalline monomer.

An alternative model for a crystalline copolymer is the "defect crystal" model proposed by Eby [74] where the minor comonomers are incorporated into the crystalline lattice as defects. These defects decrease the melting point and heat of fusion by disrupting the intermolecular bonding within the crystalline lattice. This theory is functionally similar to the Flory theory in that it predicts a linear relationship between the reciprocal melting point and the mole fraction of the comonomer, but in addition a parameter is provided that can be interpreted as the degree to which the comonomer disrupts the crystalline lattice. According to this model the efficacy of a particular comonomer to depress the melting point is to a first approximation proportional to the molar volume of the comonomer. The slope of the reciprocal melting point versus comonomer mole fraction curve is no longer inversely proportional to the heat of fusion of the homopolymer, as in the Flory theory. This slope can be interpreted as an estimate of the degree to which the comonomer disrupts the lattice.

We will now review several studies on the melting behavior of acrylic copolymers and compare the results to the predictions based on the above two theories. Slade [75] determined the melting point of a series of acrylonitrile-vinyl acetate (AN/VA) copolymers using differential thermal analysis (DTA). The VA content of the copolymer covered the range from 0 to 38.5% by weight. The melting point of the homopolymer could not be directly determined due to thermal degradation, but the extrapolated value was 322°C. The heat of fusion of the homopolymer calculated from the Flory theory, (7.3), was 573 cal/mol, which is much lower than the values obtained from analysis of the gel melting points. The melting behavior of acrylonitrile-styrene (AN/S), and acrylonitrile-isobutylene (AN/I) copolymers were studied by Berger et al. [76] by DSC. The calculated heat of fusion for the PAN homopolymer, again using the Flory theory, was found to be dependent on the particular comonomer. Values based on the AN/S and AN/I copolymers were 650 and 1260 cal/g, respectively. The authors suggested that the disparity in heat of fusion values might be interpreted as support for the Eby defect model. Recall that large differences in the calculated heat of fusions are not expected from the Flory equation. The molar volume of the styrene comonomer is significantly larger than that of isobutylene, and if the Eby model were more appropriate for PAN, then one would expect the former comonomer to disrupt the crystalline lattice to a greater degree, thus resulting in a lower calculated value of the heat of fusion.* Direct measurement of the experimental heat of fusion of the copolymers was given as additional support for the defect model; at equal molar concentrations the heat of fusion was smaller for the styrene comonomer. The experimental heat of fusion is the energy represented by the integrated area under the DSC endotherm. The heat of fusion of the AN/I copolymer was two and one-half times that of the AN/S copolymer at equal comonomer mole fraction.

A novel thermal analytical technique was developed by Frushour [57, 77, 78] that utilizes water to depress the melting point of AN copolymers. This allows melting and melt crystallization studies to be carried out on the PAN homopolymer and copolymers below the temperature where the exothermic cyclization reaction begins. Information obtained by this techniques gives additional insight into the morphology of PAN homopolymer and copolymers. In this technique polymer or fiber is mixed with water and sealed in a high-pressure DSC capsule [57]. As the water content is increased, the polymer melting point will decrease continuously until a critical water concentration is reached where a pure water phase is formed [77]. The maximum melting point depression will have occurred at this point and excess water simply goes into the pure water phase.

This critical water concentration is dependent upon the acrylonitrile level of the polymer. For PAN homopolymer it is 33% water, based on polymer weight, and at this concentration and for all higher concentrations the PAN melting point remains at 185°C. This represents a 135°C depression in the melting point, assuming a dry polymer melting point of 320°C. The region of excess water is

---

*The defect model would predict that at equal mole fractions, the larger comonomer would give a larger melting point depression, and hence a larger slope when the reciprocal melting point term (left side of 7.3) is plotted against mole fraction comonomer. In (7.3), the ΔH_m is inversely proportional to this slope. Of course, if the slopes for different comonomers differ, then the Flory theory is invalid.

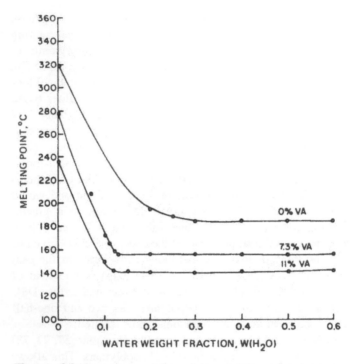

Figure 14 Dependence of the polymer melting point on water content for acrylonitrile/vinyl acetate copolymers [77].

referred to as the constant melting plateau. Incorporation of comonomers decreases both the critical water concentration required to reach the constant melting plateau and the corresponding plateau melting point. This can be seen in Figure 14 where the melting point is plotted against water concentration for the homopolymer and two AN/VA copolymers.

It is the presence of the constant melting plateau that makes routine analysis of the melting and crystallization behavior so straightforward because if excess water is used, then the measurement becomes sensitive only to the polymer structure. The role of the water, then, is simply to reduce the melting temperature to a range where thermal degradation is no longer a serious problem.

The melting endotherms that can be obtained by this procedure are shown in Figure 15. They are well shaped and the peak areas $\Delta H_{f(exp.)}$ can be easily and reproducibly measured. The technique was used to study the melting behavior of copolymers, and higher-order copolymers [78]. One can see from Figure 15 that the effect of comonomer incorporation on reduction of the melting point and experimental heat of fusion (endotherm area) is quite apparent. It turns out that the dependence of the reciprocal melting point on comonomer mole fraction is

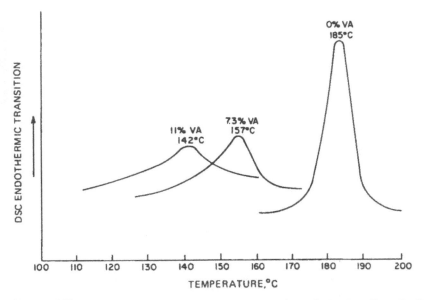

Figure 15 Melting endotherms by differential scanning calorimetry of acrylonitrile/ vinyl acetate copolymers. The polymers were mixed with water (one part polymer to two parts water) and sealed in a high-pressure DSC capsule [77].

identical for dry polymers and polymers in the constant melting plateau region. In Figure 16 the reciprocal melting point is plotted against mole fraction of comonomer for a number of different copolymers both in the dry state and with excess water. Two salient features should be noted. First, the slopes of the curves for the dry and wet polymers are essentially identical, and secondly, the magnitude of the slope increases as the size of the comonomer becomes larger. The first observation means that for virtually any acrylic polymer the dry polymer T_m can be calculated from the wet polymer T_m, thus circumventing the problem of thermal degradation.

The correlation between the comonomer size and the slope was established by calculating the molar volume of the comonomer side chain and comparing this to the slope. These values are given in Table 3 along with the results of the least-squares correlation between the molar volumes and the slopes. This reasonably strong correlation was taken as justification for interpreting the value of the slope for each comonomer as a parameter that measured the degree to which the comonomer disrupted the crystalline lattice. This parameter was termed the melting point depression constant. The correlation between slope and comonomer volume was interpreted as support for the Eby model of the

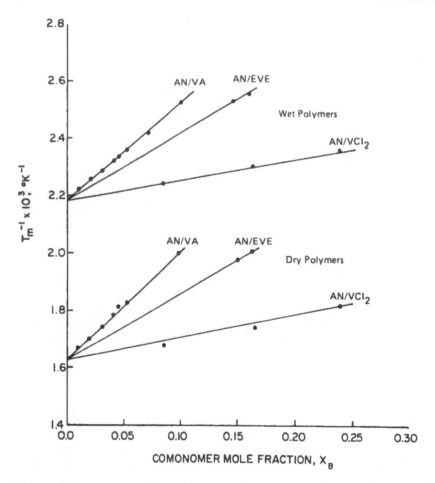

Figure 16 A comparison of the dependence of the wet and dry-polymer melting points on comonomer type and mole fraction: VA (vinyl acetate), EVE (ethyl vinyl ether), and VCl$_2$ (vinyl chloride) [77].

copolymer where the minor comonomers become incorporated in the crystalline domains as defects.*

---

*This does not imply that the defect model is generally more suited to crystalline copolymers. The Flory model works well in other cases, such as polyesters. The success of the defect model for acrylics may be attributed to the dominating effect of the strong dipolar interactions. These bonds probably develop as soon as the acrylic polymer polymerizes, in the case of polymer, or precipitates, in the case of the fiber. There is little opportunity for the neighboring chains to move about and allow the long sequences of AN units to align and crystallize.

Table 3 Melting Point Depression Constants for Acrylic Copolymers

	Wet-polymer $K_B \times 10^3$ (K^{-1})	Dry-polymer $K_B \times 10^3$ (K^{-1})	Molar volume of comonomer side-chain[a]
Methyl acrylate	3.370	—	36 cm³/mol
Vinyl acetate	3.343	3.923	36
Ethyl vinylether	2.20	2.40	39
Vinyl bromide	0.909	—	11
Vinylidene chloride	0.744	0.710	15
Vinyl chloride	0.510	—	7.5

[a]Volume of R side chain on monomer (CH_2=CHR). Volumes calculated from tabulated values of the molar volumes of common functional groups. The correlation between the K_B. values and the side-chain molar volumes is

$K_B = 8.5 \times 10^{-5} \,(M.V.) - 1.5 \times 10^{-4} r^2 = 0.83$

Source: Data by Frushour [78].

The melting point depression constants can be used in a generalized melting point equation for acrylic copolymers of any order.

$$\frac{1}{T_m} - \frac{1}{T_m^\circ} = \sum_{i=1}^{n-1} K_i X_i \tag{7.4}$$

where T_m and T_m° are the melting points of the copolymer and homopolymer, respectively, K_i and X_i are the melting point depression constants and mole fraction of the ith comonomer and n is the order of the polymer ($n = 2$ for a copolymer, 3 for a terpolymer, etc.).

Using this equation, the melting point of numerous terpolymers and tetrapolymers could be accurately predicted [78].

The wet-polymer DSC technique was shown to be much more sensitive than x-ray diffraction to the structural changes brought about by the incorporation of comonomers into the PAN chain [57]. It was postulated by Frushour [57] that the melting point and experimental heat of fusion directly measured the regularity and strength of the dipolar bonding network that stabilized the two-dimensional laterally bonded structure. DSC and x-ray were proposed as complimentary techniques that were sensitive to two different levels of order within the structure. This conclusion was based on results obtained when a series of AN/VA copolymers were analyzed by the two techniques. This comparison is summarized in Table 4. The structurally sensitive x-ray diffraction parameters (half-width of the intense [200] equatorial reflection and the crystalline index) and DSC parameters (wet-polymer melting point and experimental heat of

Table 4 Comparison of the Response of Thermal and X-ray Parameters to Vinyl Acetate Comonomer Level

Vinyl acetate (mole fraction)	Wet-polymer $(T_m \, ^\circ C)$	Wet-polymer $\Delta H(\text{exp})$ (cal/mol AN)[a]	Half-width [200] reflection	Crystal-linity index
0	185°C	1150	1.42	0.39
0.0092	177	860	1.4	0.40
0.0294	166	550	1.4	0.37
0.0404	158	500	1.5	0.37
0.0509	153	420	1.6	0.37
0.0708	142	300	1.8	0.38
0.0981	125	205	2.4	0.33
0.130	108	70	3.1	0.29
0.150	102	50	4.1	0.20
0.178	None	None	4.6	0.12
0.240	None	None	8.0	—

[a]$\Delta H_{(\text{exp})}$ is the experimentally determined heat of fusion.
Source: Data by Frushour [57].

fusion, $\Delta H_{f(\text{exp.})}$) for a VA range of zero to 0.24 mole fraction, or 34% by weight, are given. The half-width data in Table 4 also appears in Figure 11, and was discussed briefly in Section IC. The [200] reflection is the intense equatorial reflection attributed to the interchain spacing in the crystalline phase, and the crystalline index is defined as the integrated diffractometer scattering intensity under the sharp equatorial reflections divided by the total scattering intensity. Very little change in the two x-ray parameters was detected until a VA level of approximately 0.08 mole fraction (12% by weight) was reached, but by this point very significant reductions in both T_m and $\Delta H_{f(\text{exp.})}$ had occurred. Once $\Delta H_{f(\text{exp.})}$ had been reduced below about 150 cal/mole·AN, which required a VA mole fraction of approximately 0.12 (18% by weight), disorder in the structure as measured by x-ray became apparent. The distribution of the interchain distances became broader and the crystalline index began to decrease. The laterally bonded crystalline phase is assumed to be stabilized by the intermolecular dipolar bonding, as originally proposed by Bohn et al. [16], though the well-defined melting endotherms may indicate that this bonding is more regular in nature than had been previously suspected. The VA comonomers are incorporated into the crystalline phase as defects at the expense of the dipolar bonds. The crystalline structure is sufficiently imperfect to tolerate a loading of up to 12% of VA by weight before significant disruption on the scale measured by the x-ray technique is detected. These results are not specific to VA. However, the molar concentration required to effect a given level of crystalline disruption should increase as the molar volume of the comonomer decreases.

E. Glass Transition and Dynamic-Mechanical Properties

1. The Glass Transition

PAN and the fiber forming acrylic copolymers do have glass transitions, but with some unusual characteristics. The glass transition, as stated previously, is the temperature range over which a glassy polymer becomes rubbery. Polymer chain segmental mobility, or chain Brownian motion, becomes activated and for typical glassy polymers, such as atactic polystyrene, the modulus of the glass decreases by about 3½ decades over a 15°C range. The temperature range is a function of the polydispersity of the polymer. Narrow molecular weight fractions will have correspondingly sharper transitions. Other changes that occur at the glass transition are pronounced increases in specific volume, heat capacity, and diffusion rate of absorbed molecules, and all of these can be attributed to the increasing amplitude of chain segmental mobility. The glass transition is a very important thermal property of a fiber, because key physical and tensile properties will change as the fiber is heated through the transition. The fiber modulus will decrease and the fiber will become more extensible. In dyeing operations it is necessary to be above the glass transition of the wet fiber so that the dye molecules can diffuse into the fiber and reach the dye sites.

The glass transition temperature, T_g, can be determined in a number of ways. In general, one monitors the change in a physical property as a function of temperature, and then defines T_g as either the beginning or midpoint of the change in the property. Each technique will give a distinct value, because the glass transition is a viscoelastic phenomenon [79] and the measured T_g will be a function of the heating rate and measurement frequency. The physical properties that are most often monitored to measure T_g include thermal expansion coefficient, heat capacity, diffusion coefficient, and dynamic modulus. T_g's based on the first three measurements are referred to as static measurements because the polymer sample is at rest during the measurement. The measured value of T_g will increase with the heating rate. In dynamic measurements, the modulus or dielectric constant are measured many times per second as the sample is heated, and the observed T_g will increase with the measurement frequency.

In typical semicrystalline polymers and fibers the glass transition will be associated with the amorphous domains, and the temperature and magnitude of the physical property changes at the transition will reflect changes in the morphology. If, for glass example, the crystallinity of the fiber is changed, then the amorphous content will move in the opposite direction. Processes such as annealing and drawing that increase the crystalline volume fraction will decrease the amorphous fraction and this will be reflected in a decrease in the change in heat capacity, modulus, and volume during the glass transition.

A glass transition is observed for PAN, but it cannot be unequivocally associated with amorphous domains (at least to the degree possible with more conventional semicrystalline polymer and fibers). However, this is consistent with the ambiguous nature of the PAN morphology. Kimmel and Andrews have summarized measurements of the PAN T_g up to the year 1965 [80]. The values fall into two ranges. For static measurements distinct changes in the temperature dependence of the measured property is observed around 85 to 95°C. Higher values of T_g, 105 to 140°C, are obtained using dynamic-mechanical and dielectric measurements.

2. Glass Transition Studies on Copolymers; Effect of Water

We will now review the measurements of T_g and the influence of comonomers and water content. The latter two factors are especially important in understanding and modifying the dyeing behavior of these fibers.

T_g values of 85 and 87°C, based on the linear expansion coefficient, were reported by Bohn et al. [16] and Howard [81]. The former group also showed that the interchain spacing of the crystalline lattice increased at approximately the same temperature (see Section IIIA) and concluded that the PAN structure must be single phased. In conventional semicrystalline polymers the crystalline parameters of the crystalline phase do not change significantly when the amorphous phase undergoes the glass transition. Howard determined the T_g for a series of (AN/VA) copolymers by measuring the linear expansion coefficient of polymer discs. The T_g for the PAN homopolymer was determined to be 87°C. Upon incorporation of VA the value of T_g remained constant until the VA level exceeded 27 wt % and then T_g began to decrease and approach the literature value of 30°C for the pure polyvinyl acetate. In Figure 17 the reciprocal of the T_g is plotted against weight fraction VA. At VA levels of 27% and greater the data fit a straight line in accordance with the classical behavior described by the Loshack and Fox equation [82]. The constant T_g region was attributed to the interfering effect of the crystalline phase, and extrapolation of the line to zero VA gives a value of 107°C, which was interpreted as the T_g for completely amorphous PAN.

Differential scanning calorimetric (DSC) measurements give a similar dependence of T_g on VA level [83]. In Figure 18 the DSC scans of the glass transition region are shown for a series of AN/VA copolymers. The T_g is taken to be the midpoint of the inflection in the change in heat capacity. This value is 100°C for the PAN homopolymer, and remains unchanged until the VA level exceeds 11%, and only then does T_g begin to decrease. One would expect DSC to give higher T_g values than volumetric measurements because of the higher heating rate. The salient point here is that both thermal expansion coefficient and heat capacity measurements show the PAN T_g to be initially insensitive to incorporation of the VA comonomer. The DSC also gives quantitative values of the change in heat capacity that occurs as the polymer is heated through the glass

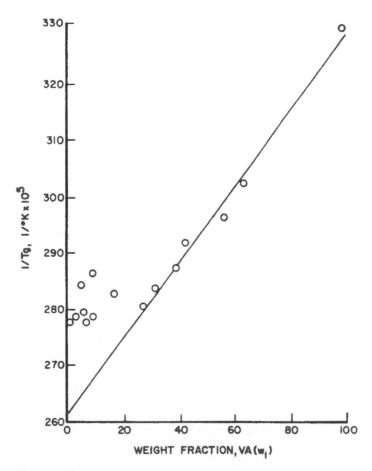

Figure 17 Reciprocal of glass-transition versus composition of acrylonitrile/vinyl acetate copolymers. The glass transition was derived from measurements of the linear expansion coefficient [81].

transition. The DSC scans in Figure 17 show that the magnitude of the heat capacity change increases upon incorporation of VA, and is a more sensitive indicator of change than the midpoint temperature.

The effect of water on the glass transition of acrylic polymers and fibers is quite pronounced. In Figure 19 the DSC curves of an AN/VA copolymer containing 7% VA are compared in the dry and wet state [83]. Water lowers T_g from 100 to 64°C. It is not surprising that water would plasticize the acrylic structure. Water has a high dielectric constant and the water molecules should be effective in decreasing the magnitude of the dipolar interactions between

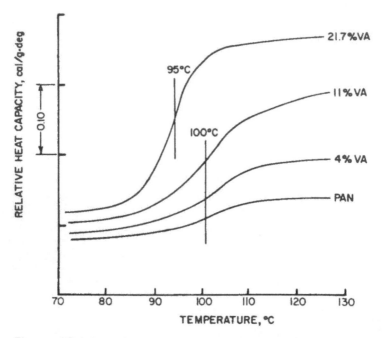

Figure 18 Differential scanning calorimetric determinations of glass transition temperatures for acrylonitrile/vinyl acetate copolymers. The scanning rate was 20°C/min [83].

nitrile groups. The water does not, however, appear to affect significantly the magnitude of the heat capacity change. This suggests that the water affects only those chain segments that participate in the glass transition of the dry polymer, i.e., if the water increased the magnitude of chain segmental mobility, then one might expect an increase in the heat capacity change as well. The large plasticization effect of water has important technological consequences for acrylic fibers. It means that when these fibers are placed in very hot water, such as conditions of dyeing, they will become highly plasticized and soften.

Several experimental approaches have been applied to determination of the fiber T_g under hot-wet conditions [84–87]. Aiken et al. [84] compared the T_g of a commercial acrylic yarn in the dry state and in water using dynamic-mechanical analysis [88], and observed a reduction from 92 to 72°C. Bell and Murayama [85] observed that the T_g of a commercial AN/VA copolymer decreased from 128°C when dry to 80°C in a 100% relative humidity atmosphere. Gur-Arieh and Ingamells [86] related the extension in length of Acrilan filaments to T_g reduction and showed a shift from 90°C in air to 57°C in water. Finally, Hori et al. [87] used DSC to show that the T_g of four kinds of acrylic fibers decreased

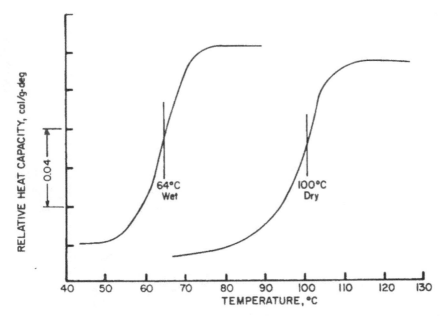

Figure 19 Influence of water on the glass transition temperature of an acrylonitrile/ vinyl acetate copolymer containing 7% vinyl acetate. The curves were obtained using a differential scanning calorimeter at a heating rate of 10°C/min [83].

with increasing water content and approached an almost constant value for all four fibers.

3. Dynamic-Mechanical Analysis of PAN Homopolymer and Copolymers

Dynamic-mechanical measurements of polymers are very sensitive to molecular transitions such as the glass transition and the more localized transitions that are found at temperatures below T_g [88, 89]. In dynamic-mechanical analysis, a material is subjected to a periodic strain of small amplitude while being heated at a programmed rate and the resulting periodic stress in the material is recorded [88, 89]. A dynamic modulus can be calculated continuously from the stress and strain. If the material is perfectly elastic, the applied strain and measured stress will be in-phase. In viscoelastic materials such as polymers, the stress in the material will lag behind the strain by a phase angle delta [δ] and the dynamic modulus can be resolved into two orthogonal components: the real modulus and the loss modulus. The real modulus is a measure of the recoverable strain energy and the loss modulus is a measure of the strain energy that is lost as heat due to viscous dissipation. Most commercial equipment will record both the dynamic modulus and the tangent of the phase angle, or tan δ. The real and loss moduli are then calculated from these quantities.

When a polymer begins to undergo a molecular transition such as the glass transition, the real modulus will decrease and the tan δ will go through a maximum. This represents a sudden absorption of energy as the polymer transition becomes activated. Dynamic-mechanical measurements are useful because a wide spectrum of transitions can be detected. The glass transition is by far the most prominent transition. Less prominent transitions are found in many glassy polymers at temperatures below T_g; these arise from rather specific main-chain or pendant side-group molecular motions. The combination of the real modulus and the tan δ is especially beneficial, because tan δ is able to resolve distinct transitions that underlie the inflection in the real modulus. The presence of these transitions can contribute to the impact strength of glassy polymers. Transitions are also observed in crystalline polymers that can be attributed to molecular motions in the crystalline phase.

Analysis of the dynamic-mechanical properties of acrylic polymers has been the subject of numerous investigations [59, 90–98]. In most of these studies, the measurements were made using a Rheovibron, which utilizes the forced-resonance principle [88]. The instrument is manufactured by Toyo Measuring Co., Ltd., of Tokyo and is based on the design of Takayanagi [99]. The specimen is mounted between two pinch clamps that are attached to metal rods. One rod is connected to an electromagnetic driver that applies a sinusoidal strain to the sample and the other rod is connected to a force transducer that measures the stress. The sample itself is contained within an oven that allows the temperature to be raised at a programmed rate. Heating rates on the order of 1–2°C/min are commonly used.

The dynamic-mechanical spectrum of a PAN fiber obtained by Minami [90] is shown in Figure 20. The real modulus at room temperature is approximately 4 × 10^9 MPa. The modulus begins to decrease near 75°C and drops to 2 × 10^8 MPa, which is a decrease of slightly over one order of magnitude, and corresponds to the onset of tht glass transition. The subsequent increase above 200°C arises from cross linking due to thermal degradation and is not pertinent to this discussion. The T_g based on the midpoint of the decrease in real modulus is approximately 115°C. This is higher than the values obtained using thermal expansion coefficient (85°C) or DSC (100°C), and reflects the temperature shift characteristic of dynamic techniques mentioned earlier.

Most glassy polymers such as polystyrene exhibit a much larger decrease in the real modulus upon being heated through the glass transition temperature. A decrease of approximately three to four orders of magnitude would be considered typical behavior [100]. Kenyon and Rayford [97] interpret the small drop in the real modulus as evidence of crystallinity in PAN. Using an empirical relationship developed by Nielsen [101] between the decrease in modulus of a

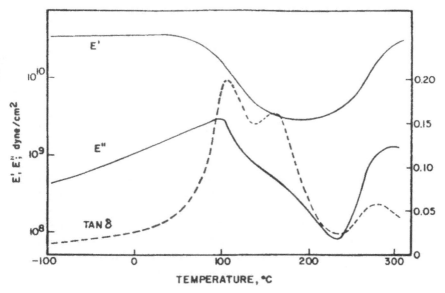

Figure 20 Dynamic-mechanical properties of undrawn polyacrylonitrile fiber measured with a Rheovibron at 110 Hz [90].

polymer and the degree of crystallinity, a crystalline fraction of 0.47 was calculated. Recall that the increase in heat capacity of PAN upon being heated through the glass transition was significantly lower than that commonly observed for glassy polymers. The magnitudes of the changes in both the real modulus and heat capacity are undoubtedly manifestations of the same structural feature, i.e., the presence of a laterally ordered crystalline phase.

The tan δ curve of PAN in Figure 20 has two distinct transitions appearing near 110°C and 160°C at a measurement frequency of 110 Hz. The presence of two transitions in the glass transition region is not at all apparent from the real modulus alone, and thus demonstrates the superior resolving power of tan δ. We will use the convention of Minazi and refer to the 110°C and 160°C transitions as α_{II} and α_{I}, respectively. The interpretation of these two peaks has not yet been entirely resolved. Some investigators have taken their appearance as evidence of a double glass transition for this polymer [58–60, 98, 102]. The PAN structure is assumed to have two phases both essentially amorphous, but with different levels of intermolecular bonding. It is assumed that both phases are capable of undergoing a glass transition, though at different temperatures, and the more ordered phase should have the higher T_g. Padhye and Karandikar [98] treated PAN fiber in various solvents and then observed the effects of these solvents on the tan δ peaks. They found that aromatic solvents (phenol, aniline, and resorcinol) seem to make the lower transition disappear and reduce the temperature for the higher one, while nonaromatic solvents (methanol, amylamine, dimethylamine, ethylene glycol, and acetonitrile) shift the lower-temperature transition lower and the higher-temperature transition even higher. It is argued that the aromatic solvents penetrate only the loosely bond phase associated with the lower T_g and plasticize that phase thus lowering tan δ peak temperature. The nonaromatic solvents are able to penetrate the more highly bound phase, or at least the more accessible regions, and this removes the lower temperature side of the tan δ peak envelope, thus shifting the remainder of the peak to higher temperatures.

An alternative explanation put forth mainly by Minami and co-workers [90, 91, 94, 35, 36] associate the lower temperature transition, α_{II}, with an unspecified motion within the laterally bonded crystalline phase, and assigns α_I, the higher temperature transition, to the glass transition of a lesser ordered or amorphous phase. The basis for these assignments is the response of the two transitions to structural changes introduced first by changing the level of chain orientation imparted by the fiber-spinning process and second, by incorporation of comonomers.

The degree of polymer chain orientation is controlled through the drawing and relaxation stages of the wet-spinning process. The freshly coagulated fiber has very little chain orientation, but a high degree of orientation is required to obtain an acceptable level of tensile strength. After the coagulation step is completed, the fiber is removed from the coagulation bath, washed, and then heated by steam or boiling water and drawn between rolls to develop the required orientation. The drawing process sometimes leaves the fiber in an excessively brittle state. By shrinking or relaxing the fiber in saturated steam, some of the molecular orientation can be eliminated and internal strains removed to reduce the brittleness.

The response of the PAN tan δ curve to drawing and relaxation can be seen in Figures 21 and 22, respectively. The magnitude of α_I, decreases upon drawing and, this is reversed upon subsequent relaxation. In other words, the magnitude of α_I is essentially controlled by the degree of chain orientation. This behavior is typical of the glass transition in semicrystalline fibers such as polyester when the degree of orientation is varied [90, 55]. An increase in orientation causes the chains in the amorphous phase to pack more closely thus restricting the degree of chain mobility that becomes available in the glass transition. The magnitude of α_{II} is not significantly affected by the drawing and relaxation processes. This is consistent with the assignment of the relaxation to a motion in the crystalline phase. It is well known that the degree of crystallinity in acrylic fibers is largely independent of drawing and relaxation process downstream of the initial coagulation step.

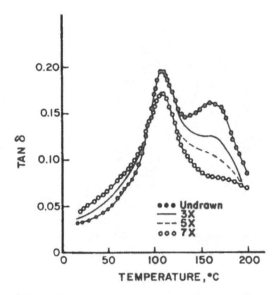

Figure 21 Temperature dependence of tan δ for polyacrylonitrile fibers with different draw ratios. Fibers were drawn in boiling water in the wet-spinning process (94).

The disruption of the acrylic structure caused by the incorporation of comonomers has a pronounced effect on the dynamic-mechanical properties [90], as well as the melting behavior that was discussed previously. The drop in real modulus observed upon heating the polymer through the glass transition increases with comonomer level. At a vinyl acetate level of 24.6%, the decrease in real modulus is approximately three orders of magnitude and the copolymer now behaves more like a typical amorphous polymer. Recall from the previous discussion of the melting behavior that the experimental heat of fusion approaches zero at this level of comonomer. Therefore, the dynamic-mechanical and thermal properties data indicate that the laterally bonded crystalline lattice has been completely disrupted at this point.

The two tan δ peaks respond differently to comonomer incorporation. The lower-temperature transition, α_{II}, which Minami [90] ascribes to a molecular motion in the crystalline phase, changes little while α_I, the apparent relaxation for the glass transition, decreases in temperature and increases in magnitude. This is consistent with the crystal defect model in which the comonomer disrupts the dipolar bonding in the lattice thereby making more polymer chain segments available to participate in the glass transition.

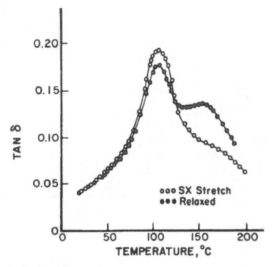

Figure 22 Temperature dependence of tan δ from polyacrylonitrile fibers before and after relaxation (shrinkage) in saturated steam at 130°C [94].

II. DETERMINATION OF MOLECULAR WEIGHT

A. Introduction

The polymers from which typical acrylic fibers are manufactured have number average molecular weights (M_n) of 40,000–60,000 Daltons, which means a degree of polymerization of approximately 1000 repeat units. The weight average molecular weight (M_w) is typically 90,000–140,000 Daltons. Free radical addition polymerization yields a distribution of molecular weights and a simple indicator of the breadth of this distribution is the ratio of M_w to M_n. This ratio is commonly referred to as the polydispersity index, Q. The polydispersity of the acrylic polymers is a function of the method of polymerization and typical values for free-radical polymerization range from 2 to 3, though some commercial processes can apparently yield higher values. Specific studies of M_n, M_w, and molecular weight distribution of acrylic polymers will be covered in this section after a discussion of the methods for their determination.

The molecular weight of the polymer has a strong influence on many if not all of the key acrylic fiber properties, and therefore accurate measurement and control is essential for producing a high-quality product. The molecular weight must be low enough that the polymer is readily soluble in spinning solvents, yet high enough to give fibers of good physical properties. Polymers with a very high molecular weight fraction may form insoluble microgels in the spinning solution. Fiber dyeability is another property that depends upon molecular weight, since most acrylic fibers derive all or a large portion of their dyeability from sulfonate and sulfate initiator fragments at the polymer chain ends. The latter is an inverse function of M_n. The number of sulfate/sulfonate groups per chain can range from one to two depending on the initiator system and chain transfer agents employed. This is discussed in detail in Chapter 3. Fiber physical properties such as the common tensile properties and durability are also affected to various degrees by the molecular weight distribution.

B. Methods for Determining Polymer Molecular Weights

1. Molecular Weight Averages and the Distribution

Most polymeric materials are comprised of mixtures of molecules of different sizes. The distribution of molecular weights is caused by the statistical nature of the polymerization process [103]. It is useful to define various molecular weight averages, and specific experimental techniques are available for their determination. These averages are defined in terms of the molecular weight, M_i, and the number of moles, n_i, or the weight, w_i, of the component molecules. Two important molecular weight averages are the number average, M_n, and the weight average, M_w and they are defined by the equations given below [104]:

Number-average molecular weight

$$M_n = \frac{\Sigma n_i M_i}{\Sigma n_i} = \frac{\Sigma w_i}{\Sigma (w_i/M_i)} \tag{7.5}$$

Weight-average molecular weight

$$M_w = \frac{\Sigma n_i M^2_i}{\Sigma n_i M_i} = \frac{\Sigma w_i M_i}{\Sigma w_i} \tag{7.6}$$

The M_n is equal to the molecular weight of the polymer chain of average length, and in its calculation all chains are given equal weighting. The M_w is more of an indicator of how the mass of the polymer is distributed across the molecular weight distribution.

2. Absolute Methods for Determining Molecular Weight

Prior to the development of size exclusion chromatography molecular weights were routinely determined by absolute methods. By absolute we mean the direct determination of a single molecular weight average from an specific experimental measurement on a polymer in a solution. The experimental result need not be calibrated against polymers of known molecular weight.

The M_n can be obtained from the colligative properties of polymer solutions. Colligative properties are those that depend on the number of species present rather than their kind. Common colligative properties used to measure molecular weight are vapor pressure lowering, ebulliometry (boiling point elevation), cryoscopy (freezing-point depression) and membrane osmometry. Common to all of these techniques is that the magnitude of the effect measured on a polymer solution is directly related to the number of polymer molecules dissolved in the solution. This information together with the weight fraction of polymer in the solution gives M_n. The relations between the colligative properties and molecular weight for infinitely dilute solutions rest upon the fact that the activity of the solute in a solution becomes equal to its mole fraction at small solute concentration. The activity of the solvent must equal its mole fraction under these conditions, and it follows that the depression of the activity of the solvent by the solute is equal to the mole fraction of the solute [105].

Most of the M_n determinations on acrylic polymers have utilized membrane osmometry. In this technique two compartments of an osmometer are separated by a semipermeable membrane, through which, ideally, only solvent molecules can penetrate. One compartment contains only solvent and the other contains the polymer solution. The solvent activity will be different in the two compartments and the thermodynamic drive to reach equilibrium will result in transfer of solvent molecules through the membrane into the compartment containing the dissolved polymer. This will increase the hydrostatic pressure in this compartment, and this pressure can be measured using several methods, depending upon the design of the osmometer. The success of the method depends upon the performance of the membrane to allow passage of only the solvent molecules. In practice, all membranes will allow passage of very low molecular weight polymer, so one must determine the lower limit for measurement of M_n for the various types of membrane. In general, osmometry can be used successfully for M_n values ranging from 10,000 to 50,000 Daltons.

Determination of M_n by osmometry follows from a generalization of the van't Hoff law of osmotic pressure [106]. In the limit of zero concentration,

$$\lim_{c_2 \to 0} \left(\frac{\pi}{c_2} \right) = \frac{RT}{M} \tag{7.7}$$

where π is the measured osmotic pressure, c is the solute concentration, R is the gas constant, T is temperature in Kelvin, and M is the solute molecular weight.

This is the van't Hoff equation. It is usual to plot π/c versus c and from the straight line one determines the intercept at $c = 0$. This intercept is the reduced osmotic pressure and is set equal to the right-hand side of (7.7), thus allowing the solute molecular weight to be calculated.

For a mixture of monodisperse polymer chains, each with concentration c_i and molecular weight M_i,

$$\pi = RT \frac{c_2}{M} = RT \sum \left(\frac{c_i}{M_i} \right) \tag{7.8}$$

from (7.7). Since $c_i = n_i M_i$, we can write

$$\frac{\pi}{c_2} = RT \left(\frac{\Sigma n_i}{\Sigma n_i M_i} \right) = \frac{RT}{M_n} \tag{7.9}$$

Thus, the reduced osmotic pressure (π/c) measures M_n.

At low concentrations the slope of π/c_2 versus c_2 is linear, but curvature develops at higher concentration. This curvature can be treated by extensions of the van't Hoff equation to include the nonideal effects. Similar extensions can be made for any of the colligative property equations for molecular weight, and are known as the polymer virial equations [106, 107]. The coefficients in the virial equations contain information about parameters such as chain branching, polymer dimensions, and the degree of polymer-solvent interaction.

The M_w of polydisperse randomly coiled polymers is typically measured from light scattering measurements, though sedimentation velocity in an ultracentrifuge can also be used [103–105]. Sedimentation measurements are more suited for compact protein molecules than for random-coil polymers, where extended conformations increase deviations from ideality and the chances of mechanical entanglement of the polymer chains exists. Very long times on the order of days are sometimes required for accurate M_w measurements using the sedimentation velocity methods. Light scattering is not without its problems, but is generally reliable if a good solvent is found, the solution is not highly colored, and if all particulate matter can be removed by filtration.

In the classical light scattering apparatus, light from a mercury-arc passes through a lens, a polarizer, and a monochromatizing filter. It then strikes either a calibrated reference standard or a glass cell containing a carefully filtered polymer solution. Scattered light from the cell is viewed by a photomultiplier tube after passing through slits and another polarizer. The scattered light intensity is measured as a function of scattering angle for a range of polymer concentrations, and then extrapolated to both zero concentration and zero scattering angle using the method of Zimm [108] to give the M_w, the second virial coefficient, and the radius of gyration of the polymer molecule.

The apparatus must be capable of measuring the intensity of the light scattered at very low angles with respect to the incident beam that has passed through the solution. The replacement of the mercury arc light source with a laser beam has improved the accuracy and greatly simplified the method.

In practice, polymer molecular weights on the order of 10,000 to 10,000,000 Daltons can readily be measured. The method cannot be easily applied to copolymers owing to the difference in refractive index between comonomer units. Branched polymers can be measured without restriction.

3. Relative Methods for Determining Molecular Weight
We now discuss relative methods for determination of molecular weights. These methods require calibration against known molecular weight standards, but are very useful due to their simplicity and power.

The simplest relative method for measuring molecular weight is viscosity. It is employed by most acrylic fiber producers as a control method for polymer quality. Solution viscosity is basically a measure of the size or extension in space of polymer molecules, and is empirically related to molecular weight for linear polymers. Measurements of solution viscosity are made by comparing the time required for a specified volume of polymer solution to flow through a capillary tube with the corresponding time for the pure solvent. This ratio ($t_{(polymer\ solution)}/t_{(solvent)}$) is referred to as the relative viscosity or viscosity ratio, η_r. Other commonly used viscosity quantities are the specific viscosity, η_{sp}, which is equal to η_r-1, and the reduced viscosity, η_{red}, which is equal to η_{sp} divided by the solution concentration, c [105].

The intrinsic viscosity, $[\eta]$, is obtained by plotting the reduced viscosity, η_{sp}/c, as a function of c and extrapolating to zero concentration. It is a function of the solvent. The $[\eta]$ is typically obtained using both the Huggins [109] and Kraemer [110] equations, which are given below.

Huggins equation

$$\frac{\eta_{sp}}{c} = [\eta] + k'[\eta]^2 c \qquad (7.10)$$

where k' is a constant for a series of polymers of different molecular weights.

Kraemer equation

$$\frac{\ln \eta_r}{c} = [\eta] + k''[\eta]^2 c \qquad (7.11)$$

In almost all cases k' + k'' = 1/2. The latter equation is commonly used as an internal check on the quality of the data. Either (7.10) or (7.11) gives the intrinsic viscosity, $[\eta]$, which is empirically related to the molecular weight by the Mark-Houwink-Sakurada equation [111]:

$$[\eta] = K'M^\alpha \qquad (7.12)$$

where K' and α are constants determined from a double logarithmic plot of intrinsic viscosity versus molecular weight.

For randomly coiled polymers the exponent α varies from 0.5 in a theta solvent to a maximum of 1.0, but typically lies between 0.6 and 0.8. Values of K' range between 0.5 and 5×10^{-4} deciliters/gram. Both K' and α are functions of the solvent as well as the polymer. The relationship is valid only for linear polymers. One use of the Mark-Houwink-Sakurada equation is to relate the intrinsic viscosity to the viscosity-average molecular weight, M_v:

$$M_v = \frac{\Sigma n_i M_i^{1+\alpha}}{\Sigma n_i M_i} = \frac{\Sigma w_i M_i^{\alpha}}{\Sigma w_i} \qquad (7.13)$$

We see that M_v depends on α as well as on the distribution of molecular species. M_v is often used in place of M_w since it is generally only 10–20% lower than M_w, and is much easier to determine. The two averages become equal when α is 1. The α and K' parameters in the Mark-Houwink-Sakurada equation can be easily determined if fractionated and reasonably monodisperse polymers are available, i.e., where $M_n \cong M_v \cong M_w$. The molecular weights can then be determined using any absolute method and the log of these MW's is plotted against the log of the intrinsic viscosity. From the slope and intercept one can determine α and K', respectively.

With the advent of size-exclusion chromatography (SEC) coupled with sensitive detectors, a common practice has been the direct determination of molecular weight distribution. When the distribution in known, then any of the molecular weight averages can be calculated. One can determine the polydispersity index (M_w/M_n) directly from the distribution rather than through the absolute methods described previously. Size exclusion chromatography, which encompasses the original gel permeation method by Moore [112], is a separation method for polymers that has become widely used for estimating molecular weight distributions. The separation is accomplished by passing a polymer solution down through a column packed with beads of a rigid porous "gel" such as highly crosslinked porous polystyrene. As the dissolved polymer molecules flow past the porous beads, they will diffuse into the internal pore structure of the gel to an extent depending on their size and the pore-size distribution of the gel. The amount of time that a given polymer molecule will spend diffusing in and out of the pores will depend on its size, and size is a function of molecular weight. Therefore, different molecular weight species are eluted from the column in order of their molecular size, i.e., the largest molecules are the first to elute.

An important development in size-exclusion chromatography was the establishment of universal calibration methods by Grubisic et al. [113] They recognized that the product of the intrinsic viscosity times the molecular weight for narrow moledular weight fractions was a measure of the hydrodynamic volume. Narrow molecular weight fractions covering a wide range in molecular weight were obtained for nine different polymers including linear and then star, comb, and ladder types of branched polymers. The retention time was obtained for each fraction and then log $[\eta] \cdot$MW was plotted as a function of retention time on the column. All points fell on a single line, which proved that the separation was based on hydrodynamic volume.

This meant that a column could be calibrated using narrow molecular weight fractions of any polymer provided that [η] was known for those fractions, and then using that calibration curve the molecular weight distribution and averages of other polymers could be calculated from their respective SEC chromatograms [114, 115]. Narrow molecular weight fractions of polystyrene or polyethylene oxide are often used to generate the calibration curve. As we shall see, the validity of using universal calibration with polar polymers such as PAN has been questioned because of so-called secondary affects associated with the interactions of the polymer with the column packing.

Various types of detectors are used in SEC including refractive index, ultra-violet/visible absorption, and low-angle laser light scattering (LALS). The latter deserves special mention because it is a critical component in the SEC-LALS liquid chromatography approach that is commonly used in many industrial polymer characterization laboratories. High-performance instruments have opened a significant performance gap between classical and modern SEC, so that analysis can be routinely performed in minutes instead of hours. The new instruments operate under high pressure, which has been made possible by use of more rigid column packings. The high pressure allows more rapid elution. Coupling a LALS detector to the SEC apparatus allows the molecular weight to be continuously determined during the separation process. This provides a rather complete description of the molecular weight distribution from which any of the molecular weight averages can be calculated, and universal calibration becomes unnecessary [115].

C. Molecular Weight Studies on Acrylic Polymers

1. Specific Issues with Acrylic Polymers

Obtaining accurate molecular weight determinations on acrylic polymers has proved very challenging and the values of fundamental parameters such as the Mark-Houwink-Sakurada, α and K' constants obtained by different researchers are somewhat scattered. Several factors are responsible for the experimental difficulties. First, PAN is not soluble in the types of solvents commonly used in the classical absolute weight measurements, and therefore it became necessary to learn how to get good measurements in solvents such as DMF and ethylene carbonate. Light scattering measurements on PAN were particularly difficult in the early work because of fluorescence from the yellow chromophores in the polymer. Obtaining narrow molecular weight fractions by fractional precipitation, a technique successfully used for typical amorphous polymers such as polystyrene also proved to be difficult.* These fractions are used to determine

---

*Fractionation is a process for the separation of a chemically homogeneous polymer specimen into components (called "fractions"), which differ in molecular weight and have narrower molecular weight distributions than the parent polymer. Fractionation involves the adjustment of the solution

the Mark-Houwink-Sakurada constants, and if they are not reasonably narrow, then the constants will be incorrect. This problem was initially addressed by directly preparing PAN polymers with narrow molecular weight distributions by polymerization in solution to less than 10% conversion. More efficient fractionation methods were later developed.

The type of end groups can affect the molecular weight values obtained from those methods that depend on the hydrodynamic viscosity, e.g., viscometry and SEC. As stated in the introduction, commercial acrylic polymers are generally made using an initiator system that results in the chains being terminated with ionic sulfate and sulfonate groups. In addition, some acrylic fiber producers add additional dyesites by copolymerization of an vinyl monomer with a pendant sulfonate group. The presence of these ionic groups affects the shape of the SEC chromatogram and causes unrealistically low retention times, as though the polymer chains had a much larger hydrodynamic volume than would be expected from the intrinsic viscosity. This behavior has been attributed to the clustering or association of chains such that they cannot easily diffuse into the pores of the column packing. It is now known that addition of small amounts of salts such as LiBr to the solvent neutralizes the effects of the charges and causes the polymer to behave more normally. This problem is circumvented by using acrylic polymers prepared by solution polymerization with free-radical initiator systems that do not produce ionizable end-groups.

2. Viscometric and Absolute Measurements of PAN Molecular Weight

The Mark-Houwink-Sakurada K' and α constants relating M_v to the intrinsic viscosity have been determined by a number of investigators and are listed in Table 5 [116]. This table gives these constants for a wide range of temperatures and solvents. The absolute methods used in each determination as well as the molecular weight range of the fractions studied and the method of preparation of the fractions is included. The values of α fall well within the range expected for randomly coiled polymers, i.e., 0.5 to 1.0.

Krigbaum and Kotliar [124] carefully reviewed much of the early work on determination of M_n and M_w, and made their own measurements on polymers prepared both by solution and aqueous polymerization. Osmometry was used to determine M_n, and both light scattering and sedimentation velocity were used to determine M_w. The solution polymers were prepared in DMF using an azo-bis-isobutyronitrile initiator and the aqueous polymers were made using a persul-

---

conditions so that two liquid phases are maintained in equilibrium, removal of one phase and then adjusting solution conditions (either by addition of a nonsolvent or reduction in solution temperature) to obtain a second separated phase, and so on. With amorphous polymers the separation is governed by the decrease in solubility that occurs with increasing molecular weight. Complications can occur if the polymer to be fractionated is crystalline, because the reduction in solvent power can induce precipitation of the crystalline polymer.

Table 5 Mark-Houwink-Sakurada Constants for Determination of M_V for Polyacrylonitrile in Various Solvents

Temp. (°C)	$K' \times 10^3$ (mL/g)	α	Mol. wt. range MW $\times 10^{-4}$	Method[a]	Prep.[b] F,P	Ref.
			-Butyrolactone			
20	34.3	0.730	4–40	LV(LS)	F	117
30	57.2	0.67	4–30	SA	F	118
30	34.2	0.70	6–30	SA	F	118
30	40.0	0.69	15–53	LS	P	119
50	28.7	0.74	4–40	LS	F	117
			Dimethylformamide			
20	17.7	0.78	7–30	LS	F	120
25	16.6	0.81	5–27	SD	F	121
25	24.3	0.75	3–25	LS	P	122
25	39.2	0.75	3–100	OS	P	123
			Deionized Dimethylformamide			
25	15.5	0.80	3–10	LS,SD	F,P	124
25	39.6	0.75	4–30	OS	P	125
25	44.3	0.70	2–20	LS	P	125
25	69.8	0.65	8–140	LS	P	126
30	29.6	0.74	4–30	SA	F	118
30	33.5	0.72	16–48	LS	P	119
35	27.8	0.76	3–58	DV	F	127
35	31.7	0.746	9–76	LS	F	117
50	30.0	0.752	4–102	LV	F	117
			Dimethylacetamide			
20	30.7	0.761	2–40	LV	F	117
			Acrylonitrile-Methylacrylate Copolymer in Dimethylformamide			
25	21.3	0.74	5–53	—	—	128

[a]Methods: osmometry (OS), light scattering (LS), limiting viscosity number-molecular weight relationship (LV), Archibald's method for approaching sedimentation equilibrium (SA), sedimentation and diffusion (SD), diffusion and viscosity (DV)
[b]Preparation of samples: fractionation (F), polymerization (P).
Source: Data by Kurata et al. [116].

fate-bisulfite redox initiator system, which resulted in ionizable sulfate and sulfonate end groups. For the solution polymers, reliable results could be obtained using distilled DMF, but for the aqueous polymers it was necessary to deionize the solutions and remove all traces of water. A comparison of the M_n

and M_w values indicate that the polydispersity index, Q, is 2.1 for the solution polymers and 2.8 for those made by the aqueous process.

3. Measurements of Molecular Weight of PAN Using Size Exclusion Chromatography

We now consider the investigations done using size-exclusion chromatography to determine the molecular weight distribution and various molecular weight averages of PAN. The importance of neutralizing the effects of the pendant ionic groups by addition of salts such as LiBr in order to obtain undistorted chromatograms has been described by Cha [129]. SEC analysis along with molecular weight determinations using osmometry and light scattering were done on two different polymers; a PAN homopolymer containing no pendant sulfonate groups and a PAN copolymer containing 0.2% sulfur corresponding to an average of about four sulfonate groups per average chain. The intrinsic viscosity, [η], of both polymers decreased when LiBr was added to the DMF at the 0.1 molar level, but the decrease for the sulfonate-containing copolymer (S-copolymer) was much more pronounced indicating considerable hydrodynamic contraction. The polydispersity index, Q, was 2.21 and 2.46, respectively, for the PAN homopolymer and the S-copolymer.

The SEC chromatograms of these two polymers in DMF were quite different even though both the M_n and M_w values were similar. The elution of the S-copolymer begins much earlier, which means that it has a larger hydrodynamic volume in spite of the similarity in molecular weight. Multiple peaks were seen in the S-copolymer that were attributed to different distributions of sulfonate groups in the polymer chain, thus leading to a distribution of molecular configurations in solution. Switching to DMF with 0.1 molar LiBr effectively swamps out the ionic effects. The retention times are increased and the multiple peaks are replaced by a single peak. The weight average size of the S-copolymer in 0.1 molar LiBr-DMF was 5.3 times smaller than that in DMF. The conclusion from this work is that valid SEC chromatograms of ionic polymers such as the S-copolymer require the presence of an electrolyte such as LiBr that is effective in screening the pendant ionic groups.

Coppola et al. [130] studied effect of LiBr on the viscosity and SEC behavior of several polymers in DMF. The polymers included PAN (prepared with and without sulfonate groups), polystyrene, and polyvinyl acetate. They concluded that all of these polymers contract upon addition of LiBr and refer to this as a general salting out effect. This contraction lowers the intrinsic viscosity by approximately 20–30% for all of the polymers, and, most importantly, was not any larger for the sulfonate-containing PAN. They contend that this contraction alone cannot account for the strong effect seen in the SEC behavior of the sulfonate-containing PAN (their experimental results confirm those reported by Cha [129]), and conclude that some type of association of the charged PAN must exist in DMF solutions absent of LiBr. This association causes the polymer to elute very quickly, as though it had a very large size. The LiBr apparently breaks up this association and allows more normal SEC elution times. The authors were unable to explain why such an association phenomenon would not also affect the intrinsic viscosity.

Specific approaches to generating molecular weight averages and molecular weight distributions from SEC chromatograms is given by Mori [131, 132], Kenyon and Mottus [133], and Azuma et al. [134]. Mori [131] emphasizes that SEC calibration with polystyrene/DMF solutions can lead to problems because of poor solubility and possible partitioning of this polymer onto the column packing. DMF is used because it is a good solvent for polar polymers such as PAN, and calibration with polyethylene oxide standards is recommended. The validity of applying universal calibration methods to PAN is discussed. In order for this approach to be successful it is necessary to be sure that the elution time is dependent only on the hydrodynamic volume, even after suppressing any associations by addition of salts such as LiBr. It is also necessary to have accurate Mark-Houwink K' and a parameters. In an earlier paper Mori [132] developed an alternative method for preparing a calibration curve for a polymer where only broad molecular weight distribution samples were available. The proposed method requires neither prior knowledge of the Mark-Houwink parameters nor the assumption of any formula for the calibration curve. This was applied to PAN and polyvinyl pyrrolidone using polyethylene oxide standards. The hydrodynamic volume, ($[\eta]\cdot MW$) of PAN polymers was calculated and shown not to fall on the hydrodynamic volume–elution time curve for the polyethylene oxide standards, whereas with the polyvinyl pyrrolidone polymers, the agreement was excellent. This suggested that secondary effects perhaps related to interactions with the column packing invalidate the universal calibration approach for PAN.

Kenyon and Mottus [133] used SEC to determine the molecular weight and molecular weight distribution of an acrylonitrile/ethyl vinyl ether copolymer. Even though this acrylic copolymer contains no pendant ionic groups, they still observed multiple peaks in the SEC chromatogram when DMF was used as the solvent. Addition of LiBr gave single peaks and increased the retention times, in manner similar to previous observations with polymers that contained ionic groups. The authors speculated that PAN polymers can form associations with the solvents that are eliminated by addition of LiBr. They were unsuccessful in applying universal calibration to PAN, and recommended calibration based on a PAN polymer with a known molecular weight distribution.

The influence of polymerization conditions on the molecular weight distribution of commercial AN-vinyl acetate copolymer (93:7 in weight) was reported by Ito and Okada [135]. The copolymer was prepared by a continuous process in aqueous medium using the persulfate-bisulfite redox initiator system, with the water to monomer ratio being the principle process variable under study. The latter controls the structure of the polymer particle, and as this ratio is decreased from 4.0 to 1.5, the polymer particles become more compact thus resulting in polymer filter cakes having lower water contents and requiring less energy for the drying process. It was important to understand how this process change would affect the molecular weight distribution because the latter is directly related to the number of dye sites in these polymers (see Section IIA).

The molecular weight distribution's were evaluated by the M_v/M_n ratios estimated from viscosity and osmotic pressure measurements, and by M_w/M_n ratios estimated using size-exclusion chromatography. It was determined that the universal calibration approach was not suitable for these polymers, and therefore the SEC curves were calibrated using fractions prepared by fractionating one of the copolymers included in the study. All of the copolymers showed broad molecular weight distribution's with the polydispersity index, M_w/M_n varying from 4.4 to 5.11. These differences were not considered significant with respect to variation in dye site content and therefore the water to monomer ratio could be optimized without concern for dyeability and other properties related to the molecular weight distribution.

REFERENCES

1. Alexander, L. E., *X-ray Diffraction Methods* in *Polymer Science*, Wiley-Interscience, New York, 1969, pp. 12–21.
2. Samuels, R. J., *Structured Polymer Properties*, Wiley, New York, 1974.
3. Parrini, P., and Crespi, G., *Encyclopedia of Polymer Science and Technology*, vol. 13, Interscience, New York, 1970, p. 89.
4. Natta, G., Corradini, P., and Ganis, P., *J. Polym. Sci.*, **58**; 1191 (1962).
5. Schaefer, J., *Macromolecules*, **4**; 105 (1971).
6. Bovey, F. A., and Winslow, F. H., *Macromolecules*, An *Introduction to Polymer Science*, Academic Press, New York, 1979, p. 223.
7. Pasquon, I., and Giannini, U., Stereoregular linear polymers, *Encyclopedia of Polymer Science and Engineering*, 2nd ed., vol. 15, Wiley, New York, 1989, p. 632.
8. Chiang, R., *J. Polym. Sci. A*, **3**; 2019 (1965).
9. Chiang, R., Rhodes, J. H., and Holland, V. F., *J. Polym. Sci. A*, **3**; 479 (1965).
10. Van Krevelan, D. W., *Properties of Polymers*, Elsevier, New York, 1972, p. 43.
11. Billmeyer, F. W., Jr., *Textbook of Polymer Science*, 3rd. ed., Wiley, New York, 1984, p. 153–154.
12. Hinrici-Olivé, G., and Olivé, S., *Adv. Polym. Sci.*, **32**, 128, (1979).
13. Hopfinger, A. J., *Conformational Properties of Macromolecules*, Academic Press, New York, 1973.
14. Krigbaum, W. R., and Tokita, N., *J. Polym. Sci.*, **43**; 467 (1960).
15. de Daubney, R., Bunn, C. W., and Brown, C. J., *Proc. R. Soc. (London)*, **226**; 531 (1954).
16. Bohn, C. R., Schaefgen, J. R., and Statton, W. O., *J. Polym. Sci.*, **55**; 531 (1961).
17. Saum, M., *J. Polym. Sci.*, **42**; 57 (1960).
18. Delmartino, R. N., *J. Polym. Sci.*, **28**; 1805 (1983).

19. Geil, P. H., *Polymer Single Crystals*, Interscience, New York, 1963, p. 21.
20. Holland, V. G., Mitchell, S. B., Hunter, W. L., and Lindenmeyer, P. H., *J. Polym. Sci.*, **62**; 145 (1962).
21. Klement, J. J., and Geil, P. H., *J. Polym. Sci. A-2*, **6**; 1381 (1968).
22. Patel, G. N., and Patel, R. D., *J. Polym. Sci. A-2*, **8**; 47 (1970).
23. Gohil, R. M., Patel, K. C., and Patel, R. D., *Makromol. Chem.*, **169**; 291 (1973).
24. Gohil, R. M., Patel, K. C., and Patel, R. D., *Colloid Polym, Sci.*, **254**; 559 (1976).
25. Kumamaru, F., Kajiyama, T., and Takayanagi, M., *J. Crystal Growth*, **48**; 202 (1980).
26. Colvin, B. G., and Storr, P., *Eur. Polym. J.*, **10**; 337 (1974).
27. Corridini, P., Natta, G., Ganis, P., and Temussi, P. A., *J. Polym. Sci. C*, **16**; 2477 (1967).
28. Takahashin, M., and Nukushima, Y., *J. Polym. Sci.*, **56**; 519 (1962).
29. Lindenmeyer, P. H., and Hoseman, R., *J. Appl. Phys.*, **34**; 42 (1963).
30. Bell, J. P., and Dumbleton, J. H., *Textile Res. J.*, **41**; 196 (1971).
31. Hinrichsen, G., *J. Polym. Sci.*, **38**; 303 (1972).
32. Gupta, A. K., and Singhal, R. P., *J. Polym. Sci.: Polym. Phys. Ed.*, **21**; 2243 (1983).
33. Matta, V. K., Mathur, R. B., and Bahl, O. P., *Carbon*, **28(1)**; 241 (1990).
34. Bahl, O. P., Mathu, R. B., and Dhami, T. L., *Mat. Sci. Eng.*, **13**; 105 (1985).
35. Imai, Y., Minami, S., Yoshihara, T., Joh, Y., and Sato, H., *J. Polym, Sci. B, Polym. Lett.*, **8**; 281 (1970).
36. Joh, Y., *J. Polym. Sci., Polym. Chem, Ed.*, **17**; 4051 (1979).
37. Hinrichsen, G., and Orth, H., *Polym. Lett.*, **9**; 529 (1971).
38. Hinrichsen, G., *Angew Makromol. Chem.*, **20**; 121 (1974).
39. Jain, M. K., and Abhiraman, A., *J. Mater. Sci.*, **18**: 1979 (1983).
40. Warner, S. B., Uhlmann, D., and Peebles, L., *J. Mater. Sci.*, **10**; 758 (1975).
41. Warner, S. B., *Polym. Lett.*, **16**; 287 (1978).
42. Prevorsek, D. C., Butler, R. H., Kwon, Y. D., Lamb, G. E. R., and Sharma, R. K., *Textile Res. J.*, **43** (Feb. 1975).
43. Fischer, E. W., and Fakirov, S., *J. Mater. Sci.*, **11**; 1041 (1976).
44. Gupta, A. K., Chand, N., Singh, R., and Mansinga, A., *Eur. Polym. J.*, **15**; 129 (1979).
45. Gupta, A. K., and Chand, N., *Eur. Polym. J.*, **15**; 899 (1979).
46. Gupta, A. K., and Chand, N., *J. Polym, Sci. Polym. Phys. Ed.*, **18**; 1125 (1980).
47. Gupta, A. K., Singhal, R. P., and Agarwal, V. K., *Polymer*, **22**; 285 (1981).
48. Gupta, A. K., and Singhal, R. P., *J. Appl. Polym. Sci.*, **26**; 3599 (1981).
49. Gupta, A. K., and Maiti, A. K., *J. Appl. Polym. Sci.*, **27**; 2409 (1982).
50. Gupta, A. K., Singhal, R. P., and Maiti, A. K., *J. Appl. Polym. Sci.*, **27**; 4101 (1982).
51. Gupta, A. K., Singhal, R. P., and Bajaj, P., *J. Appl. Polym. Sci.*, **28**; 1167 (1983).
52. Gupta, A. K., Singhal, R. P., and Agarwal, V. K., *J. Appl. Polym. Sci.*, **28**; 2745 (1983).
53. Williams, M. L., Landel, R. F., and Ferry, J. D., *J. Am. Chem. Soc.*, **77**; 3701 (1955).
54. Bueche, F., *Physical Properties of Polymers*, Interscience, New York, 1962, Chap. 6–9.
55. Dumbleton, J. H., and Murayama, T., *Kolloid-Z. Z Polym.*, **220**; 41 (1967).
56. Kulshreshtha, A. K., Garg, V. N., and Sharma, Y. N., *J. Appl. Polym. Sci.*, **31**; 1413 (1986).
57. Frushour, B. G., *Polym. Bull.*, **4**; 305 (1981).

58. Andrews, R. D., Miyachi, K., and Doshi, R. S., *J. Macromol. Sci.-Phys. B*, **9**; 281 (1974).
59. Miyachi, R., and Andrews, R. D., *J. Appl. Polym. Sci. Appl. Polym. Symp.*, **35**; 127 (1974).
60. Andrews, R. D., Yen, R. C., and Changs, P., *J. Macromol. Sci.-Phys. B*, **19**; 729 (1981).
61. Grobelny, J., Tekely, P., and Turska, E., *Polymer*, **22**; 1649 (1981).
62. Turska, E., and Grobelny, J., *Eur. Polym. J*, **19**; 985 (1983).
63. Grobelny, J., Sokol, M., and Turska, E., *Polymer*, **25**; 1414 (1984).
64. Sokol, M., Grobelny, J., and Turska, E., *Polymer*, **28**; 843 (1987).
65. Grobelny, J., Sokol, M., and Turska, E., *Eur. Polym. J.*, **24**; 1195 (1988).
66. Grassie, N., and McGuchan, R., *Eur. Polym. J.*, **6**; 1277 (1970).
67. Shalaby, S. W., Thermoplastic polymers, in *Thermal Characterization of Polymeric Materials*, (E. Turi, ed.), Academic Press, New York, 1981.
68. Flory, P. J., *Principles of Polymer Chemistry*, Cornell University Press, Ithaca, NY, 1953, p. 568.
69. Mandelkern, L., *Chem. Rev.*, **56**; 903 (1956).
70. Flory, P. J., Garrett, R. R., Newman, S., and Mandelkern, L., *J. Polym. Sci.*, **12**; 94 (1954).
71. Krigbaum, W. R., *J. Polym. Sci.*, **28**; 213 (1958).
72. Flory, P. J., *Trans. Faraday Soc.*, **51**; 848 (1955).
73. Wunderlich, B., *Macromolecular Physics*, vol. 3, Academic Press, New York, 1980, p. 275.
74. Eby, R. K., *J. Appl. Phys.*, **34**; 2442 (1963).
75. Slade, P. S., *Thermochim. Acta*, **1**; 4459 (1970).
76. Berger, W., Heller, A., and Adler, H-J., *Faserforsung Textiltechnic*, **24**; 484 (1973).
77. Frushour, B. G., *Polym. Bull.*, **7**; 1 (1982).
78. Frushour, B. G., *Polym. Bull.*, **11**; 375 (1984).
79. Ferry, J. D., *Viscoelastic Properties of Polymers*, 2nd ed., Wiley, New York, 1969.
80. Kimmel, R. M., and Andrews, R. D., *J. Appl. Phys.*, **36**; 3063 (1965).
81. Howard, W. H., *J. Appl. Polym. Sci.*, **5**; 303 (1961).
82. Loshack, S., and Fox, T. G., *Bull. Am. Phys. Soc.*, **1**; 123 (1956).
83. Frushour, B. G., and Knorr, R. S., Acrylic fibers, in *Handbook of Fiber Science: Vol. IV, Chemistry* (M. Lewin and E. M. Pearce, eds.) Marcel Dekker, New York, 1985.
84. Aiken, D., Burkinshaw, S. M., Cox, R., Catherall, J., Litchfield, R. E., Price, D. M., and Todd, N. G., *J. Appl. Polym. Sci. Appl. Polym. Symp.*, **47**; 263 (1991).
85. Bell, J. F., and Murayama, T., *J. Appl. Polym. Sci.*, **12**; 1795 (1968).
86. Gur-Arieh, Z., and Ingamells, W. C., *J. Soc. Dyers Colour.*, **90**; 8 (1974).
87. Hori, T., Khang, H., Shimuzu, T., and Zollinger, H., *Text. Res. J.*, **58**: 227 (1988).
88. Murayama, T., *Dynamic Mechanical Analysis of Polymeric Materials*, Elsevier, New York, 1978.
89. McCrum, N. G., Read, B. E., and Williams, G., Molecular theories of relaxation, in *Anelastic and Dielectric Effects in Polymeric Solids*, Wiley, New York, 1967, Chap. 5.
90. Minami, S., *J. Appl. Polym. Sci.: Appl. Polym. Symp.*, **25**; 145 (1974).
91. Minami, S., Yamamori, H., and Sato, H., *Prog. Polym. Phys. Jpn.*, **14**; 379 (1971).
92. Schneider, X., and Wolf, K., *Kolloid-Z*, **234**; 149 (1953).

93. Cotton, G. R., and Schneider, W. C., *Kolloid-Z*, **192**; 16 (1963).
94. Minami, A., Yoshihara, T., and Sato, H., *Kobunshi Kagaku* (English ed.), **1**; 125 (1972).
95. Okajima, S., Ikeda, M., and Takeuchi, T., *J. Polymer Sci. A-1*, **6**; 195 (1968).
96. Minami, S., Sakurai, T., Yoshihara, T., and Sato, H., *Prog. Polym. Phys. (Jpn.)*, **14**; 385 (1972).
97. Keyon, A. S., and Rayford, M. J., *J. Appl. Polym. Sci.*, **23**; 717 (1979).
98. Padhye, M. R., and Karandikar, A. V., *J. Appl. Polym. Sci.*, **33**; 1675 (1987).
99. Takayanagi, M., *Proceedings of the Fourth International Conference on Rheology*, *Part I*, Interscience, New York, 1965, p. 61.
100. Nielsen, L. E., *Mechanical Properties of Polymers*, Van Nostrand Reinhold, New York, 1962, p. 163.
101. Nielsen, L. E., *J. Appl. Phys.*, **25**; 1209 (1954).
102. Andrews, R. D., and Kinmel, R. M., *Polym. Lett.*, **3**; 167 (1965).
103. Peebles, L. H., *Molecular Weight Distributions in Polymers*, Wiley-Interscience, New York, 1971.
104. Cooper, A. R., Molecular weight averages and distribution functions, in *Determination of Molecular Weight*, Cooper, A. R., ed. Wiley, New York, 1989, chap. 1.
105. Billmeyer, Jr., F. W., Measurement of molecular weight and size, in *Textbook of Polymer Science*, 3rd ed., Wiley, New York, 1984, chap. 8.
106. Rudin, A., *The Elements of Polymer Science and Engineering*, Academic Press, New York, 1982, chap. 3.
107. Stockmayer, W. H., and Casassa, E. F., *J. Chem. Phys*, **20**; 1560 (1952).
108. Zimm, B. H., *J. Chem Phys.*, **16**; 1093 (1948).
109. Huggins, M. L., *J. Am. Chem. Soc.*, **64**; 2716 (1942).
110. Kraemer, E. O., *Ind. Eng. Chem.*, **30**; 1200 (1938).
111. Kamide, K., and Saito, M., Viscometric determination of molecular weight, in *Determination of Molecular Weight*, (A. R. Cooper, ed.) Wiley, New York, 1989, chap. 8.
112. Moore, J. C., *J. Polym. Sci.*, *A*, **2**; 835 (1964).
113. Grubisic, Z., Rempp, P., and Benoit, H., *J. Polym. Sci.*, *B*, **5**; 753 (1967).
114 Weiss, A. R., and Cohn-Ginsberg, E., *J. Polym. Sci.*, *B*, **7**; 379 (1969).
115. Yau, W. W., Kirkland, J. J., and Bly, D. D., *Modern Size-Exclusion Liguid Chromatography*, Wiley, New York, 1979, p. 285.
116. Brandrup, J., and Immergut, E. H., *Polymer Handbook*. 3rd ed. Wiley, New York, 1989.
117. Fujisaki, Y., and Kobayashi, H., *Kobunsi Kagaku (Chem. High Polym. (Tokyo)*, **19**; 73, 81 (1962).
118. Inagaki, H., Hayashi, K., and Matsuo, T., *Makromol. Chem.*, **84**; 80 (1965).
119. Shibukawa, T., Sone, M., Uchida, A., and Iwahori, K., *J. Polym. Sci.*, *A-1*, **6**; 147 (1968).
120. Scholtan, W., and Marzolph, H., *Makromol. Chem.*, **57**; 52 (1962).
121. Bisschops, J., *J. Polym. Sci.*, **17**; 81 (1955).
122. Cleland, R. L., and Stockmayer, W. H., *J. Polym, Sci.*, **17**; 473 (1955).
123. Onyon, P. F., *J. Polym. Sci.*, **22**; 13 (1956).
124. Krigbaum, W. R., and Kotliar, A. M., *J. Polym. Sci.*, **32**; 323 (1958).
125. Onyon, P. F., *J. Polym. Sci.*, **37**; 315 (1959).
126. Peebles, L. H., *J. Polym. Sci.*, *A*, **3**; 361 (1965).
127. Kobayashi, H., *J. Polym. Sci.*, **39**; 369 (1959).

128. Kamide, K., Miyazaki, Y., and Kobayashi, H., *Polym. J.*, **14**; 591 (1982).
129. Cha, C. Y., *J. Polym. Sci., B*, **7**; 343 (1969).
130. Coppola, G., Fabbri, P., and Palleri, B., *J. Appl. Polym. Sci.*, **16**; 2829 (1972).
131. Mori, S., *Anal. Chem.*, **55**; 2414 (1983).
132. Mori, S., *Anal. Chem.*, **53**; 1813 (1981).
133. Kenyon, A. S., and Mottus, E. H., *J. Appl. Polym. Sci.; Appl. Polym. Symp.*, **25**; 57 (1974).
134. Azuma, C., Dias, M., and Mano, E., *Macromol. Chem., Macromol. Symp.*, **2**; 169 (1986).
135. Ito, S., and Okada, C., *Sen-I-Gakkaishi*, **42**; T-619 (1986).

8

Thermal and Actinic Degradation

Gary Wentworth

The C.P. Hall Company
Chicago, Illinois

I. INTRODUCTION

Exposure to elements of the environment can cause degradation of acrylic fibers, affecting both physical properties and aesthetics. Deterioration of tensile strength or loss of elongation leading to embrittlement is undesirable in most fiber applications. Reduction of gloss, yellowing, or fading of color are of concern where appearance is important.

This chapter deals with the effects of heat and actinic (i.e., ultraviolet and visible range) radiation on acrylic fibers. The controlled use of high temperature to form carbon fibers from acrylics [1, 2] is beyond the scope of this discussion, except insofar as it may offer clues to degradation and stabilization mechanisms under less extreme conditions.

While there are commonalities among polymeric materials such as fibers, coatings and plastics with respect to degradation due to environmental factors, fibers are particularly susceptible to degradation because of the large surface areas involved [3]. To understand the causes of fiber degradation it is necessary to consider thermal, oxidative, and photochemical processes, alone and acting in concert on the polymers involved.

II. THERMAL/OXIDATIVE

A. Mechanism

Acrylic fibers have good color and heat stability at temperatures below about 115°C [4]. Still, thermal effects, in combination with air (oxygen) can be important in fiber processing where they can cause discoloration (yellowing). And in some end uses the combination of even mild heat with oxygen can lead to discoloration over a period of time. Grassie et al. [5–10] used both thermal analysis (DTA and TGA) and IR spectroscopy to study thermal degradation of polyacrylonitrile (PAN) under a wide variety of conditions. Isothermal aging below 200°C in nitrogen resulted in "polymerization" of nitrile groups as judged by changes in the infrared spectrum.

The loss of the C≡N absorption in the infrared was accompanied by a change in color, deepening from yellow to gold to tan with concomitant appearance of new bands at 1640–1610 cm^{-1} and 1580 cm^{-1} attributed to C=N. Further heating above 350°C led to formation of aromatic structures along with some chain scission and volatilization of low molecular weight heterocycles, HCN and NH$_3$.

Finally, continued heating above 900°C gave loss of nitrogen and formation of graphite. Similar reactions occurred in the presence of oxygen, except that loss of volatiles was markedly reduced, a finding that would later have great importance for the formation of graphite fibers from acrylics.*

Based on differences in degree of degradation of PAN and polymethacrylonitrile, Grassie [7] argued that the α-hydrogen on PAN plays an important role in intramolecular condensation of nitrile groups and suggested a mechanism of intramolecular hydrogen transfer to account for initiation and termination at frequent intervals along the chain. The result would be a series of blocks of conjugated, cyclic imine structures along each PAN chain. Involvement of the α-hydrogen was also indicated by the observation [11] that α-deutero PAN discolored less on heating than the PAN control, a qualitative indication of a deuterium isotope effect.

While the conditions used by Grassie and co-workers to study PAN color-forming reactions are much harsher than would be found in typical textile end

---

*In the field of graphite fiber formation, treatment of PAN with oxygen at intermediate temperatures prior to graphitization at >1000°C is said to "stabilize" the structure against undesirable loss of volatiles. Considerable work [1, 2, 12, 14] has been directed to characterization and control of this reaction. It is discussed further in Chapter 12, on industrial end uses.

uses, inferences can be drawn about the effects of polymer structure, additives, and stabilizers on fiber discoloration under milder conditions. It was known empirically that the tendency of an acrylic fiber to yellow is dependent not only on conditions but also on comonomer choice and content. Grassie and others showed that polymerization of nitrile groups can be initiated at lower temperatures, perhaps by an ionic mechanism, when carboxylic acids (and, to a lesser extent, amides) have been incorporated in the polymer structure [8, 14].

Both acidic (phenols, zinc chloride) and basic (thiourea, guanidines, potassium cyanide) additives accelerate the reaction [6]. α-Substitution (methacrylonitrile, vinylidene cyanide) changes the course of the reaction to favor fragmentation by depolymerization [7]. Acrylate and methacrylate monomers [9] have a diluent effect, while styrene was said to inhibit the reaction because the stability of the styryl radical favored fragmentation pathways [9].

It is also likely that since styrene is known to give largely alternating structures in copolymerization with acrylonitrile, long sequences of conjugated nitrile-derived structures are statistically less likely. (Styrene-acrylonitrile polymers containing 30–60% styrene are the basis for commercially viable thermoplastics which can melt cleanly to give nearly colorless molded parts.)

Vinyl acetate, acrylic acid and acrylamide [8, 15] accelerate the thermal degradation reaction while maleic anhydride and maleimide strongly inhibit it [15]. Incorporation of vinyl chloride, vinylidene chloride, and α-chloroacrylonitrile changes the course of reaction to dehydrohalogenation [16, 17] and reduces volatilization through crosslinking reactions which prevent fragmentation.

In a comprehensive paper, Friedlander et al. [18] attributed color development to formation of partially hydrogenated naphthyridine rings produced by linkage of adjacent nitrile units, some of these being oxidized to nitrones. They

found multiple pathways to formation of color in PAN (Figure 1) and showed that discoloration is a very complex process which can be affected by the conditions of polymer synthesis as well as effects of atmospheric oxygen and heat. Formation of a β-ketonitrile unit (a) during polymerization in aqueous media (path A on the following page) was confirmed by the strong UV absorption at 275 nm which could be used to quantify the compound photometrically [19].

This structure (a) was also shown to be responsible for fluorescence exhibited by PAN (PAN prepared by ionic catalysis does not fluoresce).

Ketonitriles were shown to form from the action of oxygen on the polymer (path B on the following page).

Intermediate formation of the hydroperoxide could be determined iodometrically. Under the conditions studied, no loss of molecular weight was observed which appeared to rule out oxidation at the α-carbon (path C) which, according

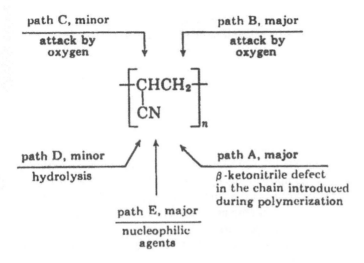

Figure 1 Mechanism of thermal degradation of PAN. (Reproduced with permission from Ref. 18, Copyright 1968, American Chemical Society.)

Path A

Path B

to the authors, is expected to lead to fragmentation. The tautomeric form of these keto units could then play a role initiating nitrile unit condensation to form colored species.

The amount of discoloration in PAN heated at 145°C could be related directly to the concentration of β-ketonitrile units in the polymer [18]. Path E (Figure 1) is the nucleophilic attack on the nitrile group leading to naphthyridine structures. This can be accelerated by β-keto units (as above) or by carboxylic acids [8], formed from hydrolysis of nitrile units (path D) during synthesis in aqueous media. (The hydrolysis can be controlled by suitable choice of polymerization conditions, particularly pH [33].)

Nucleophilic species in general accelerate the color-forming reaction. Thus, thiourea, potassium cyanide, bases, etc., greatly enhance color formation.

Isotopic tracer techniques have been used to confirm earlier infrared spectroscopy assignments of structure in thermally degraded PAN [20]. Degradation of poly (d_3-acrylonitrile), poly (1-^{13}C-acrylonitrile), and poly (^{15}N-acrylonitrile) in the presence of $^{18}O_2$ permitted unambiguous assignment of infrared bands because of shifts in absorption frequency and/or intensity due to isotope effects which could be picked up using Fourier transform techniques (see Figures 2 and 3). By following these shifts as a function of both oxygen concentration and heating time of the labeled PANs, the authors confirm that the thermooxidative degradation proceeds with time from nitriles to imines to fused pyridine/ dihydropyridine structures. The observation that PAN degraded in air and in $^{18}O_2$ gives the same spectrum (Figure 2, curves A and B) was used to rule out formation of nitrone structures [18] as important in the color forming reaction.

Grasselli and co-workers [21] studied thermal degradation of PAN under a variety of conditions. In dimethylformamide solution at 180°C, color deepened with time. In the UV spectrum, wavelength cutoffs also increased with time, but upon dilution all samples showed a single absorption band at 352 nm (Figure 4). This means that the deepening of color is due to an increase in concentration of

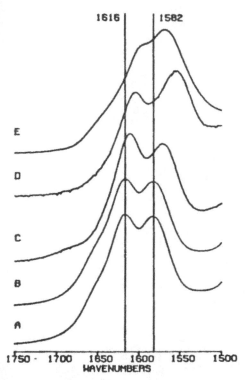

Figure 2 Infrared absorbance spectra from 1750 to 1500 cm^{-1} of labeled PANs degraded at 235° for 6 hr with CN/O$_2$ = 1300: (A) PAN, (B) PAN with ^{18}O$_2$, (C) ^{15}N-PAN, (D) 1-^{13}C-PAN, and (E) d$_3$-PAN. (Reproduced with permission from Ref. 20, Copyright 1984, John Wiley and Sons, Inc.)

chromophores not to a lengthening of the conjugated sequence (which would cause a shift to longer wavelength). Thus, the nitrile condensation reaction proceeds to a certain number of rings and stops. The authors estimate the sequence length at three to five rings. Other workers [22, 23] have argued that relatively short sequences are to be expected as a consequence of tacticity, i.e., isotactic placement of nitrile groups is preferred for intramolecular ring closure. Intermolecular crosslinking was thought to accompany the intramolecular reaction. Further support for this concept comes from the report that PAN with enhanced isotacticity [24] showed an enhanced rate of thermal degradation [25, 26]. PAN was compared to its α-deuterated counterpart in thermal degradation of powders in the form of KBr pellets [21]. Rates of disappearance of nitrile groups were identical. The lack of a deuterium isotope effect would seem to rule out Grassie's hydrogen transfer mechanism [7] of thermal polymerization

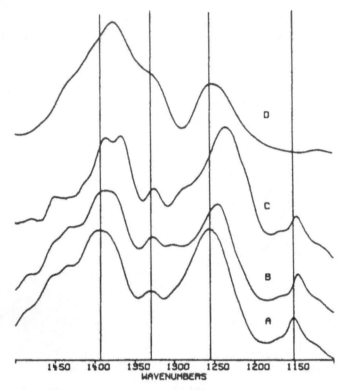

Figure 3 Infrared absorbance spectra from 1500 to 1100 cm⁻¹ of labeled PANs degraded at 235° for 6 hr with $CN/O_2 = 1300$. (A) PAN, (B) ^{15}N-PAN, (C) 1-^{13}C-PAN, and (D) d_3-PAN. (Reproduced with permission from Ref. 20, Copyright 1984, John Wiley and Sons, Inc.)

(however, note that earlier workers [11] reported the opposite result). The color of PAN film was more intense when degraded in air than in nitrogen [21]. This was attributed to carbonyl structures showing infrared absorbance at 1680 cm⁻¹ which could cause bathochromic shifts, broadening the UV absorbance bands, and deepening the color. ^{13}C NMR ruled out the presence of nitrone structures [18] which would have been expected to result in a peak at about 200 ppm.

Recently, Tverskoi et al. have reported ESR evidence for the existence of extended, cyclic, conjugated sequences in thermally degraded PAN by relating the band width of the ESR signal to the degree of conjugation in the system [27]. Comparing calculated bandwidths to those observed with a series of model compounds, they concluded that thermally degraded PAN contains a minimum of 8–12 conjugated, pyridine-type structures.

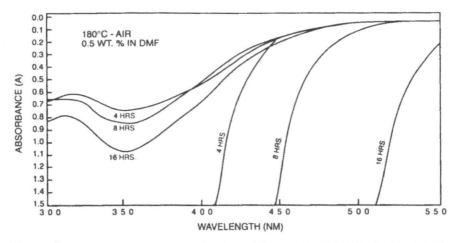

Figure 4 Ultraviolet spectra of solution degradation of 0.5wt% PAN in DMF at 180°C in air. 4 hr-diluted 1:5 with DMF, 8 hr-diluted 1:10 with DMF, 16 hr-diluted 1:200 with DMF. (Reproduced with permission from Ref. 21, Copyright 1985, Pergamon Press Ltd.)

B. Control of Thermal Degradation

Discoloration of PAN fibers can be controlled by choice of polymer synthesis technique, comonomer structure and a wide variety of additives. Synthesis of PAN in a solvent such as ethylene carbonate/propylene carbonate, rather than the common commercial practice of aqueous redox polymerization, creates a polymer without the defects which are a major initiator of color formation [8, 28]. Low conversion of monomer also improves thermal stability [29], since this minimizes the chance of polymer radicals entering into side reactions. Reaction temperatures above 50°C which lead to polymers with a greater propensity for color formation are avoided by using a redox initiator such as persulfate/bisulfate. Polymerization using azobis (isobutyronitrile) (AIBN) as initiator leads to lower color than using benzoyl peroxide, presumably due to formation of catalytic oxidative species in the latter case [18]. Additives which have been used effectively during suspension polymerization include reducing agents such as sulfinic acids [30] or organophosphorus compounds [31]. Sulfates, carbonates, and nitrates [32] have been effective in solution polymerization.

Treatment of acrylic fibers, yarns, or fabric with thiosemicarbazide solutions improves thermal stability [34] Impregnating wet-spun fibers in the gel state with alkaline earth metal salts [35] or with sodium sulfide [36] has been claimed to be effective. Additives to the spinning "dope" have been cited [37–41]. Treatment of dry fiber with a number of agents has also been claimed [42–44]. Finally, fiber whiteness can be recovered by use of bleaching agents [43]; sodium chlorite (Textone) is used commercially for this purpose. Yellowness can be masked using fluorescent brighteners which absorb ultraviolet radiation and reemit the energy in the blue region of visible light. Brighteners may be added to the spin dope in dry spinning, incorporated in the uncollapsed fiber during processing or added to the yarn or fabric by a converter. Some typical optical brighteners are shown below [45]. Additions of blue dyes and pigments in the part-per-million range are used for a similar purpose [46–50]. Since these reduce the total reflectance, the fiber appears more gray. A review on the use of stabilizers has been published [3].

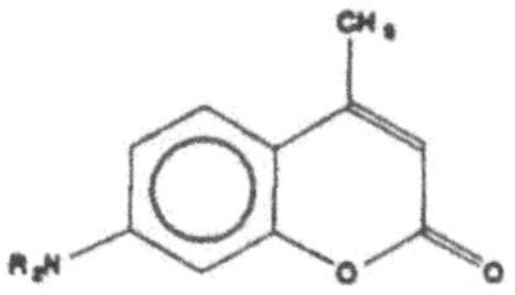

Coumarins 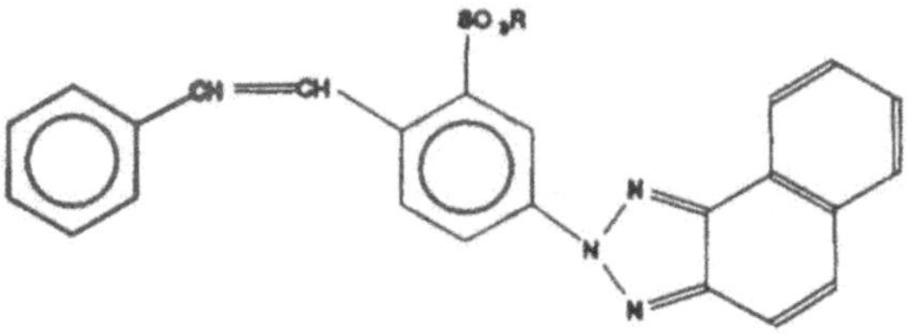   *Naphthotriazolylstilbenes*

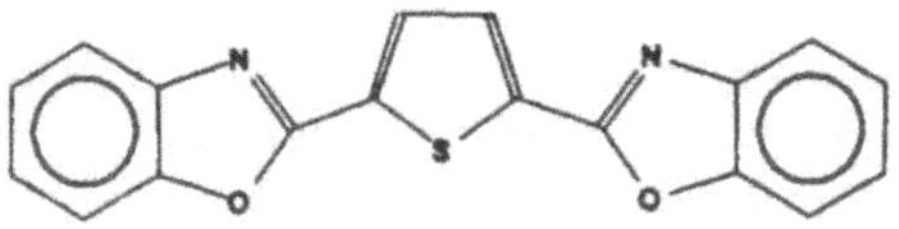

bis (benzoxazoyl) thiophene

Compared to other synthetic fibers, acrylics are more difficult to produce with good whiteness, owing to the degradation mechanisms previously reviewed. In addition, the lack of stability toward heat limits the processes which can be used by converters to produce garments and other textile products from acrylic fibers. At the consumer level, however, acrylic heat stability is not a significant factor in limiting utility of the product.

III. PHOTOOXIDATIVE

A. Exposure Conditions/Comparison with other Fibers

Weathering of acrylic fibers over a period of time can cause discoloration or deterioration of physical properties. In actual environmental exposure, a fiber is subject to the effects of actinic radiation (UV and visible light), oxygen, heat, moisture, and airborne pollutants. And of course, the magnitude of these effects can vary from location to location. Blakely researched climatic data for a number of sites around the world; it was shown that the solar energy associated with exposure in Phoenix is double that of a site in England and that atmospheric moisture and temperature vary greatly as well [51].

Acrylics are among the most resistant fibers to the effects of weathering and this has led to niche applications in awnings, tents, outdoor furniture [2] and sandbags [52]. Further discussion of this technology may be found in Chapters 6 and 11. Indeed, it can take several years before the effects of weathering on an acrylic fiber or fabric become obvious. A number of accelerated tests have been used to study weathering effects on polymeric materials. The weatherometer intensifies UV/visible light exposure while remaining close to the actual frequency distribution of sunlight (there are several variations on the theme depending on the actual light source used: carbon arc, xenon arc, sun lamps with suitable choice of phosphors). Weatherometer exposure has been coupled with environmental chambers in which humidity and temperature can be varied [2]: the QUV device used largely by the coatings industry combines light exposure with hot/cold and wet/dry cycles [53]. In recent years, accelerated outdoor exposure testing techniques have been developed which employ mirrors to focus and intensify sunlight on the polymeric objects under test [54, 55], and there is a variation in which the test item is sprayed each night to simulate dew [56, 57].

In one study, several synthetic and natural fibers were exposed to sunlight for 200 hr and loss in tenacity measured [2]. The clear superiority of acrylics is shown in Figure 5 where the acrylic fiber lost 10% in tenacity versus 25% for polyester, 35% for nylon, and even higher losses for natural fibers. In similar work, the exposure time in months required for a 50% loss in yarn strength was

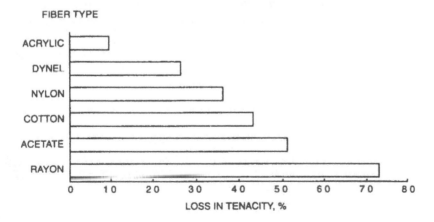

Figure 5 Loss of tenacity by fiber type after 200-hr exposure in direct sunlight. (Reproduced with permission from Ref. 2, Copyright 1985, Marcel Dekker, Inc.)

several times greater for acrylics than for olefin fibers, nylon, cotton, and wool [4]. Exposure in a lightfastness tester under varying conditions of humidity had less effect on acrylics and polyester than on viscose, cotton and nylon fibers [58], although all fibers suffered increasing losses of tensile strength with increased humidity. Temperature effects were also enhanced at high humidity. In another study, yarns from nylon, acrylic, aramid, polyester, and polypropylene fibers were exposed to natural weathering and to wet/dry cycles in the lab with and without UV irradiation [59]. Effects on physical properties were followed by measuring changes in the stress-strain curves. Although the acrylic yarn was the least affected by dry UV exposure, it was the most affected by wet/dry cycles. Oxidation has been shown to increase moisture absorption of acrylics [60]. Virgin acrylic fibers absorbed 0.5–1.0% by weight of water, whereas fibers heat-treated in air at 255°C absorbed 8% or more by weight. Since uptake of moisture is accompanied by expansion of the fiber, this could "open" the structure to further action by atmospheric oxygen and initiate a cycle of accelerating loss of properties.

Although loss of physical properties during weathering is an important consideration, deterioration of aesthetics is also of concern. Acrylic, nylon, and wool fabrics, natural and dyed yellow and blue, were exposed to natural sunlight (85 days) and to accelerated testing (xenon arc weatherometer). The acrylic fabric was the most stable to tensile strength degradation but, along with wool, was susceptible to discoloration by yellowing [61]. The presence of blue dye lessened the effect, perhaps by masking the formation of yellow color.

Synthetic polymers in the form of molded objects or coatings films are not as easily degraded by light and atmospheric conditions as fibers of the same chemical composition, because of differences in surface-to-volume ratios. For example, a cube of acrylic polymer 1 cm^3 in volume has a surface area of 6 cm^2, while extrusion into a 3-denier fiber causes the surface area to increase to 2074 cm^2 [3]. Thus, even though acrylics are generally regarded as among the most resistant of textile fibers to photooxidative degradation, it is necessary to consider ways to optimize this resistance. This can be done most effectively through an understanding of photooxidative processes.

B. Mechanism

When a polymeric material such as an acrylic fiber is exposed to actinic radiation, degradation can be caused by direct photolytic scission of the polymer chain [62]. Irradiated in a vacuum, polyacrylonitrile evolves hydrogen, methane, acrylonitrile, and hydrogen cyanide; crosslinking reactions also occur [63].

The energy of ultraviolet radiation at the earth's surface (290–400 nm) is more than sufficient to break most chemical bonds [64]. For example, 300-nm radiation corresponds to the 95-kcal/mole energy required to break an aliphatic C-H bond and 350 nm (90-kcal/mole energy) will break many C-C bonds [16].

Still, it is far more likely that degradation during weathering will result from the reaction of excited polymer molecules with atmospheric oxygen. Figure 6 is a schematic diagram of the various pathways and reaction steps that can be involved in weathering. Polymer molecules are excited by absorption of ultraviolet energy, undergoing a transition from their ground state into an excited state (a singlet state or, after a level "crossing" reaction, a triplet state). This can occur directly to the polymer, where the presence of aromatic and carbonyl groups which absorb at ultraviolet wavelengths can enhance the absorption. Or, chromophoric impurities found in most commercially prepared polymers, i.e., additives, colorants, catalyst residues [65], can absorb energy and transfer it to the polymer. In the case of PAN prepared using radical catalysts, a β-ketonitrile unit formed during polymerization [18] which was previously noted to be a major source of thermal degradation is also a likely weak link for UV energy absorption and subsequent degradation. When it does not lose its energy by radiation (fluorescence or phosphorescence) or heat transfer, the excited polymer molecule can form radicals which react with oxygen to form peroxides.

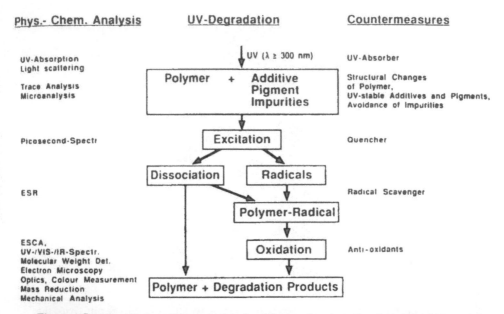

Figure 6 Individual reaction steps involved during the photochemical degradation of polymers exposed by weathering (UV degradation) together with the preventive measures and the methods of physicochemical analysis applicable in such cases. (Reproduced with permission from Ref. 65, Copyright 1991, Elsevier Sequoia S.A.)

These in turn can lead to carbonyl compounds which can be the focal point for more energy absorption or can undergo chain scission in a Norrish-type reaction [66–67], causing degradation of physical properties dependent on molecular weight.

Alexandru and Guillet [68] found that although PAN fibers retain their strength over extended periods of exposure to natural sunlight ($\lambda > 300$ nm), introduction of ketone groups in the polymer backbone (by incorporation of methyl isopropenyl ketone) led to extensive loss in tensile strength, presumably due to Norrish Type II chain scission reactions. Fibers containing methyl acrylate or vinyl acetate monomers in addition to methyl isopropenyl ketone lost even more tensile strength, perhaps because their greater molecular flexibility favored formation of cyclic Norrish transition states leading to chain scission. Since copolymerization is expected to "open up" the polyacrylonitrile structure, an increase in oxygen permeability may also be a factor.

UV irradiation at higher energy than natural sunlight can accelerate degradation processes. Stephenson [69] reported a decrease in acrylic fiber strength after 150 hr UV irradiation at 254 nm. Whether a similar loss in strength would occur at UV wavelengths corresponding to natural sunlight (>300 nm) was not mentioned, though practical end-use results would seem to indicate otherwise [2]. A fundamental study of oxidative photodegradation pathways for PAN has shown the chemistry to be complex and dependent on the energy of irradiation [70, 71]. PAN films on a silver backing were irradiated with UV light in the 250–400 nm range [71]; chemical changes in the films as well as gases produced were monitored by FTIR, while ESR was used to monitor radical formation. Use of ^{18}O air and ^{13}C PAN made some of the changes in the infrared spectrum easier to interpret. At $\lambda \geq 266$ nm, products were formed resulting from C-C bond scission in the main chain as well as cleavage of -CN and -H from the chain. At even higher energy levels ($\lambda \geq 250$ nm), formation of intermediate diradicals from $-C\equiv N$ groups and subsequent ring closure accounted for yellowish brown color and an increase in electrical conductivity. Again, the extent of such reactions occurring under the lower energy conditions of exposure to longer-wavelength, natural sunlight remains to be determined.

Figure 6 also shows techniques which can be used to follow polymer degradation processes experimentally, along with countermeasures that can be taken to prevent degradation. Specific examples will appear in subsequent discussion in this section.

Nakamura et al. reported that although IR spectra of thermally and UV degraded PAN looked much the same, UV spectra and IR dichroism showed that the chemical structures of thermal and photoproducts were different [72]. In studies on thin films, thermal degradation at 200°C was shown to lead primarily to aromatic structures, whereas photodegradation using an intense UV light source led to a high concentration of carbonyl groups. This was related to the temperature of reaction. The photoreaction takes place below the glass transition temperature (T_g), whereas the thermal reaction is above T_g and nitrile groups can align themselves to allow cyclic condensation reactions. A comparison of a absorption maxima in the UV spectra with energies computed using molecular orbital calculations led to the conclusion that blocks of cyclic imine structures were no more than 5 units long. An increase in orientation (by stretching the films) enhanced carbonyl formation by either degradation route. In other work, however, the combination of UV irradiation and elevated temperature (200°C) was said to be necessary for the cyclization of nitrile structures [73].

Application of modern instrumental methods has made it possible to define more precisely the structural changes in PAN fiber that occur due to atmospheric degradation [74]. Unfinished PAN awning fabric was exposed in a weatherometer and examined using FTIR in an ATR mode, which provides

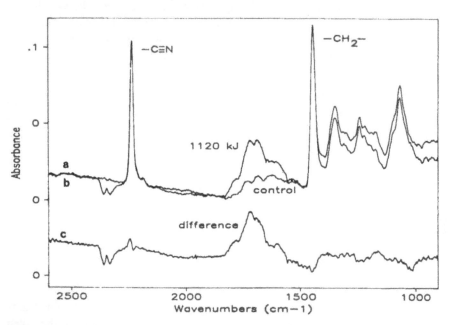

Figure 7 FTIR-ATR spectra of (a) weatherometer degraded greige PAN fiber, (b) the corresponding control greige fiber, and (c) (a) minus (b). (Reproduced with permission from Ref. 71, Copyright 1992, John Wiley and Sons, Inc.)

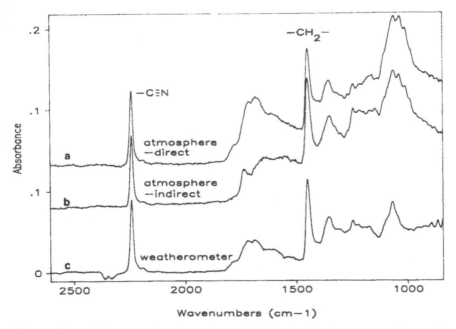

Figure 8 FTIR-ATR spectra of (a) colored PAN fiber, atmosphere-exposed but not in direct sunlight; (b) colored PAN fiber, atmosphere-exposed and in direct sunlight; and (c) weatherometer degraded (1120 kJ) greige PAN fiber. (Reproduced with permission from Ref. 74, Copyright 1992, John Wiley and Sons, Inc.)

information on surface degradation only. Figure 7 shows the computer subtraction of spectra before and after exposure; increase in the spectral bands in the 1500–1800 cm^{-1} region was shown to correlate with color development as measured by ΔE and Δb on an L-a-b scale. Commercial acrylic fibers from typical acrylonitrile-ester copolymers showed the same behavior. A comparison of weatherometer with outdoor exposure showed that exposure to direct sunlight was required for fabric to exhibit the color change and spectral characteristics of weatherometer-degraded material (Figure 8). Dyed fabric behaved the same as griege fabric and topical finishes also had no measurable effect on the chemical structures formed. Gel permeation chromatography (GPC) measurements on the bulk fabric samples showed significant reduction in molecular weight. Since FTIR-ATR results indicated a surface degradation only, a considerable amount of chain scission at the surface must have occurred to cause such a change in molecular weight of the bulk samples. Interestingly, neither ^{13}CNMR nor ESCA was sufficiently sensitive to characterize degradation structures on the fabric surface.

C. Control of Actinic Degradation

As shown in Figure 6, there are several tactics that can be taken to stabilize polymers against the effects of photodegradation. Avoiding ketone structures in the acrylic backbone [68] through judicious use of polymerization conditions [18, 27] produces an inherently more stable polymer. In general, use of a stabilizer must be matched to a specific degradation process. For example, antioxidants such as hindered phenols are not very effective at stabilizing polymers against UV light, for two reasons. First, the kinetic chain sequence is much shorter in UV initiated oxidation than in thermal oxidation and, secondly, phenols cannot prevent formation of hydroperoxide species on the polymer which lead to degradation [75]. Ionic species such as thiodipropionate esters, phosphorus acid esters, and metal complexes of dialkyldithiocarbamic and phosphoric acids do decompose hydroperoxides even at room temperature [76, 77] and therefore can be effective UV stabilizers.

$$ROCCH_2CH_2SCH_2CH_2COR' \qquad\qquad (RO)_3P$$

$$\underset{O}{\overset{\|}{}} \qquad\qquad \underset{O}{\overset{\|}{}}$$

$$\left[(RO)_2\overset{\overset{\displaystyle S}{\|}}{P}\!\!-\!\!S \right]_2 M \qquad\qquad (R_2NCS)_2 M \\ \underset{S}{\overset{\|}{}}$$

A wide variety of UV absorbers have been reported to enhance the stability of acrylics. Dye-like anthraquinone structures [78] protected fabric against degradation of color and tensile strength. Choice of dyestuff was shown to affect color fading in one study [79], though the effect was greater with natural fibers than with acrylics. In another report, however, there was no difference in degradation of yellow and blue acrylic fabric [80] exposed to either natural or artificial light. Metal chelates having an absorption peak bathochromic relative to the dyestuff help to prevent color fading on acrylics and other fibers [81]. Visually transparent, mixed oxides of iron, magnesium, tin, and zinc have been claimed to impart UV degradation resistance to acrylics by absorbing UV radiation and dissipating the captured energy thermally [82]. TiO_2 was also effective, presumably as a physical screen, but dulled the fiber. As classes of compounds, both *o*-hydroxybenzophenones and benzotriazoles protect polymeric systems by absorbing light and reemitting the absorbed energy as heat [83, 84].

The same is true of hydroxyphenyl-*s*-triazines. Care must be taken not to generalize, however, since the structurally related benzoxazoles are actually photosensitizers [83, 84].

Antioxidants can retard the rate of photooxidative processes in polymers. For acrylic fiber stabilization, these include mercaptans, thioglycerols and phenolthiazine [85, 86], phenolics [86], inorganic salts of hydrazine [87], hydrazides [88], thioureas and thiosemicarbazides [89], and polyphosphites [90]. Other spinning dope additives claimed to improve fiber stability include 1-hydroxyethane-1,1-diphosphonic acid [30] and monoesters of glycerol [91].

Nickel compounds such as nickel phenolate act as quenchers of excited states of molecules [92] and hindered amine light stabilizers (HALS) such as bis (2,2,6,6-tetramethyl-4-piperidyl) sebacate decompose hydroperoxides and are converted to nitroxyl radicals ()N-O·) which may react further with polymeric radicals [93, 94]. (Note, however, that amine groups can lead to discoloration of acrylics.)

In some instances, monomers, initiators and additives used in the polymerization process have been claimed to improve fiber stability. These include the monomers sodium methallyl sulfonate [95, 96], methacrylamidobenzene sulphonamides [97], and unsaturated carboxylic acid sulfobetaine hydrazides [97]; the catalysts sodium 1-hydroxy-1-urea sulfonate [98] and salts of sulfamic acid [99]; and the additives hexamethylenediamine sulfate [88] and metal chelates with the disodium salt of EDTA [100].

Improvement in both heat and light stability has been claimed through treatment of acrylic fiber in the gel state (during the spinning process) with metal salts of naphthalene disulfonic acid [101].

The incorporation of halogen-containing monomers to impart fire retardancy is known to increase the sensitivity of acrylic fibers to both thermal [102] and photolytic [103] degradation. Additives said to be stabilizers for halogen-containing acrylics include oxalic acid [104, 105], aluminum sulfate [106], a benzotriazole [107], combinations of organophosphites and tin compounds [108], and organotin salts of organic acids [42]. Incorporation of a phosphite ester acrylate in the polymer has also been claimed [109].

A comprehensive list of commercial UV stabilizers with chemical structures, trade names, and suppliers has been published [110].

Maleic anhydride and maleic acid have been reported as stabilizers for acrylic and modacrylic polymers [111]. The stabilization effect in high-acrylonitrile thermoplastics has been shown to be due to its action as a dienophile [112, 113]. Since in high-acrylonitrile thermoplastics the acrylonitrile content is well below the 85% minimum for acrylic fibers, it is likely that the thermal reaction of nitrile groups with each other will occur intermolecularly. Such a reaction forms conjugated imide sequences which can be interrupted with maleic anhydride in a Diels-Alder condensation. Maleimides and other strong dienophiles had the same effect.

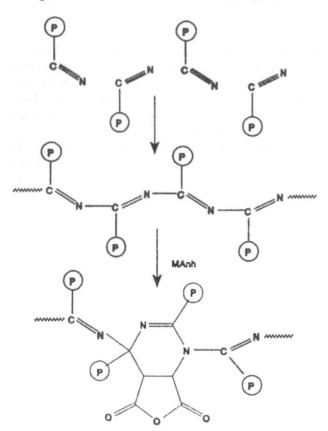

IV. FUTURE DIRECTIONS

In a mature field such as acrylic fibers it is likely that, barring serendipity, further advances will be made only through a better understanding of degradation mechanisms than is currently available. Often, a development in one field of polymer science has potential for use in a related field. One such development is the work of Gerlock and co-workers [114–120] who have been using a novel technique to quantify photoinitiation rates for a number of coatings polymers. Coatings are doped with a nitroxide free-radical scavenger and then subjected to ultraviolet light under controlled conditions of light intensity, temperature and humidity. The efficiency of nitroxide scavenging of fleeting polymeric radicals is inferred by following nitroxide radical concentration using electron spin resonance (ESR), after correcting for other quantifiable reactions of nitroxide radical. This technique, which could be applied to fibers as well as

coatings, has been shown to correlate with exterior weathering in the case of acrylic-melamine coatings. It has been used to guide the synthesis of more durable polymers [117, 118], to evaluate the influence of exposure variables in accelerated tests [117], and to examine the mechanisms of action of UV absorbing additives and hindered amine light stabilizers [119, 120]. These workers found, among other things, that the use of harsh, accelerated tests to predict resistance of polymers and stabilizers to the effects of natural sunlight can frequently lead to erroneous results.

REFERENCES

1. L. H. Peebles, Acrylonitrile polymers, degradation, *Encyclopedia of Polymer Science and Technology, Suppl. Vol. 1*, 1976, pp. 1–25.
2. B. G. Frushour and R. S. Knorr, in *Acrylic Fibers* (M. Lewin and E. M. Pierce, eds.), *Handbook of Fiber Science and Technology, Vol. IV*, 1985, pp. 335–338.
3. L. Krcma, *Kem. Ind, 28*, 247 (1979).
4. G. H. Fremon, in *Fibers from Synthetic Polymers* (R. Hill, ed.), Elsevier, 1953, Chap. 19.
5. N. Grassie and R. McGuchan, *Eur. Polym. J., 6*, 1277 (1970).
6. N. Grassie and R. McGuchan, *Eur. Polym. J., 7*, 1503 (1971).
7. N. Grassie and R. McGuchan, *Eur. Polym. J., 8*, 243 (1972).
8. N. Grassie and R. McGuchan, *Eur. Polym. J., 8*, 257 (1972).
9. N. Grassie and R. McGuchan, *Eur. Polym. J., 8*, 865 (1972).
10. N. Grassie and R. McGuchan, *Eur. Polym. J., 9*, 113 (1973).
11. G. Avery, S. K. Chadda, and R. C. Poller, *Eur. Polym. J., 19(4)*, 313 (1983).
12. G. V. Reddy and G. Radhakrishnan, *Angew. Makromol. Chem., 121*, 41 (1984).
13. T. Usami, T. Itoh, H. Ohtani, and S. Tsuge, *Macromolecules, 23*, 2460 (1990).
14. T. Takata, I. Hiroi, and M. Taniyama, *J. Polym. Sci., A-2*, 1567 (1964).

15. J. Runge and W. Nelles, *Faserforsch. Textiltech.*, *21*(3), 105 (1970).
16. N. Grassie and R. McGuchan, *Eur. Polym. J.*, *9*, 507 (1973).
17. M. Tomescu, I. Demetrescu, and E. Segal, *Rev. Roum. Chim.*, *26*(5), 793 (1981).
18. H. N. Friedlander, L. H. Peebles, Jr., J. Brandrup, and J. R. Kirby, *Macromolecules*, *1*, 79 (1968).
19. J. R. Kirby, J. Brandrup, and L. H. Peebles, Jr., *Macromolecules*, *1*, 53 (1968).
20. J. J. Rafalko, *J. Polm. Sci., Polym. Phys. Ed.*, 22, 1211 (1984).
21. H. S. Fochler, J. R. Mooney, L. E. Ball, R. D. Boyer, and J. G. Grasselli, *Spectrochim. Acta, 41A,* 271 (1985).
22. M. M. Coleman, G. T. Sivy, P. C. Painter, R. W. Snyder, and B. Gordon, *Carbon, 21,* 255 (1983).
23. G. T. Sivy, B. Gordon, and M. M. Coleman, *Carbon, 21,* 573 (1983).
24. D. M. White, *J. Am. Chem. Soc.*, *82*, 5678 (1960).
25. N. A. Kubasova, D. S. Dinh, M. A. Geiderikh, and M. V. Shishkina, *Vysokomol Soedin., A13,* 162 (1971).
26. M. A. Geiderikh, D. S. Dinh, B. E. Davydov and G. P. Karpacheva, *Vysokomol Soedin., A15,* 1239 (1973).
27. V. S. Tverskoi, V. M. Samoilov, and A. S. Kotosonov, *Vysokomol. Soedin. Ser. B, 28* (7), 525 (1986).
28. H. S. Bach and R. S. Knorr, Acrylic fibers, *Encyclopedia of Polymer Science and Engineering* (2nd ed.), Wiley-Interscience, New York, 1985.
29. R. H. Peters and R. H. Still, *Applied Fiber Science, Vol. 2*, (F. Happey, ed.), Academic Press, 1979, Chap. 10.
30. Y. Joh and T. Sugimori, U.S. Patent 3,813,372 (May 28, 1974), to Mitsubishi Rayon Co., Ltd.
31. Brit. Pat. 1,360,669 (December 5, 1970), to Bayer.
32. J. R. Kirby, U.S. Patent 3,784,511 (January 1, 1974), to Monsanto Co.
33. S. Hamada et al., U.S. Patent 3,660,527 (May 1, 1972), to Toray Industries, Inc.
34. T. Shibukawa, Jpn. Patent 03,180 (1971), to Exlan Co.
35. E. Seidel, J. Aurich, H. Ebeling, and W. Berger, E. Ger. Patent 227,169, Al (September 11, 1985), to VEB Chemiefaserwerk "Friedrich Engels".
36. U.S. Patent 3,682,004 (January 20, 1969), to Courtaulds.
37. L. A. Yasnikov, E. A. Pakshver, and Yu. V. Glazkovskii, *Khim. Volokna, 14*(5), 11 (1972).
38. Ger. Patents 1,494,047 and 1,494,048 (November 2, 1962), to Bayer A.G.
39. Ger. Patent 1,946,008 (October 28, 1968), to Engels.
40. Brit. Patent 1,345,255 (March 25, 1970), to Bayer.
41. Brit. Patent 1,243,292 (September 13, 1968), to Japan Exlan.
42. Z. Adamski, *Przegl Wlok., 25* (3), 159 (1971).
43. A. H. Bruner and T. B. Truscott, U.S. Patent 3,558,765 (January 26, 1971), to Monsanto Co.
44. G. Palethorpe, U.S. Patent 3,642,628 (February 15, 1972), to Monsanto Co.
45. N. Teranishi, H. Iwada, and H. Tamura, Jpn. Patent 39,056 (1970), to Mitsubishi Rayon Co., Ltd.
46. B. Ranby and J. F. Rabek, *Photodegradation, Photo-oxidation and Photostabilization of Polymers*, Wiley-Interscience, 1975, pp. 365–369.
47. Y. Yamanchi, *Senryo To Yakuhin, 15*(23), 75 (1970).

48. Y. Sakai and H. Kato, *Sen'i Kako*, 25 (2), 88 (1973).
49. G. Eigenmann, L. Kaiser, and C. Luethi, U.S. Patent 3,849,155 (November 19, 1974), to Ciba-Geigy.
50. Brit. Patent 1,195,143 (September 2, 1967), to Nippon Kayaku KK.
51. R. R. Blakey, *Prog. Org. Coatings, 13*, 287 (1985).
52. V. J. Bagdon, *Textile Res. J.*, 546 (1971).
53. *Annual Book of ASTM Standards* (G53), Vol. 14.02,1098 (1987).
54. *Annual Book of ASTM Standards* (ASTM EE-838-81), Part 41 (1982).
55. C. R. Caryl and W. E. Helmick, U.S. Patent 2,945,417 (July 19, 1960).
56. G. A. Zerlaut and J. S. Robbins, *Ind. Lackierbetr., 53* (10), 378 (1985).
57. C. R. Caryl, *ASTM Bull, 243*, 55 (1960).
58. C. D. Shah and R. Srinivasan, *J. Tert. Inst., 69*, 151 (1978).
59. M. W. Webb, *J. Text. Inst., 75*, 219 (1984).
60. S. B. Warner, L. H. Peebles, Jr., and D. R. Uhlmann, *J. Mater. Sci., 14*, 2764 (1979).
61. S. Urabe and M. Miyajima, *Kenkyu Kiyo-Tokyo Kasei Daigaku, 23*, 179 (1983).
62. H. H. G. Jellinek and I. J. Bastien, *Can. J. Chem., 39*, 2056 (1961).
63. C. V. Stephenson, J. C. Lacey, Jr., and W. S. Wilcox, *J. Polym. Sci., 55*, 477 (1961).
64. B. Ranby and J. F. Rabek, *Photodegradation, Photo-oxidation and Photostabilization of Polymers*, Wiley-Interscience, Great Britain, 1975, p. 424.
65. G. Kämpf, K. Sommer, and E. Ziringiebl, *Prog. Org. Coatings, 19*, 69 (1991).
66. P. I. Plooard and J. E. Guillet, *Macromolecules, 5*, 405 (1972).
67. R. O. Kan, *Organic Photochemistry*, McGraw-Hill, New York, 1966.
68. L. Alexandru and J. E. Guillet, *J. Polym. Sci. Polym. Chem. Ed, 13*, 483 (1975).
69. C. V. Stephenson, B. C. Moses, and W. S. Wilcox, *J. Polym. Sci., 55*, 451 (1961).
70. J. F. McKellar and N. S. Allen, *Photochemistry of Man-made Polymers*, Applied Science, London, 1979.
71. C. A. Sergides, A. R. Chughtai, and D. M. Smith, *Macromolecules, 19*(5), 1448 (1986).
72. R. Nakamura, H. Yoshida, and M. Hida, *Sen-i Gakkaishi, 39*(10), T415 (1983).
73. T-C. Chung, Y. Schlesinger, S. Etemad, A. G. Macdiarmid, and A. J. Heeger, *J. Polym. Sci. Polym. Phys. Ed., 22*, 1239 (1984).
74. H. E. Howell and A. S. Patil, *J. Appl. Polym. Sci., 44*, 1523 (1992).
75. G. Scott, *Atmospheric Oxidation and Antioxidants*, Elsevier, Amsterdam, 1965.
76. J. D. Holdsworth, G. Scott and D. Williams, *J. Chem. Soc.*, 4692 (1964).
77. G. Scott, *Chem. Brit., 9*, 267 (1973).
78. M. R. Benedict and H. L. Needles, *Can. Textile J.*, 91 (1972).
79. N. Grewal and B. Balakrishnaiah, *Manmade Textiles in India, 29* (3), 137 (1986).
80. S. Urabe and M. Miyajima, *Kenkyu-Kiyo Tokyo Kasei Daigaku, 23*(2), 179 (1983).
81. W. F. Smith and G. A. Reynolds, Brit. Patent 1,496,506 (December 30, 1977), to Eastman Kodak.
82. A. S. Patil, F. G. Crouch and W. E. Streetman, Eur. Patent 328,119 (September 2, 1989), to BASF.
83. H. J. Heller, *Eur. Polym. J., Suppl*, 105 (1969).
84. G. R. Lappin, *Encyclopedia of Polymer Science and Technology*, Vol. 14, Wiley-Interscience, New York, 1971, p. 14.
85. Brit. Patent 1,360,669 (December 5, 1970), to Bayer A.G.
86. Brit. Patent 1,349,669 (September 28, 1970), to Mitsubishi Rayon.

87. S. Hamada et al., U.S. Patent 3,660,527 (May 1, 1972), to Toray Ind., Inc.
88. S. Nakao, N. Numata, and T. Yamamoto, Jap. Patent 29,872 (September 13, 1973); 29,873 (September 13, 1973) and 29,874 (September 13, 1973), to Kanebo Co., Ltd.
89. H. Logemann et al., Ger. Offen. 1,494,047 and 1,494,048 (November 2, 1962), to Bayer A.G.
90. W. Schmidt et al., Brit. Patent 1,345,255 (March 25, 1970), to Bayer A.G.
91. H. Sekiguchi et al., U.S. Patent 3,681,275 (August 1, 1972), to Japan Exlan Co., Ltd.
92. K. B. Chakraborty and G. Scott, Chem. Ind. (London), 30, 237 (1978).
93. G. Berner and M. Rembold, Organic Coatings, Marcel Dekker, New York, 1984, Chap. 4.
94. J. G. Calvert and J. N. Pitts, Photochemistry, Wiley, New York, 1967.
95. K. Nakao et al., Jap. Kokai Pat. 73,979 (September 1, 1973); to Kanagefuchi Co., Ltd.
96. Z. I. Burlyuk and A. N. Reshetova, Proizvod. Sin. Volok., 144 (1971).
97. Brit. Patent 1,206,484 (December 5, 1966) and 1,178,913 (February 2, 1966), to Bayer A.G.
98. T. Ohfuka et al., Ger. Offen. 1,720,202 (September 21, 1972), to Ashai Chem. Ind. Co., Ltd.
99. S. Matsumura and C. Kenemitsu, Jap. Patent 26,333 (July 30, 1971), to Teijin, Ltd.
100. N. Yokouchi et al., Jap. Patent 38,561 (November 13, 1971), to Mitsubishi Rayon Co., Ltd.
101. P. A. Ucci, U.S. Patent 3,281,260 (October 25, 1966), to Monsanto Co.
102. M. Dunke, Ger. Patent 70,985 (January 20, 1970).
103. S. Kumazawa and A. Yoshida, Jap. Kokai Patent 75,899 (October 12, 1973), to Kanegafuchi, Ltd.
104. F. Takemoto el al., (Jap. Kokai Patent 20,318 (July 15, 1987), to Mitsubishi Rayon Co., Ltd.
105. M. Fujimatsu and T. Shibata, Jap. Kokai Patent 80,666 (March 20, 1990), to Japan Exlan Co., Ltd.
106. M. Fujimatsu and M. Miura, Jap. Kokai Patent 10,158 (January 19, 1987), to Japan Exlan Co., Ltd.
107. F. Takemoto et al., Jap. Kokai Patent 126,913 (May 30, 1988), to Mitsubishi Rayon Co., Ltd.
108. I. Nakanome et al., Ger. Offen. 1,946,330 (August 13, 1990), to Japan Exlan Co., Ltd.
109. B. Huber and H. J. Kleiner, Ger. Offen. 2,449,468 (April 29, 1976), to Hoechst A.G.
110. P. J. Shirmann and M. Dexter, in Handbook of Coatings Additives, (L. J. Calbo, ed.), Marcel Dekker, New York, 1987, pp. 262–268.
111. M. Dunke, Ger. Offen. 1,946,008 (May 14, 1970), to VEB Chemiefaserwerk "Friedrich Engels".
112. B. A. Marien, J. Polym. Sci. Polym. Chem. Ed., 17, 425 (1979).
113. B. A. Marien, J. Polym. Sci. Polym. Chem. Ed., 17, 435 (1979).
114. J. L. Gerlock, Anal. Chem., 55, 1520 (1983).
115. J. L. Gerlock and D. R. Bauer, J. Polym. Sci. Polym. Lett. Ed., 22, 447 (1984).
116. J. L. Gerlock, D. F. Mielewski, and D. R. Bauer, Poly. Deg. Stab., 20, 123 (1988).
117. J. L. Gerlock, D. R. Bauer, L. M. Briggs, and R. A. Dickie, J. Coat. Tech. 57, 722 (1985).
118. J. L. Gerlock, D. F. Mielewski, D. R. Bauer, and K. R. Carduner, Macromolecules, 21, 1604 (1988).

119. J. L. Gerlock, D. R. Bauer, and L. M. Briggs, *Poly. Deg. Stab.*, *14*, 53 (1986); J. L Gerlock, T. Reiley, and D. R. Bauer, *ibid*, *14*, 73 (1986); D. R. Bauer and J. L. Gerlock, *ibid.*, *14*, 97 (1986).
120. D. R. Bauer, J. L. Gerlock, D. F. Mielewski, M. C. Paputa Peck, and R. O. Carter III, *Poly. Deg. Stab.*, *28*, 39 (1990).

9

Dyeing Processes

Hubert Emsermann

Miles Inc., Rock Hill, South Carolina

Reimund Foppe

Bayer AG, Leverkusen, Germany

I. INTRODUCTION

A. Fiber Types

Acrylic fibers, as first produced, consisted only of acrylonitrile monomer units polymerized into large linear molecules which were almost impossible to dye [1]. High-temperature dyeing and use of carriers and swelling agents were unsuccessful. One process, the cupro-ion dyeing method, had limited success. The fiber was treated with copper sulfate and hydroxylamine sulfate, leaving cuprous ions tightly bound to the fiber. Selected acid dyes could be applied to give rather dull colors.

Fibers of 100% polyacrylonitrile primarily have industrial application due to their excellent chemical and weathering resistance but are rarely used commercially because of difficulty in dyeing. The main reason is the rather high second-order glass transition point ($\geq 100°C$). A successful dyeing required polymer modifications. As discussed in Chapter 3, acrylonitrile is usually copolymerized with at least one other monomer in a concentration of 5–10%. Some of these contain sulfonic or carboxylic acid groups; others, amine or quaternary ammonium groups. This results in a more open structure and lowering of the glass transition point to 85–95°C which enables better dye penetration of the fiber. The introduction of either cationic or anionic groups makes it possible to produce contrasting colors from yarn or fabric produced from the two fiber types. In cross-dyeing, yarn or fabric is made from regular and acid-dyeable acrylic fibers in heather or patterned single yarns.

Cationic and acid dyes are combined in a single dyebath set with proper pH and auxiliaries. Two-color combinations or colored/white effects are most common. Fibers made from these polymers are dyeable at temperatures above 85°C without carrier.

If modacrylic fibers are being used, dyestuff selection is more critical due to the lower lightfastness properties of some dyestuffs on these fibers.

In today's market, we differentiate primarily between wet- and dry-spun fibers which are available as cationic or anionic dyeable types. The major importance, however, is with the cationic dyeable fiber; and we concentrate on this type in this chapter.

B. Dyestuffs

1. Disperse Dyestuffs

This group of dyes exhibits only limited exhaustion properties. Acrylic fibers absorb disperse dyes by a solution mechanism [2]. Disperse dyes do not form a strong bond with the acrylic fiber and transfer readily at the boil. In general, consistent level dyeings can be expected with a simple dyeing procedure. Dye solubility limits within the acrylic fiber restrict the amount of disperse dyes being absorbed. This fact, along with the absence of a strong chemical bond which causes marginal wetfastness, limits the useful depth of disperse dyes to approximately 0.5% o.w.g. total depth of all dyes combined. Selected dyestuffs of small molecular size are used only in pastel and light shades where level dyeings with cationic dyes have been difficult. Due to improved process control and better machinery, the use of disperse dyes for acrylic fibers is very small and steadily decreasing in importance.

2. Cationic Dyestuffs

Cationic dyestuffs can be classified in two large groups. The first group comprises dyestuffs with a localized charge (+), where the positive charge is localized on a single nitrogen atom. A typical representative is the Chrysoidin (C.I. basic orange 2). See Figure 1.

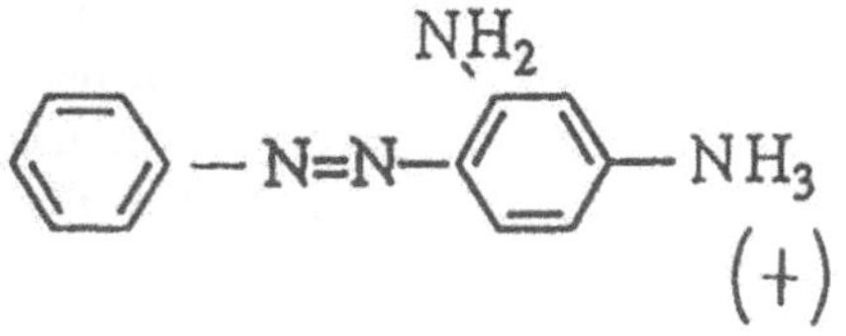

Figure 1 C. I. basic orange 2.

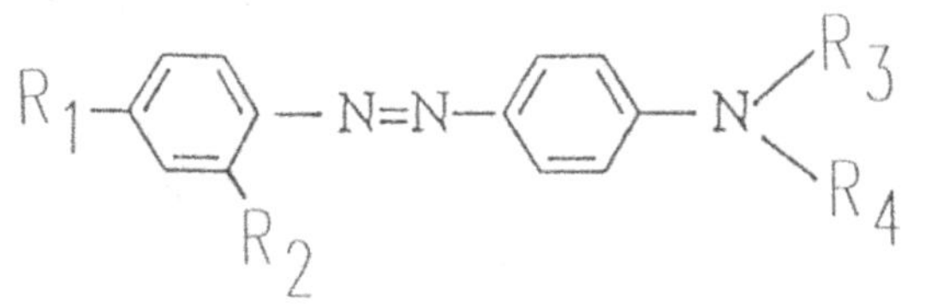

R $_1$ —⬡— N=N —⬡— N $\diagup$ R $_3$ $\diagdown$ R $_4$

R $_2$

Figure 2 Modified basic dyestuff.

The second group, called the modified basics, are dyes with a delocalized charge. Most commonly used cationic dyestuffs today are of this nature. See Figure 2.

This class of cationic dyes, with especially good lightfastness on acrylic fibers, was developed in 1954. This, together with the polymer modification, was an important event in the history of the acrylic fibers. With these dyes and earlier developed basic dyes, a full range of bright colors with good fastness properties can be obtained in conventional dyeing equipment.

C. Dyeing Mechanism

Exhaust dyeing of acrylic fibers involves three steps as shown in Figure 3.

1. Adsorption of Dyestuff on the Fiber Surface

This first step takes place very rapidly, and the concentration of dye adsorbed on the fiber surface is almost independent of the concentration of dye in the dye-

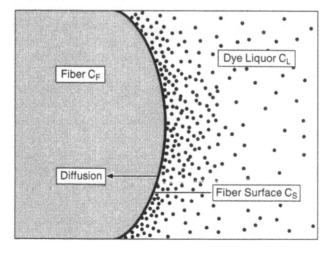

Figure 3 Exhaust mechanism of cationic dyes on acrylic fibers.

bath. The fiber will only surface dye and the dyeing proceeds from a surface layer of almost constant dyestuff concentration [3].

2. Diffusion of Dyestuff into Fiber

The speed of diffusion of dyestuff into the fiber determines the rate of dyeing. Diffusion starts above the glass transition point of the fiber (85–95°C), where the fiber molecules acquire enough energy to move, which means the fiber softens and the dyestuff is allowed to penetrate into the fiber. Below the glass transition point, the fiber is somewhat stiff and forms a barrier against the dye. At temperatures above the glass transition point, a relatively rapid dye rate which increases sharply with higher temperatures is achieved.

3. Fixation of Dyestuff within the Fiber

Fixation occurs between dye cations and anionic sites in the fiber in form of a saltlike bond. This salt formation can be mathematically described as ion exchange following stoichiometric rules.

D. Characteristic Dyestuff and Fiber Constants

The following points are of main concern in dyeing:

Dyestuff yield	Dyestuff equilibrium, exhaustion, final shade
Dyeing time and rate of temperature rise	Speed of dyeing
Levelness	Migration, compatibility of dyes, effect of retarders, migrating agents, and electrolytes

Each of these points depends on the dyeing conditions, dyestuff, and fiber properties. Due to the variety of acrylic fibers and cationic dyes, it is almost impossible to generalize a dyeing procedure. For optimum dyeing, it is necessary to have knowledge of certain basic fiber and dyestuff data which allows for calculation of optimal dyeing conditions.

1. Saturation Value (SF)

Any basic dyeable acrylic fiber contains a certain amount of anionic dyesites which are polymerized into the acrylic fiber and can bind only a stoichiometrically equal amount of cationic dyestuff. This means each fiber can take up only a limited quantity of cationic dyes. Dyestuff in excess (beyond the saturation point) will remain in the dyebath. SF actually tells how many percent of a 100% pure basic dyestuff of a molecular weight 400 (Malachite Green Crystals) can be bound.

$$p \times f \leq SF$$

where p = dyestuff in percent on weight of goods
 f = dyestuff constant (the saturation factor)
 SF = saturation value of the fiber

		SF
For example: ACRILAN 16 acrylic fiber (Monsanto)		1.3
CRESLAN acrylic fiber	(Cytec)	1.7
DRALON acrylic fiber	(Bayer AG)	2.1

The saturation value for a fiber is constant; however, the amount of dye to reach saturation varies from dye to dye. Saturation does not depend significantly on temperature (provided equilibrium is obtained) or varying pH. For most fibers, it increases only slightly with increasing pH (between pH 2–7).

Determining the saturation value (SF): Four samples of fiber to be tested are dyed in separate baths with

2.5, 3.5, 4.5, and 5.5% Astrazon Blue FRR 200
X% Acetic acid, pH 4.5
Liquor ratio 1:40

For comparison, Dralon or another acrylic fiber with a fiber saturation value (SF) of 2.1 is dyed with 4% Astrazon Blue FRR 200. After dyeing for 3–4 hr at the boil, the fiber sample is removed. A fresh sample of the same fiber is then added to this dyebath to exhaust the remaining dye. The exhaust dyeing of the unknown fiber which most closely matches the exhaust on Dralon is used to determine the saturation value (SF).

2.5% Astrazon Blue FRR 200 = SF 1.2
3.5% Astrazon Blue FRR 200 = SF 1.8
4.5% Astrazon Blue FRR 200 = SF 2.3
5.5% Astrazon Blue FRR 200 = SF 2.8

Knowledge of the fiber saturation value makes it possible to determine whether or not certain heavy shades can be dyed on a given fiber. It also prevents over saturation of the fiber with dye or retarder.

2. Fiber Dyeing Rate (V)

The fiber dyeing rate (V) is a measure of the relative rate of dyeing of acrylic fibers which, in general, is dependent on the physical structure of the fiber. There are, for example, big differences in dyeing rate between wet- and dry-spun acrylic fibers. Other manufacturing differences, such as the drawing of the fiber, will also affect the dyeing rate. A simple procedure (utilizing a cationic dyestuff of high affinity in amounts far below the fiber saturation) allows determination of this value.

Determining the fiber dyeing rate (V): A fiber of known V such as Dralon V = 1.7 is dyed in the same bath with the experimental fiber. The dyeing is made with 1% Astrazon Blue FRR 200 and acetic acid (pH 4.5) until the dyebath is almost exhausted. Depth differences can be assessed in percent to determine V.

If both fibers are equal in depth, the dyeing rate of the test fiber is equal to that of the standard fiber. A deeper shade of the unknown fiber reflects a higher dyeing rate, while a lighter shade indicates a slower rate. Data relating to (*V*) permit adjustments to be made in the starting temperature to compensate for the rate of dyeing.

3. Saturation Factor (*f*)

According to the mechanism of fixation for each dyestuff, one constant is necessary to describe the maximum quantity of dyestuff that can be taken up by any particular fiber. This constant *f* of each individual dyestuff is provided by the dyestuff manufacturer. It ranges from 0.15–1.20 for commercial dyestuffs. It depends upon the molecular weight of a dyestuff and the content of pure dye in a commercial dyestuff (E).

$$f = \frac{400}{M} \times E$$

where *f* is the saturation factor and *M* is the molecular weight. It allows the determination of the relative saturation of any acrylic fiber. If the percentage of the dye is multiplied by the individual *f* factor, the dyestuff will exhaust if the product $p \times f$ is less than SF.

Example: Dralon   SF = 2.1

Formula A	*p*	*f*	$p \times f$
Basic yellow 28	1.50	0.26	0.39
Basic violet 16	0.50	0.50	0.25
			0.64

pf < SF. Then the formula exhausts.

Formula B	*p*	*f*	$p \times f$
Basic violet 16	1.00	0.50	0.50
Basic green 4	1.50	1.00	1.50
Basic orange 30	1.00	0.50	0.50
			2.50

pf > SF. Therefore, the formula will not exhaust completely.

4. *K*-Value

The *K*-value is a constant which describes the dyeing behavior of a cationic dye in combination with other cationic dyes. This is one of the most important constants. In the dyeing of acrylics with basic dyestuffs, levelness can be one of the dyer's major problems. Leveling problems can be traced partially to the differing exhaust rates of individual dyes when dyed in combinations. Formulas consisting of dyestuffs with different combination constants require longer dyeing time and higher amounts of retarder than combinations of dyestuffs with similar combination constants. In addition, if a shade is continually changing during the dyeing cycle, it makes it difficult to control the process.

Several approaches have been used to determine the combination properties. One of them is the half dyeing time t_{70} (0.5), which is the dyeing time of single dyestuffs at which 70% of a given dyestuff at a concentration of 0.5 times the fiber saturation exhausts. This constant is helpful in determination of necessary data for steam fixation times in continuous dyeing processes. However, the half dyeing time does not describe the exhaustion behavior of dyestuffs in combinations. Dyes having the same single strike rate may exhaust totally differently in combination. If this happens, leveling problems can occur, resulting in cast differences, which are more easily detected by the human eye than differences in depth of shade. Therefore, when dyeing acrylic fibers, it is necessary to have specific information concerning the behavior of cationic dyestuffs in combination.

This infomation is given with the combination constant or K-value which can easily be determined in the average mill laboratory. Compatibility is assessed against one of two defined 1–5 scales, a yellow-red and a blue scale. The compatibility value of the dye is determined from its dyeing behavior in combination with each of the standard dyes in the relevant scale [4].

1. The K-values range from 0.5 to 5 for different dyestuffs.
2. Dyestuffs with equal or similar K-values exhaust on tone in every proportion.
3. A dyestuff with a lower K-value exhausts prior to a dyestuff with a higher K-value in the same combination. The difference in dye uptake increases with the difference in K.
4. The effect of cationic retarders and electrolyte is more pronounced on dyestuffs with higher K.
5. The K-values are valid regardless of
   acrylic fiber used
   concentration of dyestuff
   rate of temperature increase
   the presence of cationic retarders
6. Only the overall rate of exhaustion may differ depending upon changes in above mentioned parameters.
7. The K-values are, however, not valid when anionic retarder systems are used.
8. The K-values are not related to the rate of dyeing but only to the order of exhaustion of the dyestuffs in any combination.

The practical application of this is obvious. It is undesirable to have one dyestuff striking before the others and, likewise, it is undesirable for one to lay back in the dye bath. The first condition can cause leveling problems; the second, problems in reproduction of shade. For reproducible level dyeings, the difference between the K-values of dyestuffs selected for a combination should be as small as possible, generally not more than one unit. If leveling problems are expected, dyestuff combinations with a higher K-value should be selected.

Determining the K-value: The K-value is determined by dyeing the unknown in combination with a series of dyes of known K-values and different hue. The following dyes and concentrations are recommended. (The concentrations recommended are approximately 1/2 standard depth.)

K-value	Blue scale	
1	0.27% Astrazon Blue FRR 200	(Bayer AG)
2	1.35% Astrazon Blue 5GL 200	
3	0.60% Astrazon Blue 3RL 200	
4	0.60% Sevron Blue ER	(Crompton and Knowles)
5	1.20% Astrazon Blue FGLN 200	

K-value	Yellow-red scale
1.5	0.25% Aizen Cathilon Orange GLH (Hodogaya Chem. Co. Ltd.)
2	0.50% Astrazon Yellow GRL
3	0.30% Astrazon Golden Yellow GL 200
4	0.75% Sevron Yellow 3RL
5	0.37% Astrazon Red F3BL

To determine the K-value of an unknown dyestuff, prepare five dyebaths as follows:

X%	Test dye	(where X is 1/2 standard depth)
Y%	Standard dye	(see above for recommendations)
2%	Acetic acid 56%	
1%	Sodium acetate	
	Liquor Ratio 1:40, 95°C	

Prepare 30 acrylic skeins or pieces of fabric (5 g each). Enter five skeins into the five dyebaths at 95°C. Dye until all five dyeings are approximately equal in depth. The dyeing time varies from K-value to K-value, and very often it takes only 30–60 sec for the K-1 dyeing. Remove and enter five fresh skeins and dye until all five dyeings are approximately equal in depth. Repeat this cycle three more times.

If, at the end of the fifth cycle, the dyebath is not exhausted, the sixth set is added, and dyeing is continued until the bath is completely exhausted.

After dyeing, the dyed samples are rinsed, dried, and mounted in the same order in which they were dyed. (Levelness of the dyeings is immaterial)

The results are evaluated by examining the dyed samples for on-tone shade throughout the sequence. The K-value assigned to the dye under test is that of the standard dye with which it exhausts on-tone throughout the sequence.

II. APPLICATION PROCEDURES

A. General

Acrylic fibers, in general, have good affinity for cationic dyes and can be dyed satisfactorily. However, modern business is placing more and more emphasis upon lower costs and shorter dyeing cycles, thus minimizing the range of dyes available. At the same time, this places the dyer into a position where he almost has to dye under optimum dyeing conditions in order to stay competitive.

Two factors are of prime importance when dyeing acrylic fibers: first, conditions under which the dyestuff strikes onto the fiber; and second, the time required for the dye to exhaust to achieve a satisfactory dyeing.

The desirable situation is to obtain as quickly as possible dye exhaustion consistent with good levelness. It is hardly necessary to point out that the preferred technique for the best dyeing success is to select dyes and dyeing conditions which will combine both slow-medium strike rate and dye transfer. However, in the case of acrylic fibers, once a conventional dyestuff has become a part of the fiber, it remains fixed with little or no transfer, except under high-temperature ($\geq 110°C$) conditions. Therefore, a controlled exhaustion is essential for uniform dyestuff absorption and optimum levelness.

Exceptions to this are the migrating cationic dyes, which are smaller in molecule size than the conventional dyes and have higher mobility within the fiber due to their chemical constitution. The migrating cationic dyes have high diffusion rates, low affinity, and good migration properties. Therefore, a more rapid dye uptake is allowed, followed by a subsequent migration phase [5]. In order to maintain the high mobility of these dyes, adequate liquor circulation and sufficient Na^+, K^+, or NH_4^+ ions are necessary. These small cations compete with the migrating cationic dyes for dyesites and improve migration.

One of the basic characteristics of acrylic fibers is that minor temperature differences can drastically influence the dyeing rate of cationic dyes onto acrylic fibers. A temperature increase of approximately $3.0°C$ will double the strike rate. In dyeing any textile material, either the dye solution must be circulated through the material, as in package dyeing, or the material is circulated through the dye solution as in a beck or jet. In ultrashort liquor ratio jets, the dye liquor is sprayed onto the fabric. The rate of circulation of the material or the dye solution, as the case may be, must be fast enough to compensate for changes in the dye concentration due to exhaustion. Since cationic dyes exhaust very rapidly on acrylic fibers at normal dyeing temperatures, something must be done to slow down the rate of exhaustion so that, with normal circulation, the concentration of dye immediately adjacent to the fiber is equal or fairly close to that in the remainder of the bath. In order to decrease the rate of exhaustion of cationic dyes, and thus help to obtain level dyeings, retarders are often used.

Taking into consideration all the various parameters above, the dyebath conditions, including both dyes and chemicals, must be adjusted so that levelness is achieved. Not only chemical factors have to be taken into consideration, but equally important are the physical factors, such as dyebath circulation, weight of goods being dyed, and the efficiency of the equipment [6]. All of the chemical and physical factors are interrelated and form a total set of dyeing conditions which must be optimized. For example, if high dyebath circulation is available, fast-striking dyes may perform well under a given set of conditions. With poor circulation or overloaded equipment, unlevel dyeings could be obtained with the same dyestuff selection and dyebath conditions. No completely scientific set of rules or data can be given or exactly followed which will always ensure satisfactory dyeings. The individual dyer or chemist must have as many facts as possible available to him or her before establishing his or her own procedure and dyestuff selection based on such data.

B. Retarders

The dyeing speed of cationic dyes can be controlled through addition of retarders, making it easier to achieve level dyeings. Available today is a range of retarders which are needed in many cases because uniformity of the dyebath temperatures and the liquor circulation conditions are not always ideal. If it were possible to maintain complete dyebath temperature uniformity and control, the need for retarders would be greatly minimized. The principal types of retarders in use over the years have been cationic retarders. Anionic retarders are used for special applications.

1. Cationic Retarders

Cationic retarders act much the same way as dyestuffs and could even be called colorless dyestuffs. Cationic retarders compete with the dyes at the fiber surface and are used to reduce the influence of the exhaustion on the surface concentration of the dye. They have a K-value and f factor and, for optimum results, the K-value of the retarder should be lower than that of the dyestuff. An example, Astragal PAN, is shown in Figure 4.

Dyestuffs of low affinity are sometimes better controlled with retarders of low effectiveness, such as sodium sulfate, whereby the large numbers of sodium cations available compete with the dyestuffs for available sites. Practice has

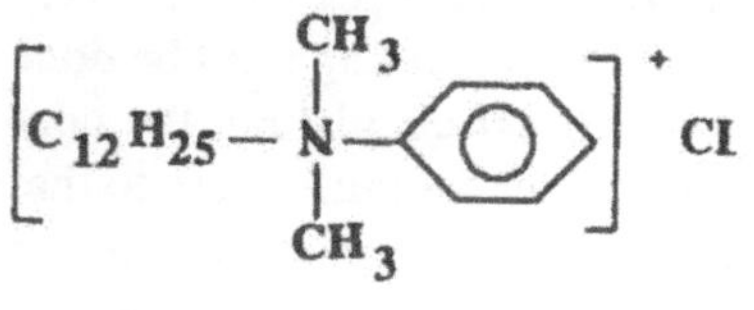

$$\left[C_{12}H_{25} - \overset{\overset{\displaystyle CH_3}{|}}{\underset{\underset{\displaystyle CH_3}{|}}{N}} - \left\langle \bigcirc \right\rangle \right]^{+} Cl$$

Figure 4 Astragal PAN.

proven that a combination of sodium sulfate, cationic migrating agent, and retarder give best results.

An important factor found in practice, which can be explained in theory, is that the rate of dyeing depends upon the temperature and concentration of dye and retarder; that is, the nearer to saturation of the fiber with dyestuff and retarder, the better the chances of obtaining level dyeings. Based upon that assumption, the relative saturation $S(\text{rel})$ can be defined as follows.

$$S(\text{rel}) = \frac{\text{number of cations}}{\text{number of dyesites available}} = \frac{p \times f}{SF}$$

For combinations with cationic retarders, the following applies.

$$S(\text{rel}) = \frac{p1 \times f1 + p2 \times f2 + p3 \times f3 + \cdots + p(\text{ret}) \times f(\text{ret})}{SF}$$

$S(\text{rel})$ should be ≤ 1

The saturation which should be obtained in order to produce level dyeings is dependent upon the fiber and the dyestuff. Fibers with higher dyeing rates and dyestuffs with higher affinity require a higher degree of saturation as illustrated by the analogy to a theater crowd in Figure 5.

Table 1 shows the percent relative fiber saturation on a fiber with SF = 2.1 required as the K-value changes or as the fiber dyeing rate (V) changes.

Table 2 gives an example of how to calculate the exact amount of retarder necessary to achieve a certain saturation level. Since the fiber saturation value (SF), the dye saturation factor (f), the retarder saturation value, and the percentages of dyestuffs are fixed values, the percentage of retarder (PR) is the only variable and can be calculated.

% relative fiber saturation for K-3 dyestuffs = 65% (from Table 1)
SF = 2.1 × 65% = 1.365

Using a retarder with f = 0.6,

$$PR = \frac{1.365 - 0.52}{0.6} = 1.4\% \text{ retarder}$$

These examples show how each of the variables contribute to the systematic approach of establishing a good dyeing procedure.

In this context, it should be mentioned that control units such as Colorex (Firma Original Hanau, Heraeus GmbH, Hanau, Germany) are being offered primarily for the dyeing of acrylic fibers. These units continuously measure the rate of dyestuff exhaustion in a production machine and automatically control temperature and rate of rise to ensure maximum levelness. Advantages are

Reduction in amounts of retarder being used, especially in light shades
Shorter dyeing times due to more accurate temperature control

Table 1 Percent Relative Fiber Saturation
Recommended on a Fiber with SF = 2.1

V	$K=1$	$K=3$	$K=5$
6.4	95	85	70
3.6	95	75	50
1.7	85	65	40
1.4	80	60	35

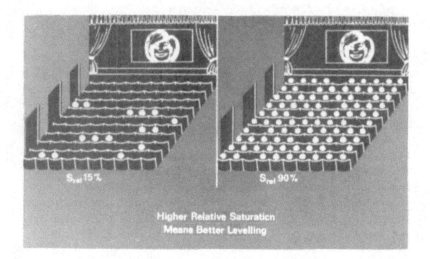

Figure 5 Relative saturation as shown by the theater analogy.

Table 2 Fiber: Dralon -- Fiber Saturation Value (SF) = 2.1
Fiber Dyeing Rate (V) = 1.7

	p	f	$p \times f$
Astrazon Golden Yellow GL 200	0.2%	0.52	0.10
Astrazon Red GTLN 200	0.4%	0.76	0.30
Astrazon Blue BRL 200	0.2%	0.60	0.12
			0.52

2. Anionic Retarders

This type of retarder is mainly used where cationic systems cannot be used due to their influence on the bulk of the yarn. They are also used successfully in piece dyeing on jet equipment in lighter depth shades and in one-bath dyeing of acrylic/cotton and acrylic/wool/nylon blends, in which cases the anionic dyestuff performs essentially as an anionic retarder. Anionic retarders do not penetrate into the fiber. They form a complex with the cationic dyestuff and the rate of absorption by the fiber is controlled by the rate at which the complex dissociates. The possibility of precipitation of cationic dyestuffs with anionic auxiliaries is overcome by using a nonionic agent which acts as a compatibilizer and antiprecipitant. In general, cationic retarder systems are used more widely than anionic systems because of their advantages in reproducibility and leveling.

K-values are not valid in an anionic retarder system, therefore K'-values have been established which will perform in the same manner in an anionic system as K-values will in a cationic system.

3. Constant-Temperature Dyeing

The objective to be achieved by this procedure is to allow the dyes to exhaust slowly and uniformly to 80% or more at a constant predetermined temperature, therefore, minimizing influence of changing dyebath temperatures. After exhaustion, the temperature is raised rapidly to the boil and held for approximately 15 min to achieve complete dye penetration and fixation.

The main concept is to dye without retarder or with a minimum of retarder at a constant temperature. The constant temperature level at which the dyeing is carried out depends upon the depth of shade, the type of dyes being used and the nature and source of the acrylic fiber to be dyed. The dyeing temperature is generally adjusted in such a manner that the bath is exhausted after dyeing. The constant dyeing temperature determined must be maintained exactly within the limits of ±1°C during the initial exhaustion process.

The whole process relies heavily on the mechanics (machine and temperature control) to obtain and maintain complete uniformity of temperature.

C. Dyeing Conditions

1. Acidity of the Dye Liquor

Basic dyes exhaust on the acrylic fiber in a pH range of 2–7. Since some dyes are unstable above a pH of 6, the entire range of dyes is preferably dyed in a weakly acid medium at a pH of 4.5–5.0. This level of acidity is most commonly obtained by addition of acetic acid or a buffer combination of acetic acid and sodium acetate. But this recommended pH range can also be established by the use of other acids or acid-producing salts. The pH of 4.5–5.0 must be stable during the whole dyeing time.

2. Dyeing Temperature/Dyeing Process

Basic dyes exhaust completely onto the acrylic fiber above temperatures of 90°C and are satisfactorily fixed at temperatures of 96–98°C. In the lower temperature range of 40–70°C, the basic dyestuff is only adsorbed onto the outer layer of the acrylic fiber, therefore, the temperature rate of rise can be rather rapid. At temperatures of 70°C and above where significant dye exhaustion takes place, the rate of rise needs to be lowered to ensure a uniform dye uptake. With addition of a retarder to the dyebath, this time can be shortened. In order to ensure complete dye penetration and good fastness properties, the dyeing time at 98°C should be 30–60 min for light to medium shades and 60–90 min for deep shades, although the dyebath may appear to have been exhausted in less time.

Pressure dyeing can offer increased dyestuff exhaustion, shorter dye cycles, and improved dye uniformity due to increased dyestuff transfer. The preferred temperature is at 105°C, whereby the dyeing process is normally completed after 20–30 min at top temperature.

3. Preparation

Most acrylic yarns and fabrics do not require scouring before dyeing. In many cases, a 40–60°C rinse is sufficient for wetout and removal of tints and/or contaminants carried over from yarn or fabric manufacturing. In case of heavier soiling, a scour should be applied using a nonionic detergent in a mild alkaline solution with tri- or tetrasodium phosphate at temperatures of 50–70°C for 10–20 min. This process is followed by a warm rinse. If a prescour is necessary, anionic detergents or ammonia (as alkali) should *not* be used since unlevel dyeings may be encountered.

4. Dissolving of the Dyes

Basic dyes are soluble in boiling water, but they are limited in their solubility. Therefore, it is necessary to be guided by the solubility limits which are given in the respective pattern cards and not to exceed them. The dyestuffs are either pasted with acetic acid and diluted with 40–50 times the amount of boiling water, or sprinkled into hot water in the mix tank. Acetic acid is added prior to addition of the dyestuff and additional boiling of the solution should be avoided. Filtering at the time of addition to the dyeing equipment is recommended.

5. Dye Formula

The dye liquor is heated to the starting temperature and auxiliaries and dyestuff are added in listed order (see Figure 6).

0.25–0.5%	Nonionic compatibilizer
0.50–1.0%	Sodium acetate
1.00–2.0%	Acetic acid (56%) pH 4.5
X%	Retarder/migrator
Y%	Dyestuff

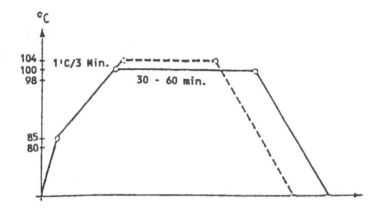

Figure 6 Procedure diagram for dyeing cycle.

D. Makeup of the Fiber

1. Tow or Cable

This is an endless filament bundle which is obtained after extrusion from the spinnerette, drawing, washing, and drying of the fiber. In this form, it is uncut and can be exhaust dyed in a long liquor ratio in a packing machine, continuous dyed by the pad-steam process, or gel dyed at the time of the fiber production when the acrylic fiber is still in a gel state. Due to economics and quality of fiber, continuous and gel dyeing are the primary dyeing methods.

2. Top

Cut or stretch-broken fibers are made into parallel fiber bundles and dyed in form of bobbins in a packing machine or sometimes on the continuous range.

3. Stock

Cut fibers are dyed either in the packing machine or continuous.

4. Piece Dyeing

Mainly dyed in tubular knitted goods for apparel end use and wovens for home furnishing. Precautions have to be taken to avoid crease and crack marks which can be a severe problem in beck dyeing. Most suitable are soft flow jet dyeing machines. Woven fabrics are preferably dyed on beam machines.

5. Fully Fashioned Articles

They are dyed on paddle or rotary drum machines.

6. High Bulk Yarn

This type is mainly used for hand-knitting or machine knit yarn. High bulk yarns consist on the average of 40% shrinkage fiber and 60% nonshrinkage fiber. This yarn type can shrink from 10% to 30% through a thermal treatment by means of either steam or hot water. It is dyed either in skeins or as preshrunk yarn in soft packages.

7. Regular Yarn

This yarn is a nonshrinkage type which is dyed on packages or warp beams.

8. Carpet Yarn

Due to the rather coarse yarn count and to avoid any flattening of the yarn, it is dyed almost exclusively in skein form. In some special cases, such as for solid carpets, the raw stock dyeing technique is used.

E. Dyeing Processes and Machines

Acrylic fibers can be dyed by either batch or continuous processes. Equipment and process are selected depending on the material (fiber, yarn, fabric, fabric construction, garment), size of dye lots, and quality requirements in the dyed fabric [7].

Machinery for dyeing must be resistant to attacks by acids, bases, other chemicals, and dyes. Type 316 stainless steel is normally used as the construction material for all parts of dyeing machines that will come in contact with dye formulations. In addition, corrosion-resistant gaskets are used, as well as nonetching glass view and light ports.

1. Batch Dyeing Processes

Batch processes are the most common method used to dye textile materials. Batch dyeing is also called exhaust dyeing because the dye gradually exhausts from a relatively large volume dyebath to the material being dyed over a relatively long period of time. Textile substrates can be dyed in batch processes in almost any stage of their assembly into a textile product including fiber, yarn, fabric or garment.

The three general types of batch dyeing machines are those in which the fabric is circulated, those in which the dyebath is circulated while the material being dyed is stationary, and those in which both the bath and the material are circulated. Fabrics and garments are commonly dyed in machines in which the fabric, or fabric and dyebath, is circulated. Fiber, yarn, and fabric can all be dyed in machines which hold the material stationary and circulate the dyebath. Jet dyeing is the best example of a machine that circulates both the fabric and the dyebath.

Becks Atmospheric becks can be used for dyeing at temperatures up to 100°C. As shown in Figure 7, a dye beck consists of a tub which contains the dyebath and a reel to move the loop of fabric through the dye liquor. The liquor to goods ratio used in becks is typically 15:1 or higher, although there are becks available operating at lower liquor ratios.

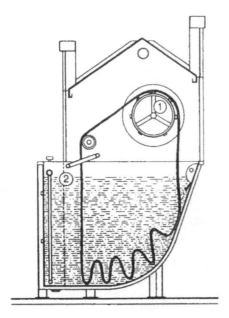

Figure 7 Dye beck: (1) reel; (2) perforated divider.

The dye beck is sometimes called a winch because of the winch mechanism used to move the fabric. The ends of the fabric piece to be dyed are sewn together to make a continuous loop. The reel pulls the fabric out of the dye liquor in the tub and over an idler roll. After leaving the reel, the fabric slides down the back wall of the beck and gradually works its way from the back toward the front of the machine. Several loops of fabric of about the same length (typically 50–100 m long) are dyed simultaneously side by side.

Chemicals and dyes used in the dyeing are added to a compartment at the front of the beck behind a baffle plate. The baffle or divider separating the compartment from the tub is perforated, allowing the added chemicals to gradually become mixed with the liquor in the tub. Live steam is injected into the compartment to heat the liquor to the required temperature.

The greatest advantages of becks are simplicity, versatility, and relatively low price. Becks subject fabrics to relatively low lengthwise tension and encourage the development of yarn crimp and fabric bulk. However, becks tend to use large amounts of water, chemicals, and energy. Becks can cause abrasion, creasing and distortion of some fabrics.

When dyeing acrylics, it is especially important that the piece length is fairly uniform. The use of a retarder is recommended, which helps to minimize piece-to-piece variation and slows down the dye uptake. Also, a conservative temperature rate of rise should be employed to ensure a level dyeing during the critical

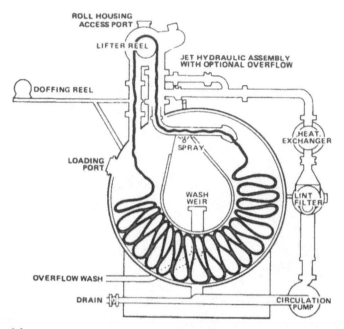

(a)

(b)

strike range. These precautions are essential since very little migration is achieved at the boil. A rather slow cooldown rate should be used to minimize crack and rope marks.

Jet Dyeing Jet dyeing machines resemble becks in that a continuous loop of fabric is circulated through the machine. However, the cloth transport mechanism is dramatically different in these two types of machines. A high-speed jet of dye liquid created by a venturi or nozzle transports the fabric through the cloth guide tube of the jet machine. A jet machine has a cloth guide tube for each loop of fabric being processed. A powerful pump circulates the liquor through a heat exchanger outside of the main vessel and back into the jet machine. The fabric travels at high velocity of 150–400 m/min while it is in the cloth guide tube. The fabric leaving the cloth guide tube enters a larger capacity cloth chamber and gradually advances back toward the cloth guide tube. Material in loop can vary from 150 to 350 kg depending on fabric weight.

Pressurizing a jet dyeing machine provides for high-temperature dyeing capability. Jet dyeing machines provide the following advantages compared to atmospheric becks:

Vigorous agitation of fabric and dye formulation in the cloth tube increases the dyeing rate, which ensures uniformity, fabric reorientation, and high energy interchange.
Rapid circulation of fabric through the machine and reorientation in the jet minimizes creasing because the fabric is not held in any one configuration very long.
Dyeing at temperatures above 100°C increases dye penetration into the fiber and improves dyestuff migration, which allows a more rapid dyeing cycle.
The lower liquor ratio used in jet dyeing allows shorter dye cycles and saves chemicals and energy.

Some disadvantages of jet dyeing machines compared to becks are as follows:

Capital and maintenance costs are higher.
The jet action tends to make fomulations foam in partially flooded jet machines.
The jet action may damage the surface of certain types of fabrics.

The first machines were partially flooded. Fully flooded machines keep the fabric completely submerged during the dye cycle. This prevents the formation of longitudinal creases which occur when the fabric is lifted from the bath in a partially flooded machine. Fully flooding the machine also prevents formation

---

Figure 8 (a) High-temperature jet piece dyeing. (b) Gaston jet piece dyeing machine. (Courtesy Gaston Country Dyeing Machine Co.)

of foam. The so-called "soft flow" machines use the same principle of a transport tube as a jet machine. These machines either eliminate the high-velocity jet or use a jet having lower velocity than that used on conventional jet dye machines. The soft flow machines are more gentle on the fabric than conventional jet machines. Jet machines offering capability of very low liquor ratios of approximately 5:1 are also available.

In jet dyeing acrylics, the loading of the fabric is very important to ensure proper fabric transportation. Load size varies for the different type jets. In cases of long fabric strands and slow turnover times, once again exact temperature control during the strike zone of the dyestuff is important for optimum levelness. Maximum dyeing temperatures are 105–108°C, which results in somewhat better dye migration as compared to the boil. Conservative cooldown rates are recommended to avoid crack and rope marks.

Package Dyeing The term "package dyeing" usually refers to dyeing of yarn which has been wound on perforated cores so that dye liquor can be forced through the package. Packages may be tubes, cheeses, or cones. Cores for dye packages may be rigid stainless steel, plastic, or paper. Plastic types are normally intended to be used either once or for limited multiple use, while stainless steel cores can be reused indefinitely. Plastic, as well as stainless steel springs, are used as compressible cores. These compressible cores allow more packages to be pressed onto the package carrier and will increase the capacity of the machine, thus lowering the liquor to goods ratio. This process also contributes to a more uniform density of the entire yarn column. Package dyeing machines may be of the vertical or horizontal spindle type.

As shown in Figure 9, the yarn packages are placed on perforated spindles on a package carrier which fits into the pressure vessel where dyeing takes place. The dye vessel is cylindrical and has domed ends. The hinged top cover, which must be lifted for loading and unloading, is secured during dyeing by bolts or a sliding ring which can be quickly locked. Most package dyeing machines are capable of dyeing temperatures up to 135°C. The number of packages may vary from as few as one in a laboratory machine to several hundred in a large production machine.

The dye liquor is pumped through the perforations in the spindles and package cores into the yarn. The flow of liquid can be either from inside to outside of the package or outside-in. In case of acrylic dyeing, it should be primarily inside out to avoid distortion of the yarn package.

A package dyeing machine has an expansion tank mounted alongside the dye vessel. The expansion tank accommodates the increased volume of dyebath resulting from thermal expansion when the bath is heated, as well as for adding the air pressure pad. Chemical and dye adds are made to the vessel through the expansion tank.

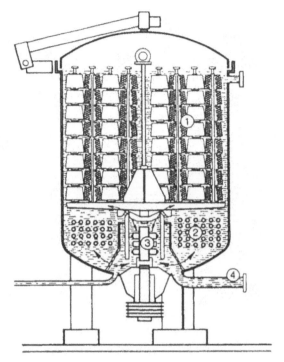

Figure 9 Package dyeing machine: (1) packages; (2) heat exchanger; (3) circulation pump; (4) drain.

A heat exchanger using high pressure steam as the heat source indirectly heats the dye liquor in a package dyeing machine. The steam coils (or heat exchanger) for heating the liquor can also be used as cooling coils after the dyeing is completed.

Liquor ratio in a package dyeing machine dyeing rigid packages is typically about 10:1 when the machine is fully loaded. Raw stock, tow, and other materials can be dyed using the same principles as package dyeing. A basket (cage) is normally used to hold these materials during the dyeing.

Advantages of package dyeing acrylic yarns versus skein dyeing are as follows [8]:

Elimination of costly skein reeling process and a general reduction at all stages of processing
Improved levelness due to a better control and more uniform liquor flow
Possibility of dyeing at temperatures above 100°C
The availability of rapid drying techniques

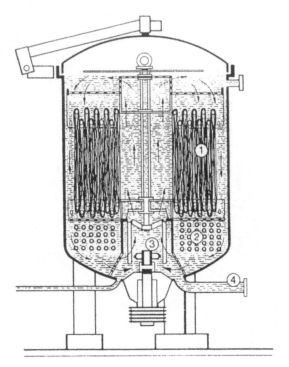

Figure 10 Skein dyeing machine: (1) skeins; (2) heat exchanger; (3) circulation pump; (4) drain.

Savings in water, effluent, energy, and dyes and chemicals due to the lower
   liquor to goods ratio employed

Beam Dyeing The principles of beam dyeing are essentially identical to
those of package dyeing. Either yarn or fabric can be beam dyed. The fabric or
yarn is wound on a perforated beam. A beam machine can be designed to hold a
single beam or multiple beams in a batch. Beam dyeing of warps for weaving is
practical in producing patterned fabrics where the warp yarn will be one color
and the filling will be another color.

Skein Dyeing In skein dyeing (also called hank dyeing), skeins of yarn are
mounted on a carrier which has rods (sticks) at the top and bottom to hold the
skeins. The skeins are suspended in the dye machine and dye liquor is gently
circulated around the hanging skeins. Perforated plates can be used at the top
and bottom of the machine to help provide uniform flow of the dye liquor.
Alternatively, the dye liquor can be pumped through perforations in the sticks so
that it cascades down over the hanging skeins. Skein dyeing produces good bulk
in the yarn because of the low tension on the yarn in the dyebath. The method is

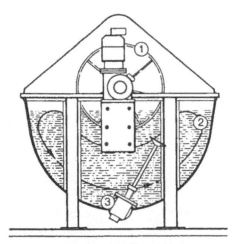

Figure 11 Paddle dyeing machine: (1) Paddle wheel; (2) garments; (3) drain.

used mainly for bulky yarns for knitted outerwear and hand-knitting. It is very important after dyeing to cool slowly to below the thermoplastic temperature to avoid yarn distortion.

Skein dyeing uses a high liquor ratio and a lot of energy. Uniform dyeing is difficult to achieve in a skein dyeing machine. Slow winding and backwinding requirements of the process make it labor intensive. Package dyeing has replaced some skein dyeing, even though the yarn bulkiness achieved in skein dyeing is usually not matched in package dyeing.

Paddle Machines and Rotary Drums Paddle machines and rotary drum machines can be used to dye textiles in many forms, but these two methods are used mostly to dye garments. Steam injection directly into the dyebath heats both of these types of machines. A schematic diagram of an overhead paddle dye machine is shown in Figure 11. The overhead paddle machine is simply a vat with a paddle having blades the full width of the machine. The blades dip a few centimeters into the vat to stir the bath and push the garments down, keeping them submerged in the dye liquor.

A rotary drum machine is a cylindrical vessel slightly larger than its internal perforated drum which holds the material to be dyed. The perforated drum is divided into several chambers, each having its own door through which it can be loaded and unloaded. The drum rotates horizontally as shown in Figure 12. Rotary drum machines are commonly used to dye hosiery.

Continuous Dyeing Continuous dyeing is often preferred because it is substantially more economical than exhaust dyeing, saving energy, water, man-hours and postcombing processes. In addition there is improved shade consis-

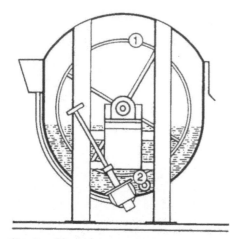

Figure 12 Rotary drum dyeing machine: (1) dyeing drum; (2) heat exchanger.

tency especially in larger lot sizes. Acrylic tow is primarily dyed continuously although there is still some dyed by the package system.

As shown in Figure 13, the dye liquor is applied mechanically onto the tow by means of a horizontal padder, and the tow is then fed into a steamer for fixation of the dyestuffs. Liquid basic dyes are preferably used due to their ease of handling. Important dyestuff properties are good buildup, steam stability, and rapid and complete fixation during steaming.

Compared with the conventional exhaust methods, the K-value is of minor importance in the pad steam process while the rate of diffusion of the dyes is of primary concern. After dyeing any unfixed dyestuff as well as auxiliaries are rinsed off and finishing chemicals such as softeners, etc., are applied in the last bath.

Gel Dyeing Dope dyeing (spin dyeing) and gel dyeing of acrylic fibers during manufacture have been increasing in popularity in recent years. Chapter

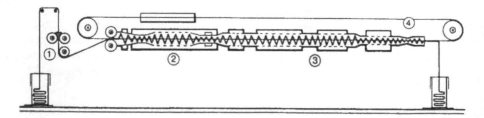

Figure 13 Continuous tow dyeing machine: (1) Padder; (2) steaming zone; (3) rinsing and finishing; (4) transport chain.

6 discussed these fiber producer dyeing processes from a manufacturing viewpoint.

In dope dyeing, pigments or special cationic dyestuff formulations are added to the polymer solution immediately before spinning. This method, which is mainly used for dry-spun fibers, has proved particularly successful for awning and outdoor fabrics. Dope-dyed black fibers are used in a wide variety of other articles too.

Nearly all producer-colored, wet-spun acrylic fibers are dyed by the gel dyeing process. This process was first used commercially by Courtaulds in 1962 as the "Neochrome Process." These days, approximately 14% of all acrylic fibers are gel-dyed. The main advantages of gel dyeing are

Integration of the dyeing process into existing spinning lines with relatively little modification of the plants

High degree of flexibility as colors can be changed quickly; batches of less than 5 tons can be dyed these days

Shades range from ultrawhite to black and include a wide range of attractive and colorful hues

The fastness levels required for apparel and home furnishing are reached by selecting suitable dyestuffs

Dyed and undyed tow have equally good processing properties

High cost-effectiveness, especially compared with batch dyeing of fibers by the exhaust method

Method Gel dyeing is performed on the spinning line after spinning of the filaments, washing and, where relevant, drawing, but before drying. At this point, the structure of acrylic fibers is very open and they have a relatively low degree of orientation, giving them high porosity and a large surface area readily accessible to dyestuffs and fluorescent whiteners. The dyeing rate in the gel dyeing process is thus far higher than when fibers are dyed after drawing and drying.

Gel dyeing basically consists of shock-type continuous dyeing of acrylic fibers while they are still in the gel state. The open structure described above means that the dyestuff is taken up by the fibers in seconds at temperatures of 30–60°C. The dyestuff is then fixed on the fiber by an air passage or by steaming as shown in Figure 14.

spinning   pre-drawing   washing   dyeing   main-drawing   finishing   drying

Figure 14 Wet spinning fiber line with gel dyeing.

It should be noted that, if fibers are dyed *before* drawing, the linear speed is lower, which leads to a prolonged immersion period in the dyebath. However, at this stage, the fiber has even higher affinity for the dye and this may result in unlevel dyeings. If fibers are dyed after drawing, spinning rates may be up to 100 m/min so the dyeing unit has to be larger to allow immersion times of 3–5 sec. It is, therefore, common to incorporate a predrawing stage before dyeing and to leave the main drawing stage until afterwards.

The rapid dye uptake means that unlevel dyeing of the hundreds of thousands of filaments in the tow cannot be ruled out. Careful selection of the dyestuffs, dyeing time, and temperature alone are not enough to ensure level dyeings. It is, therefore, advisable to use mechanical means, such as extenders, to spread the compact tow and thus ensure that the dyestuffs penetrate it more evenly. The more thinly and evenly the tow is spread when it enters the dyebath, the more level the dyeing obtained.

Other prerequisites for level dyeing are uniform drawing (predrawing) of the tow and a uniform residual solvent level. Figure 15 shows several variations of dyeing units suitable for gel dyeing. The construction of the dyeing unit is very important.

Good results are achieved if the tow is moved by mechanical means during dyeing; for example, by using channelled rollers or drums. Spray units installed above the dyebath have also proved a successful method of ensuring uniform penetration of the dyestuffs.

Specially modified washing units can also be used for dyeing, either in vertical stack washing machines or in trough washers. Continuous circulation of the liquor in the dyebath is absolutely essential.

Dyestuffs can be metered into the bath by a variety of methods. The simplest method is automatic metering of the various liquid dyes directly into the dyebath. However, constant monitoring of the bath, or of the dyed fiber, is necessary so that metering errors are picked up immediately. A more reliable method is batchwise preparation of the liquor in a formulation tank. Once the liquor has been checked, either the concentrated dyestuff mixture or a dilute solution can be metered into the bath.

Dyestuff Selection For gel dyeing, the saturation values and characteristic data for the fiber and dyestuff are the same as for exhaust dyeing. Combination indices can be used, too, but these differ from the indices used in exhaust dyeing. The values are closer together and influence shade consistency rather than levelness.

The fastness levels of gel-dyed fibers are similar to those of fibers dyed by conventional methods.

The dyestuff selection takes account of the special requirements of this process. The dyestuffs must be stable to the solvents used in the spinning

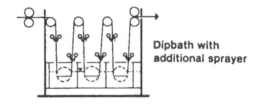

Dipbath with
additional sprayer

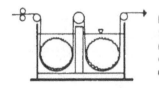

Double dyeing bath
with additional
mechanical movement
of the tow by rippled
cylinders

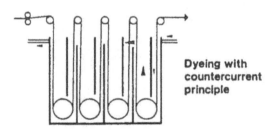

Dyeing with
countercurrent
principle

Figure 15 Gel dyeing units.

process—for example, sodium thiocyanate. Liquid dyes are almost always used
because of the large batch sizes and thus the large dyestuff additions involved.

Unlike powder formulations, liquid dyestuffs are nondusting and do not
require time-consuming pasting or dissolving processes or special dissolving
units, and they are simple to apply with automatic metering units.

A full range of liquid dyestuffs is available today and includes some dyes
developed specifically for gel dyeing. The concentrations available are the same
as the powder brands. The liquid dyestuffs can be mixed in concentrated form
and diluted with water without the aid of auxiliaries. The mixtures are stable and
have no tendency to settle or crystallize.

REFERENCES

1. P. H. Hobson, Acrylic and modacrylic fibers: development, use and potential;
Textile Chem. Colorist, 4, 37–42 (1972).

2. Dyeing and Finishing of Orlon Acrylic Fibers, Du Pont Technical Information, Bulletin OR-196 (Dec. 1978).
3. Wolfhard Beckmann, Physikalisch-chemische Eigenschaften kationischer Farbstoffe und ihre Bedeutung fuer das Faerben von Acrylfasern, *Z. Ges. Textilin.*, *9*, 603–608 (1969).
4. Charles L. Zimmerman, Compatibility of cationic dyes for acrylic fibers, AATCC Committee RA87, *Textile Chem. Colorist, 7*, 44 (1975).
5. Marshall White, Jr., F. Schlaeppi, N. E. Houser, and J. T. Larkins, Jr., Economical and energy efficient systems for the dyeing of cationic-dyeable fibers and blends, Book of Papers, AATCC National Conference, New Orleans, 1983.
6. G. R. Turner, A summary of dyeing acrylic fibers, Du Pont, *Dyes Chem. Bull.* A paper, based on this information, was given by G. R. Turner before the Canadian Association of Textile Colourists and Chemists. Quebec Section, March 17, 1972.
7. Warren S. Perkins, A review of textile dyeing processes, *Textile Chem. Colorist, 23*, 23–27 (1991).
8. J. Park, Package dyeing acrylic yarns, *Textile Chem. Colorist, 11*, 15–16, 39 (1979).

10

Apparel End Uses

Arthur Lulay

Consultant
Sun City, Arizona

I. BACKGROUND

Acrylic fibers are defined by the U.S. Federal Trade commission as having at least 85% by weight of acrylonitrile units (–CH2–CH[CN]–). Modacrylic fibers are defined as being composed of at least 35% but less than 85% acrylonitrile units by weight. Modacrylic fibers are less suitable for apparel end uses compared to acrylic fibers because, generally, they are higher priced, yellower, and have lower dimensional stability to heat. Therefore, the consumption of modacrylic fibers is very limited in apparel and reference to them will be confined to the one or two markets where they are utilized to a small extent.

In the three decades of the 1950s, 1960s and 1970s, acrylic fibers expanded rapidly into sweaters, socks, craft yarns, pile fabrics, and fleecewear. The advent of the "stretch-break/high bulk" yarn process gave acrylics the bulk, warmth, softness, and comfort desired by the consumers of natural fibers. Development of a spiral crimped bicomponent acrylic fiber with wool-like aesthetics by DuPont in the late 1950s gave impetus for penetration into craft yarns and sweaters. Development of producer-dyed fibers, acid dyeable fibers, larger size filaments, and other variants by Monsanto, Courtaulds, and others permitted entry into fake furs, other high pile fabrics and fleecewear for sweatshirts and active sportswear.

II. FIBER CHARACTERISTICS

A. Wet Versus Dry Spun

In the United States, all fiber producers except DuPont* utilize the wet spinning process to produce acrylic staple and tow. This is a more economical continuous process with higher productivity for heavier deniers, higher dyeing rate, and easier "in-line" dyeing. This process also produces a round or kidney-bean-shaped cross section in contrast to the dog-bone-shaped fiber obtained by the dry-spinning process. Typical fiber cross sections obtained from the wet- and dry-spinning processes are shown in Figure 1.

Because of its dogbone shape, the dry spun fiber has a lower bending modulus or stiffness relative to a round or kidney-bean-shaped fiber. This is shown mathematically in Figure 2. Calculations were performed for bending along the A-B axis in the assumed idealized shapes [1]. These results predict softer yarn or fabric aesthetics will be obtained at a comparable fiber denier with the dry-spinning process.

The dog-bone cross section also prevents the fibers from packing as closely in the yarn assembly leading to a less dense or bulkier product. Measurement of yarn thicknesses by Onions et al. [2], confirm this theory (Table 1). Dog-bone-shaped fibers were 30% and 15% thicker than round and kidney-bean-shaped fibers respectively in comparable yarn structures.

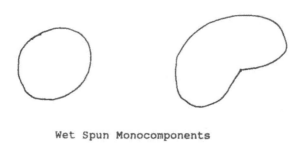

Wet Spun Monocomponents

Dry Spun
Monocomponent

Dry Spun
Bicomponent

Figure 1 Fiber shapes obtained from wet- and dry-spinning processes.

---

*DuPont ceased production of Orlon in October 1991.

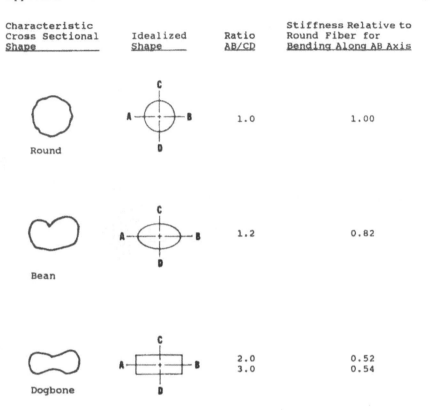

Characteristic Cross Sectional Shape	Idealized Shape	Ratio AB/CD	Stiffness Relative to Round Fiber for Bending Along AB Axis
Round		1.0	1.00
Bean		1.2	0.82
Dogbone		2.0 / 3.0	0.52 / 0.54

Figure 2 Dependency of fiber stiffness on cross section.

Studies by the author also indicate that the lower packing ability of dry-spun fibers contribute to increased comfort in garments during high-stress activities where rapid removal of perspiration from the skin is essential. It is theorized that there are larger spaces or capillaries between adjacent dog-bone fibers compared to round because of their inability to pack closely. This permits larger quantities of moisture or water to move between fibers and away from the skin. The basket sink test, one measure of water transport, confirms the superiority of dry-spun acrylics compared to wet-spun acrylics, cellulosics, polyester, and polypropylene [3]. This is shown in Figure 3, where the amount of water transferred in a gram of fiber per second is plotted against fiber type.

B. Acrylic Fiber Properties Versus Other Fibers

An attempt to qualitatively assess the properties of acrylics versus polyester, nylon, wool and cellulosics has been made [4] and is presented in Table 2. A scale of 5 (highest or best) to 1 (lowest or poorest) was used to assess properties.

Table 1 Yarn Thickness—Wet- versus Dry-Spun Acrylic Fiber

Cross section	Round	Kidney bean	Dog-bone
Spinning process	Wet	Wet	Dry
Fiber denier	3	3	3
Thickness index			
Twist 8.0 tpi	382	442	514
Twist 9.1 tpi	395	432	508
Twist 10.6 tpi	378	439	506
Twist 11.4 tpi	392	443	507
Twist 13.3 tpi	405	447	504
Average	390	441	508
Equated to round			
Cross section	100	113	130

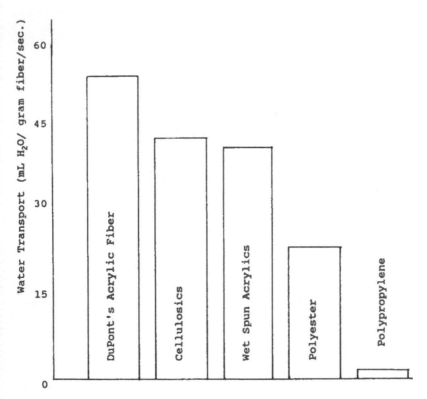

Figure 3 Water transport of various fibers, basket sink test (staple).

Table 2 Qualitative Comparisons of Textile Fibers

Properties	Poly-ester	Nylon	Acrylic	Wool	Cellu-losics
Category 1: Highly Desired					
Abras. resist. (durability)	4	5	3	3	3
Strength (durability)	4	5	3	2	3
Wash-wear performance	5	3	3	1	1
Wrinkle resistance	5	3	3	4	1
Pill resistance	3	1	3	3	5
Category 2: Somewhat Desired					
Bulk (cover)	3	3	5	4	1
Water transport (HS comfort)	3	2	5	3	4
Static resistance	2	1	4	5	5
Speed of drying	5	5	5	1	1
Category 3: Relative. Unimport.					
Resistance to burning	5	5	3	5	1
Res. to degrad. by dry heat	5	3	5	5	3
Res. to degrad. by wet heat	3	4	5	2	4
Res. to degrad. by sunlight					
Outdoors Brt. yarn or nat.	3	3	5	2	3
Beh. glass Brt. yarn or nat.	5	3	5	2	3
Resistance to insects	5	5	5	1	1
Res. to microorganisms	5	5	5	1	1
Avg. ranking, category 1	4.2	3.4	3.0	2.6	2.6
Avg. ranking, category 2	3.2	2.7	4.7	3.2	2.7
Avg. ranking, category 3	4.4	4.0	4.7	2.6	2.3
Avg. ranking, overall	4.1	3.5	4.2	2.8	2.5

The basic properties of the fibers have been translated to end-use performance. Furthermore, the importance of these performance properties to the consumer for apparel have been segregated into three categories: highly desired, somewhat desired, and relatively unimportant. Performance parameters such as abrasion and pill resistance are considered highly desirable by the consumer. On the other hand, resistance to insect attack has insignificant importance to the purchasing decision by the average consumer because, for example, damage of wool sweaters by moths is not a paramount consideration. Other important properties which consumers consider before purchasing, such as aesthetics and costs, were not included in this comparison because they are dependent upon individual preferences.

As shown in Table 2, polyester fibers have the highest performance qualities in the "highly desired" category followed by nylon, acrylics, cellulosics, and wool. Polyester fibers excel in wash-wear performance and wrinkle resistance, have very good abrasion resistance and strength, and adequate resistance to pilling. These properties account for the popularity of polyester fibers in men's and women's apparel and certain segments of home furnishings.

The outstanding abrasion resistance and strength of nylon staple channel it into military uniforms and work pants. These properties combined with good dyefastness to sunlight and adequate resilience provide additional markets for nylon staple in home furnishings, including carpets and furniture covering.

Acrylic fibers have intermediate properties in the highly desirable category, poorer than polyester or nylon but superior to wool and the cellulosics. Acrylic fibers are deficient to polyester in strength, abrasion resistance, wash-wear performance, and wrinkle resistance. The latter two properties can be improved for acrylic fibers by utilizing the greater wrinkle and wash-wear performance inherent in loose-knit constructions compared to the denser, more compact woven structures. This allowance for acrylics combined with their greater bulk, resilience, and preferred natural-like aesthetics compared to polyester, has permitted significant penetration into knit sweaters, socks, fleecewear, and craft yarns. By contrast, acrylic fiber usage in woven fabrics is small, and usually, in blends with better-performing polyester fibers or natural fibers such as cotton or wool to satisfy consumer demands.

Acrylic fibers rank highest in the "somewhat desired" and "relatively unimportant" categories. Some of the properties in these two categories are considered more important for nonapparel markets, such as good resistance to sunlight degradation for drapery or good resistance to degradation by microorganisms for sandbags. Because of this, acrylic fibers have penetrated some of these other end uses.

In summary, acrylic fibers exhibit a well-rounded combination of performance properties with no major deficiencies. However, for woven apparel, the key properties of good wash-wear performance, abrasion, and wrinkle resistance favor polyester fibers. In addition, the higher manufacturing cost of acrylic compared to polyester must be considered. This has reduced market share for acrylics in fleecewear and certain craft yarns such as those used for hooking rugs.

III. TEXTILE PROCESSING OF ACRYLIC FIBERS

A. Introduction

In general, three basic types of acrylic fibers are sold:

1. Tow: a bundle of parallel fibers ranging in total denier from 470,000 to over 1 million. Individual fibers in the bundle range from 2 to over 10 denier per filament.

2. Staple-long: fibers range in length from 3 to 6 inches for processing on the worsted and woolen systems of spinning. Fiber deniers range from 2 to over 10.
3. Staple-short: fibers are sold in lengths from 1 to 3 inches for processing on the cotton and modified American systems using either ring or open end spinning equipment. Fiber deniers range from 1 to 3 or more.

B. Worsted System of Spinning

1. Tow

In the 1950s, it was discovered that acrylic tow could be stretched with heat to impart about 25% residual shrinkage into the fibers. This high-shrinkage component was blended with fibers of low-shrinkage potential and spun into yarn on the worsted system. The low-shrinkage fibers were obtained by relaxing a portion of the stretched high-shrinkage fibers with steam. When the resultant yarn or fabric was exposed to wet heat, the high-shrinkage fibers relaxed, shrank, and moved to the center of the yarn bundle causing the low-shrinkage fibers to buckle and loop to the yarn surface. This process added bulk, loft, and softness to the yarn or fabric. Although yarn weight increased as shrinkage occurred, the increase in yarn diameter was relatively greater, adding bulk and cover to the product. This bulking mechanism is illustrated in Figure 4 for a 1/33 worsted count yarn [5]. The two photographs show the yarn structure before and after it was exposed to wet heat.

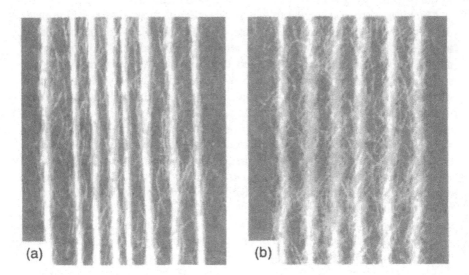

Figure 4 High-bulk yarns of Orlon (a) before and (b) after exposure to wet heat [5].

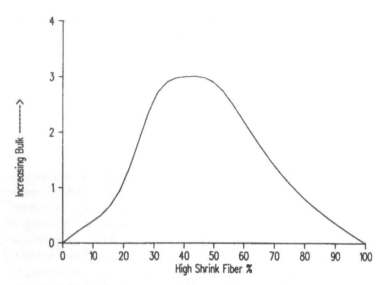

Figure 5 Bulk as a function of content of high-shrinkage fiber [5].

Further studies showed that the maximum increase in bulk was obtained when about 40% of high-shrinkage fibers were blended with 60% of relaxed or low-shrinkage fibers. This is illustrated in Figure 5, which is a plot of yarn bulk vs the percent of high-shrinkage fibers in the yarn blend (5).

Machines were constructed to handle large-denier tow bundles which made the new process economically attractive. The first machine, commercialized in the 1950s, was the Turbo Stapler. It stretches tow about 50% by passing it through electrically heated plates, and then breaking the tow randomly into sliver in a single stage using breaker bars. This produces fibers with a potential shrinkage of 25% and ranging in length from 1/2 to 6 inches with an average length of about 4 inches. After collecting the sliver, 60% by weight of the sliver is steamed relaxed in an autoclave prior to blending with the high-shrinkage sliver. This noncontinuous process introduced shrinkage and dye variations in the sliver. Currently, a J-box is attached to the Turbo Stapler which provides steam to continuously relax the sliver and avoid product variability. A schematic drawing of the Turbo Stapler without the J-box attachment is shown in Figure 6 [6]. A diagram of the J-box, called the "continuous sliver relaxer" (CSR) is shown in Figure 7 [7]. The CSR unit is also used for preparing sliver in 100% relaxed form, such as bicomponent products which rely on crimp, rather than loops, to develop bulk and cover.

True stretch-break machines, employing several stages of stretch to break the fibers were introduced to the trade in the 1970s and 1980s. The tow is repeatedly stretched in progressively shorter zones until the fibers are broken to the

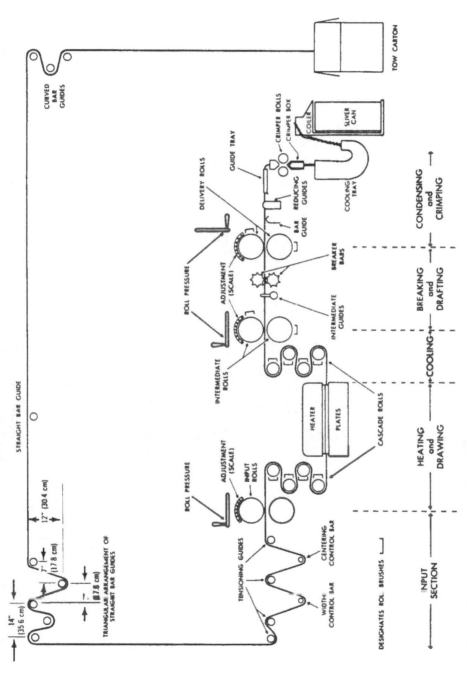

Figure 6 Schematic diagram of Turbo Stapler and creel arrangement [6].

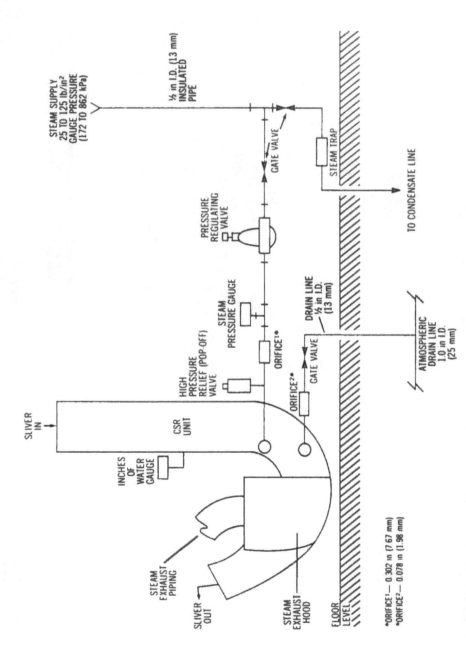

Figure 7 Continuous sliver relaxer [CSR] piping arrangement [7].

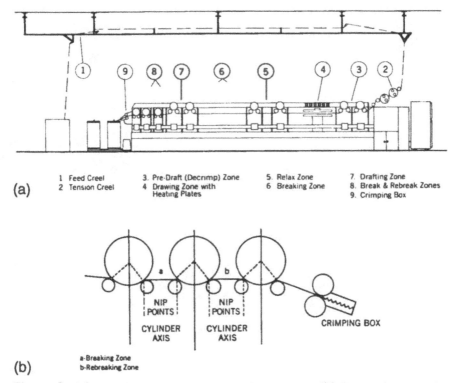

(a)

1 Feed Creel	3. Pre-Draft (Decrimp) Zone	5. Relax Zone	7. Drafting Zone
2 Tension Creel	4 Drawing Zone with Heating Plates	6 Breaking Zone	8. Break & Rebreak Zones
			9. Crimping Box

(b)

a-Breaking Zone
b-Rebreaking Zone

Figure 8 (a) Seydel model 671S stretch-break converter. (b) Ratch adjustments for Seydel converter [7].

desired length. Larger tow bundles can be processed on these machines which increases productivity compared to the Turbo Stapler. CSR units are also provided with stretch-break machines by the vendor. By 1991, most of the Turbo Staplers have been replaced by stretch-break machines; including the Seydel Models 671-S, 677, and 679, the Tematex T-19 and the Duranitre. A schematic drawing of the Seydel Model 671-S is shown in Figure 8 [7]. The CSR unit is not included in this drawing.

The slivers containing either low- or high-shrinkage fibers are subsequently blended at the rebreaker followed by pin drafting, preparation of roving, and the spinning of yarns. A typical layout of this process, from stretch-breaking to yarn spinning and winding is given in Figure 9 [7]. Potential problems and solutions to these problems occurring during stretch-breaking are given in Table 3 [7]. Potential problems and causes in downstream operations are given in Table 4 [8].

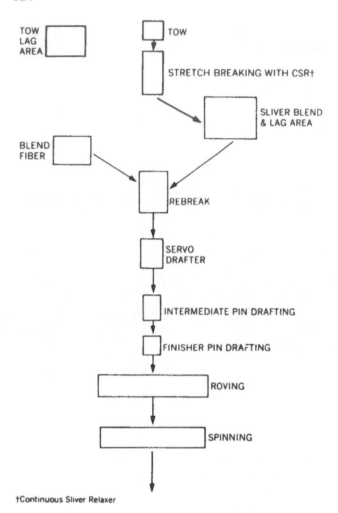

Continuous Sliver Relaxer

Figure 9 Typical processing flow-acrylic stretch/breakable tow, American worsted system [7].

2. Staple

Fibers are sold as staple in bales, similar to cotton, for processing on the worsted system of spinning. The fiber size ranges from 3 to 10 dpf or higher and can be obtained with all of the fibers cut to one length or variably cut to a range of lengths. Cut lengths vary from 2½ to 6 inches or longer.

Acrylic fibers process well on all worsted systems including the Bradford, French or modified. Typical processing sequences for 100% acrylic products or

Table 3 Problems in Stretch Breaking

Problem	Probable cause	Solution
1. Wraps on rolls 1 & 2	Excessive broken filaments in incoming tow	Reduce creel tension Clean creel bars and guides using warm water or finish
2. Poor tow tracking on roll 1	Low creel tension Improper creel alignment Improper tow carton placement Low creel height	Check and adjust for proper conditions
3. Stretch-zone breaks	Low plate temperature Dirty plates Warped plates Tow too narrow	Raise temperature Clean with fine emery cloth or sodium lactate solution Reset or replace Check creel and reset
4. Wraps on rolls 3, 4, 5, & 6	Excessive filament breakage in draw zone Inadequate cooling on rolls Cold tow Damaged or dirty rolls Roll brushes dirty or worn Tow guide spacing improper	Check and adjust heater plates Pre-condition tow 24 hr Inspect, replace, or clean water lines as needed Check water temperature Inadequate preconditioning Replace Replace Adjust
5. Wraps on 7, 8, or 9 metal rolls	Excessive static Deposits	Adjust relative humidity Clean rolls, brushes, and guides
6. Draw-zone breaks	Tow quality Machine malfunction	Adjust creeling to minimize twists or folds and optimize recommended working widths Check guides and rolls for defects Check roll pressure
7. Start-up chops/lumps	Low hydraulic pressure Undersize rubber rolls Irregular breaking	Check and reset Replace with rolls having diameters of 234 mm (1 & 2) and 240 mm (3 to 9) Check metal rolls for condensate formation—reduce cooling water flow if necessary Check where irregular breaking starts Optimize draft

Table 3 Continued

Problem	Probable cause	Solution
8. Wraps on delivery rolls rubber and metal	Excessive back pressure on crimper rolls	Reduce crimper clapper pressure
	Deposits	Remove by cleaning
9. Flaring between delivery roll and crimper on outside edge only	Worn crimper-roll bearing	Replace and realign rolls
10. Misdrafting between rolls 5, 6, and 7	Uneven feed	Adjust guides
11. Poor sliver quality	Nonuniform tow feed	Adjust creel guides
	Nicked or burred crimper surface	Buff with emery cloth
12. Excessive shrinkage variation	Cold machine start-up	Relax all sliver 1st 20 min
	Heater plate temperature out of control	Reset temperature; check setting
	Variable sliver cooling	Temperature of water supply too high
		Sliver-can temperature not to exceed 150°F (66°C)
13. Excessive variation in fiber length distribution	Tow being misfed to machine	Check creel and adjust Set as recommended
	Improper drafts	Lag at least 24 hr in
	Cold tow	conditioned area

blends with other fibers are given in Figure 10 [8]. Problems during carding or downstream processing are similar to those encountered in tow processing and are given in Table 4.

Worsted yarns produced by the staple system of spinning from monocomponent fibers do not have the bulk exhibited by yarns prepared from the stretch-break tow system. Therefore, some fiber producers supply the trade with high-

Table 4 Worsted Spinning System: Possible Causes of Processing Difficulties

Difficulty		Possible causes
Carding	Loading	1. Improper temperature and humidity (static at low humidity; "tacky" fibers at high humidity)
		2. Transfer rolls too close
		3. Insufficient opening of fiber on main cylinder
		4. Low surface speed of main cylinder
		5. High surface speed of stripper rolls relative to worker rolls
		6. Damaged wire clothing

Table 4 Continued

	Difficulty	Possible causes
	Neppy web	1. Damaged or dull card clothing
		2. "Fancy" roll needed for low denier fibers (less than 6.0 den./fil.)
	Flaky web	1. Insufficient carding (set finisher workers closer to main cylinder
		Start with last worker and work forward.)
	Web breakage	1. High tension between doffer and wind rolls
		2. Low fiber cohesion (i.e., insufficient fiber crimp)
Pin-drafting	Roll lapping	1. "Tacky" fibers due to high humidity, or static at low humidity
		2. Excessive oil or finish on fibers
		3. Damaged rolls
		4. Defective faller bars
		5. Fiber buildup
	Uneven pin-drafter sliver	1. Too few pins/inch
		2. Missing or damaged pins
Roving	Roll lapping	1. "Tacky" fibers due to high humidity, or static at low humidity
		2. Excessive oil or finish on fibers
		3. Fiber buildup on roll clearers
	High "ends down"	1. Draft too high
		2. Damaged flyers, or fiber accumulation on flyers
		3. High tension and low twist (reduce tension; increase twist, but not enough to hinder subsequent drafting)
	Uneven roving	1. Unevenness from previous processing operations
		2. Insufficient doublings
		3. Poor drafting due to improper draft, tension, or humidity
	Poor package formation	1. Spindle and flyer misaligned
		2. Improper tension
Spinning	Roll lapping	1. "Tacky" fibers due to high humidity, or static at low humidity
		2. Excessive oil or finish on fibers
	High "ends down"	1. Yarn count exceeds practical spinning limit
		2. Excessive piecing-up of roving
		3. Rough yarn-contact surfaces
		4. Tension, draft, or spindle speed too high
		5. Inadequate lubrication of rings
	Fused fibers	1. Worn or grooved travelers, pigtail guides, or rings
		2. Excessive spindle speed
	Uneven yarn	1. Unevenness from previous processing operations
		2. Poor drafting due to machine conditions
		3. Improper roll settings
	Poor package formation	1. Spindle misaligned
		2. Improper yarn tension

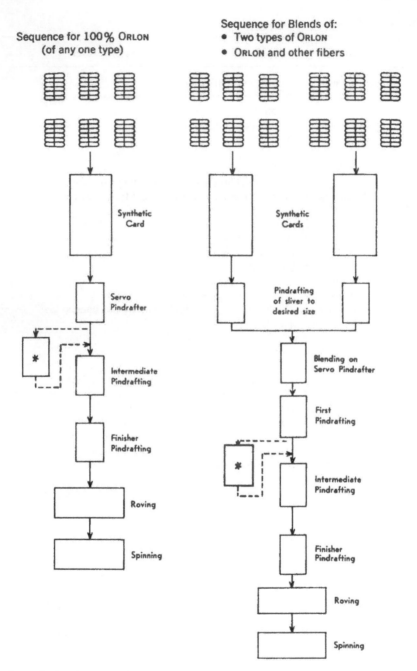

*Optional pindrafting (need determined by evenness of sliver required from the finisher pindrafting).

Figure 10 Typical flow diagrams for worsted system processing of acrylic staple [8].

shrinkage staple products which can be blended with regular staple prior to carding to achieve mixed-shrinkage, high-bulk yarns. Also, bicomponent fibers processed into yarns from staple will crimp during subsequent heat treatments to develop high-bulk and wool-like aesthetics.

3. End Uses

Yarns prepared by the worsted system of spinning, from either staple or tow, are used primarily in the following types of garments:

Sweaters of all weights
Craft yarns
Knit accessories including caps, gloves, and scarfs
Men's, women's, and children's socks
Women's dresses
Knitted sportshirts and sweatshirts and
Sleepwear

In the first four applications, acrylics are mostly used in 100% form. In the latter three end uses, acrylics are frequently blended with polyester, rayon or natural fibers.

C. Cotton System of Spinning

1. Ring Spinning

Acrylic fiber producers supply fibers ranging from 1 to 5 dpf for the cotton system of spinning. The heavier deniers are usually blended with finer deniers to aid processing and to produce a crisper hand. Fiber lengths are supplied over the range of 1 inch to 2½ or 3 inches.

A schematic drawing of the typical operations required for ring spinning is shown in Figure 11. Processing steps include opening, blending, picking, carding, drawing, roving, and yarn spinning. Temperature and relative humidity are carefully controlled in all areas of operation. Low humidity creates problems associated with static buildup and high humidity causes tackiness with the finish.

Yarns as fine as 40/1 cotton count are produced from 1.5 denier staple fibers. Based on cotton count, twist multipliers of 2.75 to 3.25 are recommended for knitting yarns, 3.00 to 3.50 for woven filling yarns, and 3.50 to 4.25 for woven warp yarns.

2. Open-End Spinning

Since the open end spinning process is relatively new, being introduced in the mid-1970s, a short description of the process is provided. The fiber feed section consists of a feed table and feed roll which nip the sliver and present it to the opening roll at a constant speed. Negative air pressure lifts the fiber from the opening roll and accelerates the fiber through a funnel-like transport channel.

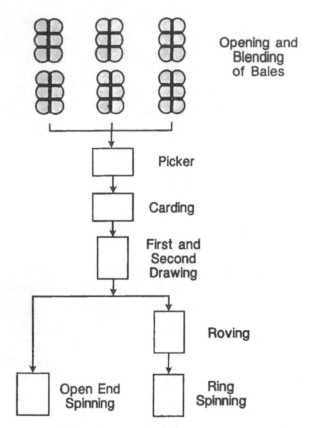

Figure 11 Typical operations required for ring- and open-end spinning.

The acceleration causes the fibers to be stretched and deposited on the rotor wall in an uncrimped state, centrifugal force causes the fibers to slide down the rotor and collect in the rotor groove. The fibers collect until the number required for the yarn count is reached. The rotation of the yarn by the rotor imparts twist to the fibers to form a yarn. Yarn is then removed at a constant speed through a funnel, commonly called a navel, and a yarn delivery tube by delivery rollers and wound on a package. This is shown diagrammatically in Figure 12 [9]. The roving step is omitted for open-end spinning. For fine count yarns, conventional two-pass drawing is recommended while one-pass drawing may be acceptable for coarser count yarns.

Open end spun yarns are more uniform than ring spun and provide 10–15% more bulk. Twist multipliers of 3.8 to 4.2 based on the cotton system are suitable for most knitting operations. However, open end spun yarns are more prone to

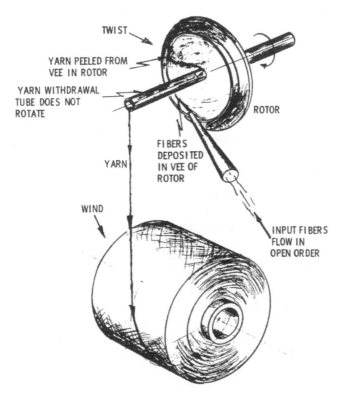

TWIST

YARN PEELED FROM
VEE IN ROTOR

YARN WITHDRAWAL
TUBE DOES NOT
ROTATE

ROTOR

FIBERS
DEPOSITED
YARN   IN VEE OF
ROTOR

WIND

INPUT FIBERS
FLOW IN
OPEN ORDER

Figure 12 Open-end spinning process. (Courtesy of P. R. Lord [9].)

pilling and higher twist multipliers may be required for some applications. Twist multipliers can be increased to 5.5 without significant loss in tensile strength.

3. Air Jet Spinning

Air jet spinning was introduced by Murata in the 1980s; it has already made a substantial impact in polyester. Its virtues are lower cost and improvement in pilling performance. It does tend to impart a harsher hand. In acrylics, programs are just now (1993) starting. Both singles yarn and plied yarn may be produced. To combat the harshness, fiber deniers of less than 2 down to microdenier are being employed.

4. End Uses

Yarns prepared by the cotton system are usually finer than yarns made by the worsted and woolen systems. Cotton spun acrylic yarns and blends with other fibers are knitted or woven into fabrics for

Women's shirts, dresses, and slacks
Men's sport and dress shirts
Brushed sweatshirts
Socks
Sleepwear
Lightweight sweaters and blouses
Men's woven slacks and suits
Sliver knit artificial fur-type fabrics for coats, hats, jackets, and slippers and
    linings for coats, jackets, gloves and robes.

D. Woolen System of Spinning

Fiber producers usually offer blends of various deniers and cut lengths for the
woolen system of spinning. Average deniers can range from 2 to 8 with cut
lengths from 2 to 3 inches. Waste products from various sources are also used in
this system.

The baled fibers go through opening, blending, picking, carding, drawing,
and spinning operations. The process is limited to coarse yarns usually heavier
than 1/20 worsted count. Yarns are less uniform than comparable worsted spun
yarns because blending and sizing passes are limited.

Yarns are used where soft aesthetics and good cover are required and durability, strength, and surface distortion are not important requirements. This
includes loop knits in 100% form and woven suitings and overcoats using
blends.

IV. KNITTING AND WEAVING

Acrylic yarns can be knit successfully on all types of weft knitting equipment.
Weft knitting is a type of knitting in which yarns run horizontally, from side to
side, across the width of the fabric. All stitches in a course (horizontal yarn) are
made by one yarn. In its simplest form, a weft knit can be made from one yarn,
such as hand knitting.

Weft knits are made on either flat knitting machines or circular knitting
machines. A special type of flat knitting is the full-fashioned knitting machine.
This machine produces the shaped garment by increasing or decreasing the
number of wales (vertical loops) per inch in the shaped area. This causes fash-
ioning marks which appear as distorted stitches. This adds prestige to the
garment. Acrylic yarns prepared by the high bulk, stretch-break principle were
very popular in piece dyed full-fashioned sweaters in the three decades preceding the 1980s.

Another special weft knitting technique is the so-called sliver knit or high-
pile knit fabric. Staple fiber, in the form of sliver, is fed into the knit structure as
the backing yarn is passing through the knitting needle. The fiber is caught in

the knit structure and is held between the stitches. After knitting, special finishing treatments are applied to produce a furlike effect. These are known as fake or man-made fur fabrics and are used in coats, coat linings, children's snow suits and hats. Some modacrylic fiber is used in this application to produce a natural guard hair effect by differential shrinkage when blended with acrylic or other fibers.

Acrylic yarns have sufficient strength to weave successfully for either warp or filling applications. Warp yarns must be slashed correctly with the proper size for good weaving efficiency. Since acrylic fibers do not have the inherent wrinkle resistance and wash-wear properties of polyester fibers, they are normally blended with polyester and other fibers for woven slacks, shirts, and suiting materials.

V. PROPERTIES OF ACRYLIC FIBERS BY END USE

A. General

Acrylic fibers convey warm, natural-like aesthetics to most fabrics as opposed to the cold plastic handle often mentioned by consumers in reference to polyester and nylon fibers. However, polyester fibers have inherent properties which produce fabrics and garments with outstanding wrinkle resistance, crease retention, and wash-wear properties. These preferred properties are developed by the ability of polyester fibers to be permanently heat-set. Acrylic fibers cannot be effectively heat-set. In addition, polyester fibers can be permanently heat-set repeatedly by exceeding the previous setting temperature. This leads to permanent pleats in garments and set textured filament yarns that can be heat-set again in fabric or garment form. In addition to natural aesthetics, high bulk is inherent to acrylic fibers because of their low specific gravity and packing factor. Also, product variants, such as bicomponent fibers or textile processing modifications, such as mixed shrinkage yarns obtained by stretch-breaking, further improve the inherent high bulk of acrylic fibers. Acrylic fibers can also be economically dyed to bright washfast colors suitable for the style changes encountered in the apparel markets.

In general, knit fabrics, because of their loose construction and the ability of the loops to move and adjust in the fabric matrix, have better wash-wear performance and wrinkle resistance compared to woven fabrics. Therefore, acrylic fibers with their poorer wash-wear performance compared to polyester fibers have gravitated to knit fabrics, while polyester fibers have captured a majority of the woven fabric markets in 100% form or in blends with rayon, cotton, wool and acrylic fibers.

In the United States, nearly 80% of acrylic fibers were shipped to apparel markets in 1989 with the remainder consumed by the home furnishings and industrial segments [10]. Of the apparel portion, as shown in Table 5, more than

Table 5 U.S. Shipments of Acrylic and Modacrylic Fiber for Apparel Markets (millions of pounds)

Year	Sweaters	Craft yarns	Socks	Pile fabrics	Fleecewear and others	Total
1975	67	85	42	71	72	337
1980	88	94	93	60	138	473
1985	76	59	62	24	145	366
1986	106	55	68	23	203	455
1987	96	45	51	21	193	406
1988	94	44	45	21	149	353
1989	91	42	42	20	142[a]	337

[a]Author's estimate of 10–20 million pounds in nonfleecewear markets.

90% of acrylic fibers were consumed in knitted fabrics. These included sweaters, socks, craft yarns, pile fabrics, and fleecewear.

B. Sweaters

1. Monocomponent Fibers

Domestic consumption of all fibers for sweaters grew at an average annual rate of 1.7% from 1976 to 1983. By 1983, acrylic fibers accounted for 72% of the total fibers consumed. During the next several years, style changes required big bulky sweaters which increased the demand for cotton and imported sweaters including the ramie/cotton blend from the Far East. The demand for acrylics peaked at 106 million pounds in 1986 (see Table 5). The recent trend back to lighter-weight sweaters plus further inroads by cotton have reduced market share for acrylic fibers.

Monocomponent acrylic fibers are prevalent in all types of sweaters, from lightweight jersey pullovers to bulky cardigan jackets. Staple and tow products, ranging from 1.5 to 10 dpf, in both bright and semidull lusters are produced to satisfy these diversified markets. The majority of yarns for sweaters are made by the worsted system of spinning. Some cotton-spun yarns prepared from 1.5 to 2.5 dpf fibers are consumed in lightweight circular knits for "cut and sew" sweaters.

Dye is imparted into the fibers by various means, mainly package, skein and piece dyeing. Other dye routes include stock and sliver dyeing and the use of producer colored fibers. Acrylics can be dyed with disperse dyes, used mainly for light and medium shades, and cationic (basic) dyes. These dyes convey good wash and lightfastness.

Sweater aesthetics can range from soft, pliant, and cashmere-like to crisp, bulky, and mohair-like by selection of the proper denier, spinning system and finishing routes. Performance properties, such as wash-wear, wrinkle resistance, durability, and dimensional stability are acceptable. Surface distortion, including pilling, is noticeable in some constructions. Fiber producers are currently working on variants which are less prone to pilling (see Chapter 6).

2. Bicomponent Fibers

To further enhance the diversity of acrylics for sweaters, fiber producers introduced product variants starting in the 1960s. These included bicomponent, acid dyeable, and producer colored fibers. These are discussed in detail in Chapter 6. The most successful bicomponent fiber for knitwear is DuPont's type 21 Orlon which is manufactured from two polymer solutions fed, side by side through each spinnerette hole. One polymer is more sensitive to water than the other. Thus one polymer of the bicomponent structure shrinks more than the other when exposed to wet heat and dried. The more shrinkable component takes the shorter path in the fiber structure causing the fibers to coil or crimp. This is a reversible change, the fibers elongate and lose crimp when wet and shrink and recrimp when dried. This behavior is similar to wool fibers and is designated water-reversible crimp.

The crimp imparts wool-like aesthetics to knitwear. By varying fiber denier from 3 to 10, wool fineness from 70 quality (lamb's wool) to 48 quality (Shetland wool) can be achieved. The crimp also increases the bulk, loft, resilience, and elasticity of knitted fabrics.

Bicomponent fibers tend to pill more compared to monocomponent fibers because of crimp development. This crimp permits greater entanglement of adjacent fibers on the fabric surface, thus creating a nucleus for pill formation. The trade has alleviated this problem by blending the bicomponent fiber with monocomponent fibers to reduce pilling propensity while retaining most of the desired wool-like properties. In addition, fiber producers have reduced the strength of the fiber in some cases so that these weaker fibers tend to break off the fabric surface prior to forming objectionable pills.

Since the bicomponent mechanism is reversible, sweaters tend to elongate when wet during laundering. The consumer must block the sweater for drying, similar to a wool garment, or tumble dry the sweater to return it to its original dimensions. This is not a serious problem in the United States where most consumers have tumble driers.

Bicomponent fibers with irreversible crimp are prepared by forming filaments using side-by-side spinning of solutions of two polymers which have major differences in shrinkage potential due to comonomer composition. Crimp is formed during the first exposure to high temperature in either the wet or dry state. This usually occurs during dyeing in the mill. Since the temperature is not exceeded by the consumer during home laundering, the crimp remains constant and does not elongate in the wet state. However, the amount of crimp developed during dyeing depends upon the restraints imposed on the fiber. As a result, package dying, where the yarn is under tension, produces low crimp and, thus, bulk. Piece and skein dyeing routes are preferred to optimize bulk development, which is lower than comparably dyed bicomponent fibers with reversible crimp. These limitations have restricted usage in the United States to specialty-type applications including blends with polyester fiber for woven apparel.

Today, most bicomponent fibers are sold as staple for processing on high productivity cards via the worsted system of spinning. Stretch-break machines, with lower productivity, are not utilized extensively for bicomponents since mixed-shrinkage yarns are not required for bicomponent bulk development.

3. Acid Dyeable Fibers

The acid dyeable acrylic fiber is used almost exclusively in yarn or fabric blends with cationic dyeable mono- or bicomponent fibers to achieve cross-dye effects for styling including heathers and stripes. A one-bath/two-step dye procedure is normally used. The fiber is sold as tow or staple for both the worsted and cotton systems of spinning. Textile processing characteristics and end use performance is similar to the cationic dyeable monocomponent fiber.

4. Producer-Colored Fibers

Producer-colored fibers are provided in the same range of deniers and cut lengths as undyed fibers. Product is provided as both staple and tow for the cotton and worsted systems of spinning. Producer-colored fibers are used to reduce dyeing and finishing costs, improve dye uniformity, and to achieve greater styling versatility through color combinations. However, early commitment by the trade to color shade is required. Stringent measures in the yarn spinning plant are required to avoid contamination and inventories are more complex than for products dyed later in the production process.

5. Other Variants

Recently, several fiber producers have introduced acrylic microfibers (0.5 to 0.9 dpf) for supersoft aesthetics in knitwear. One such product is MicroSupreme™, a 0.9-dpf product by Cytec (formerly American Cyanamid). Greater yarn strength is also obtained because of the increase in the number of fibers in a cross section of yarn. These microfibers are being spun on both ring and air jet spinning frames.

C. Craft Yarns

Most fiber requirements for craft yarns are similar to sweaters, since most consumers hand-knit craft yarns into sweaters or similar accessories such as hats, scarfs, gloves, and afghans. In general, the same fiber types are consumed in hand-knitting yarns as in sweaters. However, since the fiber is in yarn form at point of sale, dye and bulk uniformity from package to package is of utmost importance to the consumer.

Until the advent of bicomponent acrylic fibers in the 1960s, wool was the principal fiber consumed in hand knitting yarns. Its elasticity and ability to recover from stretch provided easy formation of loops on the knitting needles and uniform stitches in the fabric. Bicomponent acrylic fibers with their spiral crimp enhanced these desired wool properties at a lower cost combined with greater styling potential, resistance to damage by moths, improved dimensional stability and wash-wear properties. By 1975, about 50 million pounds of bicomponent acrylic fiber was consumed in craft yarns in 100% form or in blends. Another 30 million pounds of monocomponent acrylic fiber was used in blends with the bicomponent fiber. This 80 million pounds of acrylic fibers replaced most of the wool and significantly increased overall consumption of fibers for craft applications. The dominant United States fiber for craft yarns was Orlon (now Acrilan) 21, sold at retail under the certification marks Sayelle (100% form) and Wintuk (60–80% type 21 plus approved monocomponent). Private label blends containing less than 60% type 21 are also important in the market.

Today, acrylic fibers continue to dominate the hand-knitting yarn portion of the total craft market but shipments are down somewhat from previous years as shown in Table 5. This is caused, primarily, by the reduced discretionary time available to women working outside the home.

Other segments of the craft yarn market do not require the elasticity and recovery from stretch offered by the crimped bicomponent fibers. These areas include yarns for hooking rugs, needlepoint and crewel. These markets are dominated by cotton, rayon, and polyester.

D. Socks

United States shipments of acrylic fibers for half hose peaked in 1980 at 93 million pounds (Table 5). This accounted for 51% of total domestic fibers consumed in socks. By 1988, acrylic consumption had declined to 23%, while cotton's share had increased from 17% to 46% [11]. This trend has continued through 1991.

Acrylic fibers are found in athletic socks, both tube and with the heel knitted into the sock, anklets, men's dress and sport socks, and women's leg warmers. Most socks contain 15% to 30% of set textured filament nylon or polyester which is plaited during knitting to obtain stretch for a multisize sock. For example, men's socks prepared in this manner will fit sizes 9 through 13.

Most moderate to high-priced acrylic socks use high bulk fibers prepared by the stretch-break, mixed shrinkage principle to give bulk, loft, and softness to the product. Less expensive acrylic socks are knit from cotton spun yarns, and thus, look and feel more like a cotton sock. Most socks are piece (paddle) dyed.

Small quantities of acid dyeable acrylic fiber are used to obtain argyle color effects when piece dyed. The acid dyeable yarn is knit in combination with cationic dyeable acrylic to obtain the desired effect.

The market share for acrylic fibers could reverse itself in the future. Acrylic socks have shape and color retention when laundered that is superior to cotton. Brighter shades can be obtained and durability is comparable to cotton. Acrylics offer greater styling versatility by offering high-bulk yarns, acid dyeable, and producer-colored products. In addition, numerous tests have shown that acrylic fibers transport more moisture away from the skin compared to cotton or wool. Cotton and wool absorb and hold moisture near the skin. This advantage for acrylics increases wearer comfort and reduces formation of blisters during high-stress conditions [12].

One acrylic is available that has been specially engineered for socks: Monsanto's S-18 sold under the trade name Duraspun (formerly Acrilan II). This is a tow product suitable for Seydel stretch breaking and is claimed to wear 25% longer than conventional acrylics [13].

E. Pile Fabrics

The shipment of acrylics for pile fabrics decreased dramatically from 1975 to 1985 but has stabilized at 20 to 24 million annual pounds since that time. Polyester now has dominant market share. These high-pile fabrics are used for several products including women's coats and jackets, furlike linings for coats, hats, shoes, boots, and gloves. The majority of these pile fabrics are referred to as imitation or fake furs. They are prepared on a circular knitting machine which involves feeding staple fibers in the form of sliver into a jersey knit structure. The sliver is caught in this structure and becomes the pile yarn. This pile yarn is usually acrylic, modacrylic or polyester. The knit backing yarn is usually cotton or polypropylene. The acrylic and modacrylic fibers in the pile are monocomponent and range in denier from 3 to 10 or higher with fiber lengths below 3 inches.

Acrylic or modacrylic fibers with high shrinkage potential (20–30%) are supplied for the pile. By blending with a regular low shrinkage acrylic fiber, an animal hair simulation is obtained during a simple scouring of the fabric. The high-shrinkage fiber shrinks and becomes the underfur and the low-shrinkage fiber does not shrink appreciably and simulates the guard hair of an animal fur.

F. Fleecewear and Other Apparel

The United States shipment of acrylic fibers for fleecewear also peaked in 1986 and has decreased substantially since that time (Table 5). In 1987, acrylic fleecewear began to face serious inroads from blended polyester/cotton yarns. Consumer interest in natural fibers and increased marketing efforts by polyester fiber producers contributed to this loss of market share. Fleecewear is utilized primarily in sweatshirt products for both activewear and leisurewear.

Acrylic fibers are expected to remain competitive in dyed fleecewear where color fastness is superior to cotton and bright coloration can be achieved at lower cost than polyester. On the other hand, polyester/cotton fleecewear is whiter and does not yellow in drying after printing. Relative cost of the fibers involved will probably influence future consumption.

Acrylic fibers consumed in fleecewear usually range from 1.2 to 3 denier per filament and are cut from 1.37 to 3 inches in length for processing on either the cotton, ring, or open-end systems of spinning. Open-end spinning, because of lower manufacturing cost, is becoming the predominant route used for the low-cost fleecewear market. Both bright and semidull lusters are used in this market. Producer-colored fiber is also used, primarily in children's sweatshirts which are mostly colored.

Non fleecewear markets included in this category are single- and double-knit fabrics for sportshirts, dresses, and skirts and woven fabrics for slacks. Estimated usage for these end uses was 10 to 20 million pounds of the 142 million pounds consumed in 1989.

REFERENCES

1. duPont Co., unpublished study.
2. W. J. Onions, E. Oxtoby, and P. P. Townend, *J. Textile Inst.*, *58*, 293 (1967).
3. *Manufacturer's Journal*, Absorption Rate, Imbibition Test, EP-SAP 2301 (Basket Sink Test), June 18, 1980.
4. Author's assessment of fiber properties.
5. *Orlon Bulletin OR-183*, E. I. duPont de Nemours & Co. Inc., June 1974.
6. *Orlon Bulletin OR-194*, E. I. duPont de Nemours & Co. Inc., June 1978.
7. *Orlon Bulletin OR-199*, E. I. duPont de Nemours & Co. Inc., November 1980.
8. *Orlon Technical Bulletin 180*, E. I. duPont de Nemours & Co. Inc., December 1972.
9. P. R. Lord, *Spinning: Conversion of Fiber to Yarn Part 1*, School of Textiles, North Carolina State University, Raleigh, 1978, p. 110.
10. *Chemical Economics Handbook*, SRI International, 1990, p. 543.3000N.
11. Private communication with duPont Co.
12. G. Pontrelli, *Clothing Comfort*, (Hollies and Goldman, eds.), Ann Arbor Science, 1977, p. 71.
13. R. D. Borie, Jr., Acrilan® II: Second Generation Acrylic Fiber, Natural and Man-Made Fiber Forum, Clemson University, 1986.

11

Home Furnishings and Industrial Applications

James C. Masson

JCM Consulting
Mooresville, North Carolina

I. HOME FURNISHINGS APPLICATIONS

A. Introduction

This sector of the acrylic fiber market has undergone a marked decline in the past 20 years, particularly in the United States. The major volume loss has been in carpets, where nylon and polyester have almost completely displaced acrylic from the market. This loss has been mainly due to nylon's ease of dyeing in both beck and continuous ranges. Other markets such as blankets and draperies have suffered as well. The lone bright spot for acrylics is the upholstery market where, as noted in Chapter 2, U.S. consumption has increased from 9 to 20 million pounds in the period 1980–1991. Europe also witnessed the decline of the acrylic carpet market. However the overall importance of the home furnishings sector in Europe is much higher than in the United States. Japan has the most robust acrylic home furnishings and carpet business; it outpaces the apparel sector by several percent.

Acrylics have much to offer for home furnishings: ability to dye to brilliant shades, good dyed lightfastness, good abrasion resistance, excellent property retention in environmental exposure, high resistance to staining and good cleanability. Acrylics' weaknesses compared to other synthetics are deformation under hot-wet conditions, as in dyeing, poor high-temperature color stability, and higher cost.

Producer-dyed acrylics are now available for upholstery and carpet end uses. The problem of hot-wet deformation during dyeing is avoided, but the mill must inventory colored fabric, rather than dye to order. However, producer-dyed fibers also allow subtle heather effects through the blending of several fiber colors. For carpet styles such as Berber, which introduce small "splashes" of color on a neutral background, a producer-dyed product is ideal. Upholstery, especially monochromatic velvet styles, puts an extreme requirement for lot to lot and within lot color uniformity on the producer dyed product to eliminate streaks from light or dark ends.

A part of the overall decline in home furnishings uses of acrylics is the reduction in demand for fire-retardant modacrylics, particularly in the drapery market. Handcraft yarns could arguably be considered part of the home furnishings market, since the most popular end products are Afghans and hooked rugs, but we have chosen to count and discuss them as apparel in Chapter 10.

B. Applications

1. Upholstery

Requirements

Velvet Upholstery Velvet upholstery produced by tufting, knitting, or weaving has been an acrylic end use for many years, whereas flat woven goods (especially in the United States) are a newer end use. Velvet upholstery tends to incorporate 2 to 3 dpf bright fiber which is ring, open-end, or wrap-spun. Wrap spinning is a novel development in which an acrylic sliver is converted to yarn with a nylon or polyester monofilament wrapper yarn. For open end spinning, 3.3 dtex (3 dpf) 60 mm (2.37 inches) fiber is spun to yarns of 83–250 tex (Ernst, 1989). Ring-spun counts of 20/2 c.c. are commonly employed for upholstery velvets. In wrap spinning, a singles yarn such as 10/1 c.c. may be produced.

Using yarns produced by any of the above methods, velvet fabrics are produced by weaving using a warp of the acrylic effect yarns plus two warps of a backing fiber such as filament polyester. The filling, also filament polyester, is inserted between each of the polyester warps, while the acrylic yarn traverses the upper and lower filling yarns. Immediately after formation, a knife cuts the acrylic yarn between the upper and lower polyester structures, forming two velvet fabrics. Figure 1 shows a cross section of the fabric prior to slitting. Alternatively the acrylic yarn may be tufted into a backing made of another fiber. Both wrap-spun and ring-spun yarns are employed in tufting.

Open-end or ring-spun acrylics may utilize producer-dyed fiber or ecru, in which case the yarn is package dyed; all wrap-spun yarns employ producer-dyed acrylic. A white nylon wrapper is used in light and medium shades which gives somewhat of a frosted effect to the fabric. A black nylon wrapper is used on

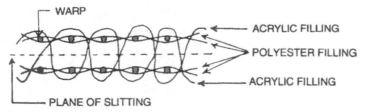

WARP

ACRYLIC FILLING

POLYESTER FILLING

ACRYLIC FILLING

PLANE OF SLITTING

Figure 1 Cross section of woven velvet prior to slitting.

dark colors to eliminate the frosting. Dye uniformity and yarn uniformity are critical since streaks from light or dark dyeing ends, or ends with different yarn weights are readily visible in velvets. The main selling point in velvets is a rich luster and pleasant hand; properly processed acrylic fabrics meet these criteria well. Most acrylic upholstery in the United States goes into the domestic furniture market, but a small amount is finding its way into the recreational vehicle and van conversion markets. Pigmented acrylic fiber would appear to be a natural for automotive upholstery because of the high lightfastness, but acrylic does not meet Detroit's requirement for heat stability.

Flat Woven Upholstery Flat woven upholstery is a newer end use for acrylic fibers, especially in the United States. Acrylics are used as effect yarns in the filling. Previously resistance to acrylics for this end use was based on perceptions of low abrasion resistance, high pilling, and high cost. In this end use, both 3 and 5 denier, bright fibers, mostly ecru, are used; they are converted to yarns by ring spinning, a typical yarn count being anywhere from 10/1 c.c. to 400 yd/lb (about 1/2 singles). The yarns are package dyed and woven as filling on Jacquard looms; warp yarns are mainly polyester and polypropylene. The acrylic may make up as little as one fourth of the total filling.

Performance In the United States, an industry standard for rating upholstery yarns is emerging. The tentative classifications are (Burleson, pers. comm.)

Cycles, Wyzenbeek abrasion test	Duty	Uses/week
<3000	Delicate	1–3
3000+	Light	3–5
9000+	Medium	4–7
15000+	Heavy	Daily

The fabric to be tested is rated at each of these levels using the rating scales shown in the table. To be acceptable for that duty, no appreciable wear at the appropriate number of cycles is required. However, the actual rating at which a mill will fail fabrics varies from mill to mill.

Rating	Velvet	Flat woven
1	Backing showing	10+ breaks
2	Slight backing visible	5–9 breaks
3	Noticeable wear	<5 breaks, noticeable wear
4	Light wear	Slight wear
5	No wear	No wear

Most commercial acrylic fabrics qualify for light or medium duty, but it is possible by fabric construction, fabric resin treatment, or fiber structural modification to prepare fabrics which meet the heavy-duty criterion.

Modacrylic yarns do not have any significant sales for upholstery in the United States. However, Teklan SRB, a modacrylic formerly produced by Courtaulds in Coventry, England has been reported (Wheeler et al., 1991) to be suitable for both domestic and contract upholstery in the British market. Table 1 shows the constructions and fabric results for domestic weight upholstery. The flammability tests are conducted with the fabric over a standard upholstery foam. Although cotton has a much lower limiting oxygen index (18 vs. 32) than the Teklan SRB, it has the ability, in combination with the modacrylic, to form a char and thus protect the foam. Passing the crib 5 test is not required for most domestic upholstery in Great Britain (Eaton, 1992). Heavier weight fabrics of Teklan FRB in combination with flame-retarded viscose rayon and either rayon or wool were shown to pass the British flame-retardance requirements for contract upholstery (Wheeler et al., 1991). They also had high abrasion resistance.

2. Draperies

Requirements Acrylic fibers have excellent potential in draperies due to their good strength retention on exposure to light (Fig. 2) (Monsanto, n.d.) compared to other fibers commonly used in window coverings. Other favorable features include low specific gravity, leading to good cover per unit weight, excellent drapability, and excellent dimensional stability to washing (Monsanto, n.d.). Despite these favorable factors, acrylic draperies have not made great inroads in the United States market, where rayon and polyester are preferred. In Europe acrylics are much more important in this market segment. More detail was given in Chapter 2.

Acrylic fibers used in draperies range in denier from 2 to 8, with the 2 to 3 denier products predominating. Bright luster is preferred for most drapery products. Yarns may be ring or open end spun in a fairly wide range of counts. Package dyeing is the preferred coloration route, although some may be beck dyed (open weave types) or continuous piece dyed. Some producer-dyed fiber is also employed. Most acrylic drapery fabric is woven onto a filament polyester warp.

Table 1 Development of Domestic Upholstery Fabrics in Teklan SRB Modacrylic

#	Weave	Construction Warp[a]	Weft	Composition cotton/Teklan	Fabric wt (g/m²)	Flammability BS 5852 Cigarette	Match	Crib 5	Abras. Resist.[b]
1	Plain	2/25 cotton	1/97 Teklan	40/60	320	Pass	Pass	Fail	23,750
2	2 × 2 Twill	2/25 cotton	1/97 Teklan	42/58	310	Pass	Pass	Fail	19,500
3	2 × 2 Twill	2/25 cotton	1/105 Teklan	39/61	329	Pass	Pass	Fail	11,000
4	2 × 2 Twill	2/25 cotton	1/105 Teklan	38/62	342	Pass	Pass	Fail	14,200
5	2 × 2 Twill	2/25 cotton	1/97 Teklan	31/69	437	Pass	Pass	Fail	16,000

[a]Yarn counts in Tex.
[b]Martindale test.
Source: Wheeler et al. (1991).

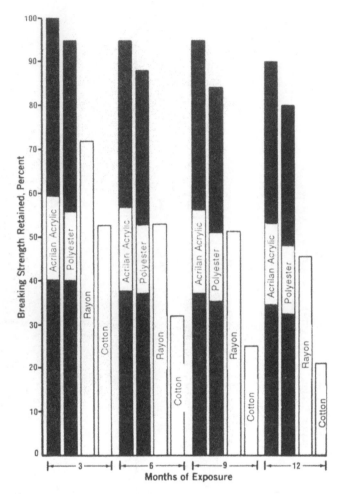

Figure 2 Comparative resistance of drapery fabrics to sunlight (under glass).

Styles include antique satins, leno and patterned jacquard constructions. Acrylic fiber is not significant in prints.

Modacrylic fibers are also used in draperies where fire codes require a flame-retardant (FR) material. In this application they compete with FR polyester and FR rayon. Construction of the fabric is critical in ensuring that the assembly of face fabric, liner, and thread meet the FR requirements. Most modacrylics are relatively dull as a result of the presence of the antimony trioxide synergist. Several modacrylics, such as Montefibre's Velicren and Monsanto's B06 maintain better luster by use of colloidal antimony pentoxide as the flame-retardancy synergist.

Performance Resistance to degradation from sunlight exposure is the most outstanding characteristic of acrylic fibers. This advantage is retained even behind glass, as draperies are typically exposed (Fig. 2) (Monsanto, n.d.). After 12 months exposure, the acrylic fabric had suffered only a 10% strength loss, polyester was second best at 20% loss, while cotton had lost 80% of its strength.

Modacrylic draperies are used for public buildings where flame retardancy is required to meet local fire codes, which in the United States are based on the National Fire Protection Association 701 regulation (NFPA, 1977). In this test, well-engineered modacrylic fabrics, or blends of modacrylic with fibers such as polyester will "pass." In actual use, drapery fabrics are combined with sheers, light blocking fabrics, etc., to form a layered composite. Testing of these layered composites showed that where char-forming fabrics such as modacrylic or cotton, or nonflammable fabric such as fiberglass was layered with a meltable fabric such as polyester or FR-polyester, the result was unexpectedly flammable in a large scale test. This result was confirmed in room tests at the Southwestern Research Institute (Gustafson, 1988).

3. Pile

Requirements An end use of long standing for acrylics has been in pile, as fake fur and coat linings. Other niche markets include robes, toys, automotive after-market seat covers, bedroom slippers, paint rollers, and golf club head covers. The main attribute for this use is the ability to be polished to present a smooth lustrous appearance and to have a luxurious hand. Some acrylic manufacturers produce specialty fibers which have only a weak, easily removed crimp so that they will polish under mild conditions. Both ecru and producer-dyed fiber are sold into this market. For many uses, a fiber of standard cross section is suitable, but specially shaped fibers with rectangular cross sections and aspect ratios (length to width) of up to 5:1 are utilized to give more exotic hand effects, simulating animal fur. The majority of the fiber is mid-denier (ca. 3) but for certain applications, blends including micro (0.9) denier have been reported used. Staple length used in pile applications is about one inch (25 mm). Polyester is a strong competitor in this market, and in the United States, has somewhat more than 50% of the volume.

Pile producers either purchase producer-dyed fiber, which may be a special merge for this end use such as Monsanto's B-37, or stock-dye ecru fiber. A heavy card sliver is then produced which is then fed to a Wildman type knitting machine together with a filament polyester or polypropylene. A tubular double-knit fabric is produced. The machine may have the ability to produce a patterned knit. The fabric is slit, then backcoated on the filament side to lock in the acrylic sliver. The acrylic side is then sheared to cut the loops and present a fabric with the tips perpendicular to the backing. Polishing with a heated roll finishes the fabric with a uniform appearance of pile lay.

It is possible to produce fabric with two levels of pile, copying the under-fur found in nature, by using a blend of fibers with different shrinkage properties, but a wet-processing step after knitting would be required to activate the shrinkage.

Performance The main attribute of pile is aesthetic: visual appearance and hand. The ability of a fiber to polish to give a uniform luster and pile lay without blotches is important to a quality pile. Certain fibers may give the appearance of more depth which is interpreted as richness of color; this is a plus in velvets as well as pile. As previously noted, the polishing operation must remove the crimp from the fiber ends to achieve the right look; polishing depths of about 1/8 inch (3 mm) are employed.

Hand is the other important attribute of pile. For many end uses, softness with some resiliency ("body") is desired. To achieve drastically different hand effects, such as the higher friction of some animal furs, shaped fibers with rectangular or triangular cross sections are required, but such fibers sell at a considerable premium and therefore are only justified for premium fabrics. Fiber blends, including fine denier, and special producer finishes are the more common ways of achieving a desirable hand.

4. Blankets

Requirements Acrylic and modacrylic fiber has been used in the blanket trade for many years. Today the competition is severe from polyester and particularly cotton, with fiber price being the major factor. The increased popularity of comforters (in effect, heavyweight bedspreads) has reduced the overall blanket demand in the United States. In addition to conventional bed blankets, another market exists for "throws," blankets of smaller size, typically 50×60 inches (127×152 cm), used in living areas.

Most blanket fiber is ecru, but a small amount of producer-dyed fiber is also used. For woven blankets, fiber deniers of 3 to 5 are generally employed for the filling yarn, either singly or as blends. Warp yarn generally uses finer denier in the range of 1.5 to 2.0. Needle-punched blankets use a batt made from 3 to 5 dpf fiber. Because this is a rather undemanding end use, and prices are low, producers may make blanket staple from products downgraded from other end uses, such as tow. Cut lengths of about 2 1/2 inches (63 mm) are common.

For woven blankets, the more expensive end of the market, the production process is shown diagrammatically in Fig. 3. A warp is made of open-end-spun yarn of about 16 to 18 c.c.; this will comprise about 18% of the final blanket. The filling yarn, also open-end-spun to about 1½ to 3 c.c., makes up the remainder. The blanket is woven typically as a 2/1 twill. Excepting those blankets made from producer-dyed fiber, the goods are then piece-dyed or in some cases pad-dyed. The dyed fabric is then "napped" to pull or break surface fibers out of the yarn and give the blanket more air-trapping ability as well as improving aesthetics. After napping, the blanket fabric is then treated to stabilize the

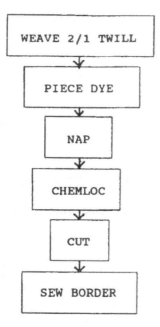

Figure 3 Woven blanket production steps.

surface and minimize pilling. It is sprayed with an aqueous solution of ethylene carbonate and then dried. As evaporation of the water proceeds, the ethylene carbonate solution is concentrated at the intersection of the filaments as shown in Fig. 4. Once the water is removed, the small residue of ethylene carbonate acts as a solvent on the acrylic fibers making them tacky enough to bond together and thus stabilize the fabric. Although some blanket manufacturers

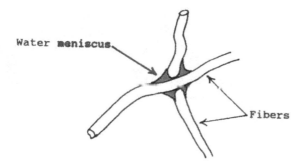

Figure 4 Blanket stabilization by Chemloc process.

have their own tradenames, this process is generally called Chemloc, the name accorded to it by Chemstand at its introduction (Biles et al., 1964). The blankets are then cut to size and the borders sewn.

The needle-punch process uses an acrylic scrim (15–20 c.c.) which is sandwiched between two acrylic webs produced on a garnet card. In the same fast operation, the assembly is needle punched. Typically the fabrics then undergo pad dyeing (unless producer dyed), napping, solvent bonding, cutting, and sewing borders as with woven blankets.

5. Carpets

Requirements As mentioned in the introduction, acrylic use in carpets, in many world areas had all but disappeared by the late 1980s. However, now they are mounting a small comeback. Three attributes are responsible for the resurgence:

Availability of producer-dyed fibers which avoid the problems of pile deformation and poor apparent value of beck-dyed acrylics.

High shrink acrylic fiber which can be used as a minor component in a blend with nylon to produce yarns with better retention of tuft definition in cut styles. This is the principle behind Monsanto's Traffic Control product.

A relatively lower priced fiber for blends with premium wool.

Since the decline of the acrylic carpet business in the late 1970s, precolored polypropylene has gained a share in several niche carpet markets, where nylon staple, either due to cost or property deficiencies, particularly pilling, was unsuccessful. These included the Berber and other textured loop styles. Acrylic fiber, however, has better properties for a carpet fiber than polypropylene. The major deficiency of polypropylene is appearance retention—the carpet assumes a matted appearance in high-traffic areas. In addition, acrylic carpets possess the better tactile response or "hand." Polypropylene does have the advantage of a 23% lower specific gravity than acrylics, which translates into lower cost. Nevertheless, polypropylene use in carpets appears to be an area that is vulnerable to acrylics, because if a mill is willing to use a precolored fiber, with the inventory problems that it creates, it ought to be willing to use a better one, if the economics are right.

As noted in Chapters 3 and 4, acrylic fibers develop their final physical properties through a relaxation process, which results in fiber shrinkage of 15–40%. This product is stable to boiling water, with residual shrinkage of less than 2%. For some end uses, however, including carpet, the relaxed product is again stretched from 20% to 40% to create a high bulk product. This high-bulk product is blended, usually at the level of 30–50% with standard product; on exposure to hot water, the high-bulk product shrinks which develops bulk in the standard, nonshrinkable portion.

In nylon carpet, the need is not to increase bulk but to preserve twist and end-point definition, particularly for styles such as "saxony," a cut pile, evenly sheared, medium-length (~16 mm) product. A small amount of a high-bulk acrylic carpet fiber (10–20%)* blended with the nylon may be used to preserve the end-point definition. The shrinkage of the acrylic, during heat setting, tends to lock in the twist and stabilize the nylon tuft structure against deformation (Wilkie and Talley, 1989). Limitations on this approach are the color "pollution" of the undyed acrylic and the flammability of the acrylic, as even FR acrylics increase the flammability of nylon due to the wick effect of the acrylic.

Process Almost all carpet yarns are now processed on the worsted system using staple of 6–7.5 inches (15–19 cm) and denier of 15–20. This produces a cleaner yarn with less tendency to "fuzz" than the woolen system yarns typical of acrylics 20 years ago. Tufting is almost universally used to convert the yarn to carpet. Only a small amount of weaving capacity, limited to very expensive grades, remains. Tufting of acrylics or blends including acrylics is the same as employed for nylon staple yarns.

In one variation in yarn spinning of acrylics, a "slub" of contrasting color is injected into a yarn of a natural or subdued base color fiber to give the "Berber" effect. Either the base fiber or the effect fiber(s) or both may be producer-dyed products. The base fiber may be a subtle blend of color and natural yarns and may contain high bulk fiber to enhance the bulk of the product. The Berber coloration process is done on the finisher card (Wallace, pers. comm.). Typically 8% to 12% of the effect fiber, as a low twist singles (2.6/1 c.c.) yarn is fed between the next-to-last worker and stripper rolls on the finisher card. Several card modifications are required to produce this special product (Wallace, pers. comm.).

Each singles yarn represents 0.5% of effect yarn in the final product; thus for 8%, 16 effect yarns must be fed. Other carpet operations in Berber yarn manufacturing are identical to those used for standard acrylic carpet yarn production on the woolen system. A typical carpet yarn produced might be a 2.00/3 c.c. with 4.25 singles "Z" twist and 2.50 "S" ply twist. Standard tufting techniques for loop carpet are employed. Typical weights of acrylic Berbers are from 35 to 55 oz/yd^2 (1186 to 1864 g/m^2).

A use of acrylic as a blending fiber, bringing special properties to the final carpet product, is illustrated by its incorporation in a blend with staple nylon. A high-shrinkage acrylic fiber is made by after-stretching a relaxed tow (1.0–1.5 MTD) up to 40% with the application of heat and subsequently recrimping the tow cold. Alternatively the stretching may be performed on a stretch-break machine such as a Seydel (Wilkie and Talley, 1989). With 40% stretch, the acrylic may have a shrinkage potential of 28%. The acrylic may be either cut

---

*More recently, the amount of acrylic fiber used to achieve the effect has been reported to be as low as 6% (Monsanto, 1975).

separately and blended with the nylon at the card, or the two fibers may be cut together, typically to 7 inch (17.5 cm) staple.

Using the modified worsted system, the blend of fibers (10–20% acrylic) is converted to a 65-grain sliver and spun on a conventional long staple carpet ring spinning frame to provide 3.50 c.c. singles with 5.0 "Z" twist; these are plied to 4.3 "S" twist. The yarns are then heat set, causing the acrylic fibers to shrink and "lock" the twist into the nylon. In Saxony (cut pile) constructions, this resulted in better appearance retention, compared to a 100% nylon control when the carpet was subjected to traffic. The 88% nylon/12% acrylic fiber blend was introduced by Monsanto as the Traffic Control Fiber System in 1990.

II. INDUSTRIAL APPLICATIONS

A. Introduction

The earliest end uses envisioned for acrylic fiber fall under the category we call industrial: convertible tops and other outdoor applications. These uses were identified as a result of acrylics' recognized superiority in stability to ultraviolet radiation, but the difficulty in dyeing the early products, plus deficiencies in physical properties, kept them out of these markets. Acrylics and modacrylics now participate in markets for outdoor use plus other niche markets where their function is reinforcement in a composite structure. Compared to apparel markets the volume is small, but the unit price is usually much more attractive. Both Japan and Europe have more significant industrial acrylic markets than the United States.

B. Applications

1. Cement Reinforcement

Requirements The discovery that certain forms of asbestos caused lung diseases (asbestosis and lung cancers) led to searches for substitutes in the various asbestos applications. The largest use of chrysotile asbestos was as a reinforcing material in cement sheets and pipe. In 1981, 75% of the world asbestos consumption of 4,726,000 metric tons was used for cement reinforcement (Caro, 1982). A wide variety of replacement fibers, including glass, PVC-coated steel, carbon, and aramid were evaluated by cement sheet producers; three—acrylic, poly(vinyl alcohol), and cellulosic—met the performance and cost requirements. Poly(vinyl alcohol) has the best performance and is the leading replacement product, but significant use of acrylics is also reported in Europe.

The performance properties necessary for a successful reinforcing fiber for cement products are

High initial modulus—minimum of 180 g/d at 1% elongation
Retention of modulus under hot/wet conditions (pressure cure products only)
Good dispersibility in the cement/water slurry
Good adhesion to the cement
Resistance to the alkaline environment over long time spans
Proper fiber diameter and length

Products from four fiber producers which meet the above criteria are offered for this use. They are Dolanit 10 and VF 11 from Hoechst, ATF 1055 from Bayer, Sekril from Courtaulds, and Ricem from Montefibre. All are PAN, with either a very small level of comonomer or none and a higher molecular weight (M_w ca. 200,000) compared to fibers for textile use. The fibers are highly oriented using stretches of 9–12x to attain maximum tenacity and modulus. The manufacturing process for acrylic reinforcing fibers is discussed in Chapter 6.

Process Cement reinforcement is the most important application for asbestos (and its substitutes), mainly for the production of flat and corrugated sheets, tiles, and pipes for pressures up to 20 kg/cm^2 (284 psi).

Asbestos cement corrugated sheets and pipes are produced with the Hatscek process on equipment similar to a paper-making machine. In this process a 10% slurry of asbestos and cement in water (approximately 88/12 cement/asbestos) is fed to a vat in which a rotating sieve cylinder picks up the solids and drains the liquid. The main function of the asbestos in this operation is the regular formation of a thin lamina of the cement, avoiding loss of the fine cement particles through the sieve. To improve this function, the asbestos fiber, normally a blend of different varieties of asbestos, needs to be opened or fibrillated on special mills to give the best possible dispersibility and uniformity.

An endless conveyer belt collects the thin lamina, which then passes over vacuum boxes which drain off the excess water and then transfers the lamina to a roller for sheet production or a mandrel for pipe production, where the lamina is wound continuously until the required thickness is reached.

The finished product is cured 15–28 days to allow the cement to reach almost complete hydration and hardness. To accelerate the process, the curing is conducted in live steam tunnels for a period of hours, then kept wet by water sprays, or by immersion in water tanks. Autoclave treatment at 160–185°C under steam pressure which forms calcium silicate by reaction of lime and silica is also used to accelerate the curing. With the autoclave process, synthetic fibers such as PAN, polyvinyl alcohol, or polypropylene cannot be used as their properties are degraded by the harsh environment.

The replacement of asbestos with synthetic fibers in the ambient cure process is complex because the fibers must provide the two main functions of asbestos: filtration and retention of "fines" during processing and reinforcement of the final cured product. For this reason, two fiber types are usually employed, a

reinforcing fiber (PAN, PVA, etc.) and a "process fiber" with a drainage factor
≥45 SR (Shopper Riegel), such as fibrillated cellulose or polyethylene fibrids
(Pulplus, DuPont), plus additives such as flocculents or fillers.

The cement composition used for a 6-mm reinforced sheet produced on a
Hatscek machine may have a composition of (Studinka and Meier, 1982)

1000 kg cement
80 kg waste paper
20 kg PAN fibers, 6 mm
15 kg aluminum sulfate
45 kg calcium hydroxide
80 g polyacrylamide (flocculent)

After compression at 250 bars (3700 psi) to give a thickness of 4.8 mm, the
sheets were cured for 28 days. The test results were

	Bending strength (N/mm^2)	Impact strength $(N/mm/mm^2)$	Sheet density (g/cm^2)
Asbestos/cement	29.2	1.8	1.76
PAN/cement	26.3	2.7	1.76

Today, only a few companies produce asbestos-free sheets and pipes using PAN
or PVA fibers on modified Hatscek machines as the recipes cost up to 30%
more. The main use of PAN fiber is for partial replacement of asbestos in sheets
to improve the productivity of the machine and the product quality and in pipes
for the total replacement of the crocidolite or blue asbestos, which is considered
most harmful and banned in many countries.

In the case of partial replacement of asbestos, PAN fiber is used at a percentage of 0.3–1.0 based on cement and to balance the cost, the asbestos is reduced
in quantities 6× the PAN fiber used. The 6-mm acrylic product is preferred for
sheet production and the 12-mm product for pipe.

Another use for PAN fiber is as an additive in mortars and plasters to reduce
cracks and microcracks during the initial stage of hardening when there is a
volume contraction and to decrease the brittleness of the final product. A typical
cement for reinforced mortar may have a composition such as (Hähne et al.,
1987)

1700 kg/m^3 aggregates (O to 8 mm)
400 kg/m^3 portland cement
20 kg/m^3 (=1.7 vol.%) Dolanite 11, 104 micron diameter, 6 mm cut length
192 L/m^3 water

The amount of water required for good workability of the slurry is related to both the fiber quantity and the cut length. An increase in fiber length from 6 to 24 mm increases the water requirement to 220 L/m³; doubling the weight (40 kg/m³) of 6 mm fibers has a similar effect.

Premixing of the fiber with the finest matrix component—cement—with the more coarse ingredients added afterwards may improve fiber distribution (Hähne, 1987). Mixers with two independent mixing appliances work best; examples are the plow share mixer of the Lodige Works (Paderborn, Germany) and the three-dimensional mixer with rotating paddles of the Zyclos Works (Vaihingen, Germany).

Performance Compared to asbestos, acrylic fibers provide lower breaking strength but higher elongation at rupture; the energy to break is increased by the substitution of acrylic fiber (Hähne, 1987). The strength retention of the homopolymer fibers in alkali at elevated temperatures (pH=13, 80°C, 24 hr) is reported to be 85% (Hähne, 1987; Anon., 1990a). At a pH of 12, which is typical of cement, no strength loss occurs at 80°C. Properties of cement composites with asbestos versus acrylic are (Akers and Chapman, 1988)

	Asbestos	Acrylic
Reinforcing fiber (vol. %)	12	4
Retention aid (vol. %)	0	16
Cement (vol. %)	88	80
Bending stress (N/mm²)	27	21
Toughness (N mm/mm²)	1.8	2.1
Density (g/cm²)	1.8	1.6

Addition of fiber to cast cements and concretes increases the ability of the structure to flex without breaking and reduces shrinkage cracking. The flexural strength increase is quantified as modulus of rupture:

$$MOR = \frac{3p_{(max)}L}{2bt^2}$$

where MOR = modulus of rupture (MPa)

$p_{(max)}$ = maximum flexural load (N)

L − span (mm)

b = width (mm)

t = thickness (mm)

The response of MOR to fiber content is shown in Fig. 5 (Studinka and Meier, 1982). Although it would be anticipated that MOR should continue to increase with fiber content, the presence of fiber increases the viscosity of the concrete so that additional water is required to have a workable mix. At

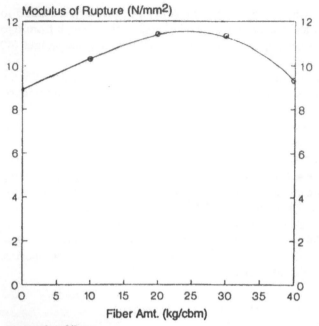

aggregates: 0/8mm
cement: 400kg/cbm PZ45F
fibers: Dolanite 11 104μm/6mm

Figure 5 Bending strength of a cement composite as a function of acrylic reinforcing fiber content.

approximately 25 kg/m^3 fiber, the adverse effect on strength of the excess water has overcome the improvement from the fiber.

Fiber length contributes to flexural strength, if the fibers are imbedded separately in the cement matrix. Length and diameter are interrelated as the tendency to form a "ball" during mixing decreases with increasing diameter, but the surface area, and thus the area to bond to the cement, increases with decreasing diameter. So finer filaments must have shorter lengths in order to be well dispersed. Early work (Koller, 1982) showed that the aspect ratio should be below 100; for acrylics, that would mean a 1-dpf fiber with a diameter of ca. 11 microns should have a length of less than 1.1 mm. In practice, the most common staple length for these products is 6 mm. The staple may be sold in the form of "chips" wherein a water soluble resin is coated over a small bundle of fibers. This allows time for the bundles to separate in the mix; then the coating dissolves and the individual fibers separate.

2. Lead Battery Plate Reinforcement

Requirements The use of fibers for battery plate reinforcing parallels the use in cement reinforcing. The purpose is to produce positive and negative plates that will not crack when drying or when they are handled during battery assembly. Later, when the battery is used, the fibers prevent cracking and flaking caused by charging, discharging, and vibration. Modacrylic fibers are generally used in this application, although it is not clear that they are superior to acrylics for this function. The most important attribute of the fiber is resistance to hydrolysis in sulfuric acid solution. Both acrylics and modacrylics rate highly in this respect; some monomeric units such as vinyl acetate or methyl acrylate may lose their pendent groups and a minority of the nitrile groups may be converted to amides, but largely the structure is intact to support the plate.

Polymer composition of the modacrylics used in battery plate reinforcing is dictated by requirements for other end uses. The two most common fibers in this application, Kanecaron and SEF, are vinyl chloride and vinylidene chloride/ vinyl bromide modified compositions respectively. The fiber has several alterations compared with that used for other modacrylic applications:

No additives are used in the spinning solution to control luster (TiO_2) or to enhance flame retardancy (Sb_2O_3) as these would dissolve in the battery paste and contaminate the cell.

The finish must allow the easy dispersion of the fiber in the paste but not degrade in the presence of acid. Silicone-based finishes appear to be preferred for this application.

Unlike those used in cement reinforcement, the fibers are not specifically engineered as high-modulus types, although since fibrillation is not a concern, the relaxation step in processing may be omitted. Thus the tenacity and modulus are somewhat higher than that of the textile end-use products.

Crimping is omitted and the fiber is sold as a tow to an intermediary processor who cuts the staple length desired by the battery manufacturer. Fibers of 2–3 denier with cut lengths of 1/16 to 1/4 inch (1.6–6.3 mm) are employed.

Process A typical application in paste manufacturing is as follows: Two pounds (0.91 kg) of modacrylic floc dispersed in 2 gal (7.7 kg) of water is added to 1900 lb (864 kg) of lead oxide (PbO) followed by 192 lb (87 kg) of water. The mass is mixed for 4 min in a Muller mixer. Then 230 lb (105 kg) of sulfuric acid is added and agitation is continued. After 25 min, 125 lb (57 kg) of water is added. Mixing is continued for several minutes until dispersion is complete. The paste is then formed into a sheet from which the individual plates are cut.

3. Awnings, Boat Covers, Outdoor Furniture, and Sandbags

Requirements Acrylics have outstanding ultraviolet stability compared to other synthetic and natural fibers. A comparison has been shown in Chapter 6, Fig. 8. Modacrylics, while not as outstanding as acrylics, do have good UV performance, plus the flame retardancy mandated for some end uses. The acrylic and modacrylic spinning processes are well suited to the addition of pigments to provide coloration in the final product. This is important as dyestuffs do not have the lightfastness to match the lifetime of the base fiber. This combination of stable polymer plus stable pigment then provides an excellent starting point for an outdoor product. This performance, however, comes at a price—compared to cotton, polypropylene, or polyester, acrylic is a premium-priced product; thus the consumer must factor in the cost per year of useful life in the buying decision. Modacrylic is another price step above; therefore its use is limited to applications where fire codes require a flame-retardant material. The production processes and pigment selection are discussed in Chapter 6.

Process A typical acrylic or modacrylic awning fabric is made from 3 dpf, 2-inch (51 mm) staple which has been ring-spun to a 18/2 cotton count. The fabric is a plain weave, usually 60 inches (152 cm) wide, with warp of 76 threads per inch (30/cm) and a filling of 30 threads/inch (12/cm). This gives a fabric weight after finishing of 9.25 oz/yd^2 (315 gm/m^2). The fabric is coated with a silicone-based finish to enhance water repellency.

Open-end spinning can also be employed to make yarns for outdoor applications, but the OE yarn is reported (Colli, 1990) to be 10–15% weaker. This in turn has an unfavorable effect on strength and elasticity of the fabric. The fabric is also reported to have a less desirable hand and reduced wear resistance. On the plus side for OE spun yarns are Uster evenness and cost.

Boat covers and outdoor furniture are typically made from the same fabric as awnings, although a lighter-weight woven—6 oz/yd^2 (204 g/m^2)—is also used.

Performance Color retention in outdoor applications is critical for an awning fabric, particularly in a commercial installation. Two types of degradation occur which lead to color change: polymer and pigment. These may or may not be interrelated. With respect to awning color stability, two climate types are most critical: hot and dry (temperature commonly 38–49°C, relative humidity <20%) with high average sunshine; and warm and humid (32–38°C relative humidity >70%) with average sunshine. In the first instance, both polymer and pigment may degrade; in the second, pigment degradation is more likely.

As shown in previous chapters, acrylic and modacrylic polymers may degrade from heat exposure. With an air temperature of 40°C in direct sun, fabric temperature may reach 77°C for dark colors and 60° for white. In tests at Tucson AZ, temperature exceeded 38°C on over 90 days in one year (Masson, unpub.) giving more than 500 hr of exposure, assuming 6 hr/day at near the maximum temperature. The polymer change expected from heat alone is yellowing due to formation of conjugated intermolecular and intramolecular double-bond structures. However, the back sides of test samples do not show a great amount of color change after this extensive exposure. The main contribution of high ambient temperature is to greatly accelerate the degradation from sunlight exposure.

Polymer yellowing, superimposed on the pigmented color (assuming no change to the pigment) results in a "muddying" of the shade. Shades such as coral (yellowish pink), teal (blue green), and white are readily rendered off-shade. Under "Florida" conditions (warm, moist) pigment degradation also occurs, usually as a fading of the shade. Pigments subject to fading under Florida conditions may be quite stable under hot, dry Arizona conditions. The AATCC 16E test (Anon., 1990b) (modified by using a water spray during the dark cycle) is a good predictor of the pigment fading problem. Despite the challenges of severe climate, a United States awning manufacturer, Glen Raven Mills, warrants its acrylic awning fabrics, Sunbrella® and Sunbrella Firesist® for five years service. In the more benign climate of Switzerland (Colli, 1990) tests of Leacril pigmented fabric indicated no color change through two years of exposure and only minimum change through five years. In contrast, cotton fabrics showed more than 50% degradation in two years and were considered degraded in three years.

In addition to color change, tensile properties are also affected by outdoor exposure. The following table shows strength retention after exposure of up to 1000 standard fade-o-meter hours (SFH) (xenon arc at 50°C) for a pigmented Acrilan acrylic awning fabric (Canaan, 1985):

	SFH exposure					
Strength (lb)	0	160	240	320	500	1000
Tear						
Warp	12.4	13.0	13.3	13.6	11.6	9.4
Filling	8.1	8.5	8.6	8.7	9.4	6.4
Tensile						
Warp	135	128	130	133	126	118
Filling	88	92	85	88	90	89

The 1000 SFH exposure is claimed to represent one year of outdoor exposure in a temperate climate. A similar study has not been completed for strength loss under the more stringent Arizona conditions, but it is expected that the strength reduction would be more severe.

The hydrophobic nature of acrylic fibers makes them difficult to wet compared to cotton. Thus, water repellency of acrylic woven awning fabrics can be achieved with a silicone finish, whereas cotton fabrics require a coating. Under the same fade-o-meter conditions as reported above for the strength tests, acrylic awning fabric loses 25% of its "spray rating" (Anon., 1990c) for water repellency at 1000 SFH, but actually increases by 58% in its hydrostatic pressure rating (Anon., 1990d). Modacrylics have poorer spray and hydrostatic ratings than acrylics in the same construction and weights because the polymer density is about 14% greater.

In the United States, awning fabric for use in regulated areas must pass NFPA-701, a vertical combustion test (NFPA, 1977). For fabrics of less than 10 oz/yd^2 (340 g/m^2), 10 samples, 5 each in the warp and fill direction, must be tested. To pass, no single sample may have a char length of more than 5.5 inches (14 cm), and the average for each direction must be less than 4.5 inches (11.4 cm). Awning fabrics from modacrylics such as Monsanto's SEF meet this stringent test provided that the water-repellent finish does not increase the flammability of the fabric.

Acrylic sandbags were first put into service by U.S. forces during the Vietnam War after testing showed their great superiority to jute burlap, cotton, and other synthetics as shown in Fig. 6 (Bagdon, 1971). In addition to the superiority with respect to light stability, acrylic bags suffer no strength loss from fungal attack when buried in soil for up to 40 weeks, whereas cotton bags failed in 2 weeks and jute in 4 (Anon., 1968). Other pluses noted by the military for acrylic bags were (Bagdon, 1971)

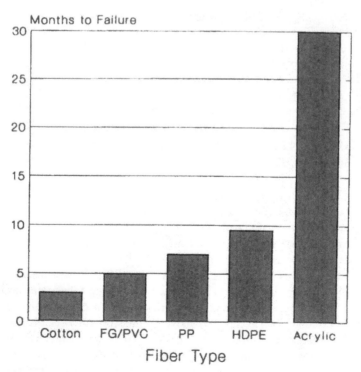

Figure 6 Lifetime of sandbags in soil burial.

Less manpower required due to the durability and consequently lower replacement rate.

Bags stack easily and do not slip.

Bags can be filled with materials other than sand without splitting.

Bags can be repaired with wire or string.

Bags meet the requirements in all tactical corps.

Although the bags cost about twice as much as polypropylene film bags, their service life expectancy of 36 months compared to 6 months for the polypropylene made them much more cost effective.

4. Oxidized PAN

Requirements PANOX is polyacrylonitrile that has been oxidized; it is a trade name for a product made by R. K. Carbon Fibres in the United Kingdom. A similar product is made by Stackpole in the United States. It is used as a flame-retardant textile material, in braking systems, and as gland packings, but primarily serves as a precursor for high-strength carbon fiber. In this section the production processes and uses will only be outlined briefly as these are subjects for entire books. Chapter 6 discussed the requirements for an acrylic fiber to be a suitable precursor for carbon fiber. If the final product is to be PANOX, these are somewhat different. Since PANOX may be processed as a stretch-broken tow or cut to staple, the acrylic precursor may be a tow of 1.5 dpf 480 K total denier. A Courtelle tow which contains itaconic acid units provides the necessary reduction in the onset temperature and the magnitude of the exotherm during the oxidation step (Morgan and Cognigni, 1990).

For ultimate carbon fiber production, high tenacity of the acrylic precursor, low level of comonomer, and smaller tow sizes are important attributes. Almost all of this production is captive—the carbon fiber producer also spins the acrylic precursor.

Process The oxidation step takes place in an air atmosphere at temperatures between 200 and 300°C. The tow passes through an oven as shown in Fig. 7 (Ko, 1991); temperature increases in each subsequent zone. Within a zone, the tow may make several traverses to attain adequate residence time. The total treatment time is 2.25 hr. The air is continually exhausted to remove volatiles such as ammonia, HCN, and other nitriles. The polymer undergoes cyclization through nitrile polymerization, cross-linking through nitrile reaction with methine hydrogens, oxidation to produce peroxides and ketones, plus scission reactions which may leave carbon-carbon double bonds (Morgan and Cognigni, 1990; Takahagi et al., 1986). On exiting the oxidation step, the product has the following properties (Morgan and Cognigni, 1990):

Composition: C = 60.3%, H = 4.1%, N = 20.7%, O = 14.5%, ash = 0.4%
Density: >1.38 g/cm^3
Limiting oxygen index: 55

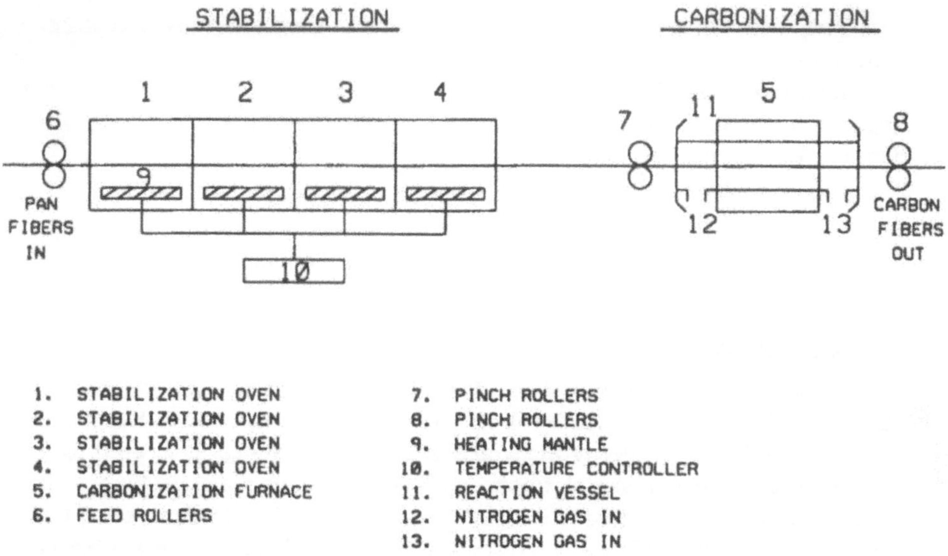
STABILIZATION
CARBONIZATION
1
2
3
4
5
6
7
8
9
10
11
12
13
PAN FIBERS IN
CARBON FIBERS OUT
1. STABILIZATION OVEN
2. STABILIZATION OVEN
3. STABILIZATION OVEN
4. STABILIZATION OVEN
5. CARBONIZATION FURNACE
6. FEED ROLLERS
7. PINCH ROLLERS
8. PINCH ROLLERS
9. HEATING MANTLE
10. TEMPERATURE CONTROLLER
11. REACTION VESSEL
12. NITROGEN GAS IN
13. NITROGEN GAS IN

Figure 7 Continuous stabilization and carbonization process.

Ko (1991) reported much higher oxygen content (20.3%) for samples stabilized at 230°C, and still higher oxygen (22.7%) and density (1.463) for samples with stepwise heating to 275°C. Heating above 275°C in air led to degradation of the stabilized polymer and poorer properties. Comparing the tensile strength and Young's modulus of stabilized AN/methyl methacrylate copolymer samples, Mikolajczyk and others (1989) found that both reached a maximum at a stabilization temperature of 210°C.

The structure of PAN stabilized at 240°C has been estimated Takahagi et al., (1986) based on x-ray photoelectron spectroscopy, Fourier transform-IR, and elemental analysis to be

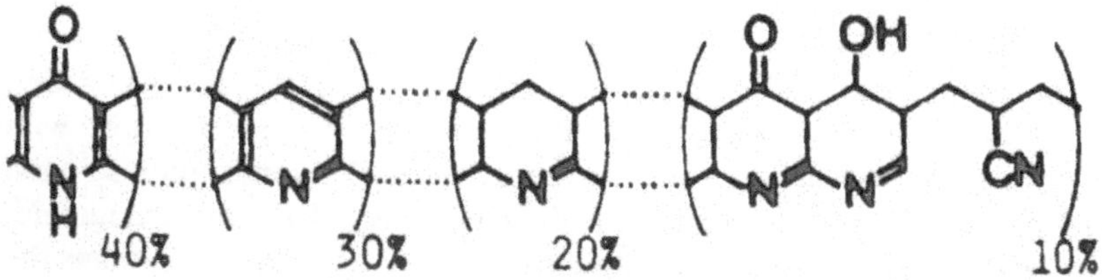
40%
30%
20%
10%

This structure would contain about 13% oxygen.

Performance Oxidized PAN can be used in 100% form; tow can be stretch broken on the Seydel system (Chapter 10), and the sliver converted to yarn on the modified worsted system. To improve abrasion resistance, the sliver may be blended with an aramid such as Nomex®; the limiting oxygen index is, however, reduced from 55 to 32 by inclusion of 10% aramid (Morgan and Cognigni, 1990). The yarns can be knit or woven on conventional equipment; PANOX can also be needle-punched. It is reported (Morgan and Cognigni, 1990; Ward, 1983) to be suitable for uses in

Clothing
  motor sports
  welding
  iron and steel manufacturing
  chemical
  petroleum
  fire, police, defense forces
Transport and furnishings
  aircraft seating and interior trim
  rail and road vehicle seating
Industrial
  gland packing (coated with ptfe)
  cable insulation
  friction linings
  fire blankets
  welding screens

The reported world tonnage in 1990 was 2000 MT (4.4×10^6 lb) (Morgan and Cognigni, 1990). One of the major limitations to its use in visible applications is that it is available only in black! For protection against intense heat, however, it offers considerably more protection than conventional fire protection textile fibers; a PANOX-based fabric is reported to maintain a barrier against a 900°C flame for more than 5 min. This is due to its inflammability plus exceptionally low thermal conductivity (Ward, 1983). In addition, it produces only a minimal release of toxic vapor and does not have afterglow. It is resistant to dilute acids and dry-cleaning solvents, but is not recommended for service in contact with strong acids or bases or oxidizing agents.

5. Carbon Fiber

Requirements Acrylic fiber is the leading precursor for carbon fiber production, although rayon and pitch are also used. The polymer and spinning requirements for the acrylic precursor are discussed in Chapter 6; the intermediate product, oxidized PAN, is discussed in Section 4. Basically, the acrylic fiber should have high tenacity and modulus, contain groups such as carboxyl which will readily start the "ladder" reaction, and have low content of metallic cations.

The oxidation of the acrylic precursor should not take place at too high a temperature. Ko (1991) found that when the stepwise oxidation extended to

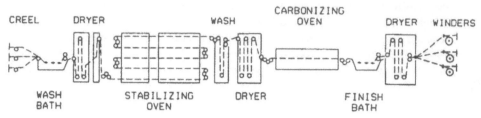

Figure 8 Carbon fiber production process.

300°C, overabsorption of oxygen occurred which led to a breakdown in the ladder structure. The final carbon fiber had 30% lower tensile strength than fiber from precursor stabilized at lower temperatures. For most carbon fiber processes, the oxidation step and the carbonization step are integrated.

Process Conversion of the oxidized PAN into carbon fiber takes place in an inert atmosphere at temperatures up to 1600°C. A diagrammatic view of the process is shown in Fig. 8 (Fourne, 1982). In this step, hydrogen, oxygen, and nitrogen are expelled from the fiber as H_2, H_2O, HCN, and other species to leave a pure carbon structure. Carbonization of PAN results in a weight loss of approximately 50%. Throughout the process, the fiber is held under tension to prevent shrinkage which would reduce the tenacity and modulus of the final product. Washing and drying sections both at the beginning of the process and after the oxidation step remove finish and by-products before carbonization. A finish is applied after the carbonization to enhance handling. A further increase in performance can be gained by a graphitization step which takes place in a relatively short oven at temperature near 2800°C in an argon atmosphere.

Performance Dramatic changes in properties occur in the carbonization step. The precursor and the final fiber are compared in the table following (Vogelsang and Boder, 1987):

Property	Ox PAN	Carbon Fiber
Density (g/cm^3)	1.4	1.8
Tensile strength (kN/mm^2) (g/den)	0.25, 2.02	3.5 ,72.9
Modulus (kN/mm^2) (g/den)	9, 22.0	240, 1510
Breaking elongation (%)	20	1.5
Thermal expansion coeff. (10^{-6}/K)	—	−0.6
Electrical resist. ($\mu\Omega$/m)	10^{14}	18

In tensile strength per unit weight, carbon fiber ranks only behind Kevlar aramid. It excels in modulus also; however, carbon fiber made from pitch can be produced with even a higher modulus (Fourne, 1982). The low elongation-to-break of the pitch-based products (0.3–0.5%) is a disadvantage.

Products are reported to be available in tow sizes ranging from 3000 to 320,000 filaments of 6.8 micron diameter (Anon., 1985) and in strengths of 1.7 to 3.5 GPa (1.7 to 3.5 kN/mm^2). The original impetus for carbon fiber was for composites for military use such as rocket casings, airplane parts, and ballistic protection. With a deemphasis on high-tech warfare, more civilian uses are being sought. Currently, sporting goods (tennis racquets, golf club shafts) and commercial aerospace are the main end uses, but potential exists in such areas as tanks for alternate automotive fuels, reinforced concrete, oil rig pressure vessels and piping, and suspension bridge cables (Reisch, 1992).

6. Filtration Fabrics

Requirements Acrylic fibers are used in filtration applications where aggressive environments, such as exposure to SO_2 and other acids, are encountered. A PAN homopolymer composition is preferred to a copolymer since the former has a lower diffusion rate and less sensitivity to moisture (Kniep, 1982). Dralon T is a fiber engineered for this end use. Both staple and filament types are sold. The newer PAN types, made specifically for cement reinforcing (Section 2), may also have application here.

Process For wet filtration under acidic conditions, woven acrylic fabrics are used, employing staple yarns. In dry filtration, in addition to woven fabrics, needled felts are employed using a scrim of filament yarn to give dimensional stability to the fabric (Anon., 1990c).

Performance Acrylics are suitable as filter fabric in mid-range applications as shown in the following table (Kniep, 1982):

Type fiber	Maximum temp. (°C)	
	Moist	Dry
Glass	260	290
Poly(tetrafluroethylene)	260	260
Aramid	180	230
Dralon T	140	140
Nylon 6, 6	120	120
copolymer acrylics	110	120
Polyester	100	150
Wool	90	110
Cotton	82	93
Polypropylene	93	107

Table 2 Physical Property Data of PAN Homopolymer Staple and Filament

	Dralon T staple	ATF 1063 staple	Dralon T filament
Denier (dtex)	2.2 (2.4)	2.2 (2.4)	802 (890)
Tenacity (g/d) (cN/tex)	3.3–4.1 (30–38)	4.9–5.4 (45–50)	4.4–5.1 (40–47)
Breaking elongation (%)	15–25	12–16	17–22
E modulus (g/d) (cn/tex)	54–65 (500–600)	87–109 (800–1000)	65–76 (600–700)
Boiling water shrinkage (%)	<5	<5	0.5–1.0
Density (g/cm)^3	1.17	1.18	1.17

Source: Kniep (1988).

Table 3 Strength Retention of Dralon T and Acrilan 16 after Exposure to Acids, Bases, and Other Chemicals

Chemical	Conc. (%)	Time (h)	Temp. (°C)	Acrilan (%)	Dralon
Acetic acid	5	10	99	97	
	40	10	21	91	
	50	15	100		Very good
	98	15	75		Very good
Acetone	100	1000	21	100	
Ammonia	28	15	20		Very good
Benzoic acid	3	10	99	94	
Chloroform	100	1000	21	100	
Ethyl ether	100	1000	21	100	
Ferric chloride	3	10	99	98	
Formic acid	5	10	99	98	
	10	15	75		Very good
	40	10	21	96	
	98	15	20		Good
Hydrochloric acid	10	10	21	100	
	10	15	75		Very good
	30	15	75		Good
	37	10	21	94	
Hydrogen peroxide[a]	0.3	10	21	100	
Hydrogen peroxide[b]	3	10	21	98	
Lactic acid	100	15	75		Very good
Nitric acid	7	10	21	100	
	10	10	21	94	
	10	15	75		Very good
	30	15	75		Satisfactory
	40	15	20		Good
Oxalic acid	5	10	21, 99	96	
Peracetic acid	2	10	21	95	
Phosphoric acid	10	15	75		Very good
Phosphoric acid	50	15	75		Good
	70	15	20		Good
Potassium hydroxide	10	15	20		Very good
	10	15	75		Good
Salicylic acid	3	10	99	98	
Sodium borate[c]	1	10	21	98	
Sodium carbonate	Dilute	15	20		Very good
	Saturated	15	100		Adequate
Sodium chlorite[d]	0.7	10	21	94	
Sodium hydroxide	10	15	20		Very good
	10	10	21	93	
	10	15	40		Good
	40	10	21	91	
	50	15	20		Good

Table 3 Continued

Chemical	Conc. (%)	Time (h)	Temp. (°C)	Acrilan (%)	Dralon
Sodium hypochlorite[a]	0.4	10	21	98	
Sulfuric acid	1	10	99	100	
	10	15	75		Very good
	30	15	75		Satisfactory
	60	15	75		Adequate

[a]pH 7
[b]pH 6
[c]pH 10
[d]pH 4
Source: Kniep (1988) and Aron. (1965).

The operating temperature should be set approximately 15°C below these maximums, to allow for short-term deviations; in this way, the fabric life is considerably extended (Kniep, 1982). The homopolymer Dralon T has a 20–30° advantage in maximum application temperature over the copolymer acrylics. Service lives of several thousand hours may be expected under favorable conditions (Kniep, 1988). It is also interesting to note that although polyester has a slight advantage over Dralon T under dry conditions, it has a 40° lower maximum under moist conditions. This is due to the difference in hydrolytic stability of the two compositions. In 48 h at pH 5 and 132°C, Dralon T retained 88% of its original strength compared to 29% for polyester (Kniep, 1988). PAN fabrics at 125°C in the presence of NO_2 and SO_2 have been reported (Weber, 1988) to degrade by backbone oxidation at the methine carbon. High water content in the gas minimizes NO_2 damage. Damage from SO_2 occurred under much higher concentrations than would be found in practice.

Physical properties of Dralon T staple and filament and the cement-reinforcing fiber ATF 1063 are shown in Table 2 (Kniep, 1988). The resistance of Dralon T and Acrilan 16 to various chemicals is shown in Table 3 (Kniep, 1988; Anon., 1965).

REFERENCES

P. J. Akers and R. J. Chapman, Diversification of the Market Applications of Acrylic Fibres, *International Man-Made Fibres Congress*, Dornbirn, Austria, 1988.

Anon., *Industrial Uses of Acrilan Acrylic Fiber*, TT-17 Monsanto Company Bulletin, August 1965, p. 6.

Anon., *Acrylics Outdoors*, Monsanto Company Bulletin, October 1968, p. 7.

Anon., *High Performance Textiles, October:*7 (1989).

Anon. *Chemiefasern/Textilindustrie, 40/92*, (March, 1990a) p. T8.

Anon., *AATCC Tech. Manual*, 65:44 (1990b).

Anon., *AATCC Tech. Manual*, 65:76 (1990c).

Anon., *AATCC Tech. Manual*, 65:219 (1990d).

V. J. Bagdon, *Textile Res. J.*, 41:546 (1971).

J. R. Biles, D. C. Nicely, and G. W. F. Sellers, U.S. Patent 3,152,919 (10/13/64) to Monsanto.

B. Burleson, Monsanto, personal communication.

D. S. Canaan, *Industrial Fabrics Product Review* (1985 Buyers Guide), p. 8.

Caro Industrial Corp. Ltd., London, 1982.

C. Colli, *Chemiefasern/Textilindustrie*, 40/92:E43 (1990).

P. M. Eaton, *Textiles* 1992, no. 1:20 (1992).

H. Ernst, *Melliand Textilber.*, 4:247 (1989).

F. Fourne, *Chemiefasern/Textilindustrie*, 32/84:433 (1982).

J. H. Gustafson, *Book of Papers, Industrial Fabrics Assn. International, Textile Technology Forum*, 1988.

Hähne, *Cana. Textile J.*, April:35 (1987).

H. Hähne, G. König, and J.-D. Wörner, in Fibre *Reinforced Cements and Concretes: Recent Developments* (R. N. Swamy and B. Barr, eds.), Elsevier, London, p. 60.

T.-H. Ko, *J. Appl. Polym. Sci.*, 42:1949 (1991).

A. Koller, *The Construction Specifier, Dec.*:44 (1982).

E. Kniep, *Chemiefasern/Textilindustrie, 32/84: June*:E39 (1982).

E. Kniep, *Chemiefasern/Textilindustrie, 38/90: Dec.*:T116 (1988).

J. C. Masson, unpublished.

T. Mikolajczyk, I. Krucinska, and K. Kamecka Jedrzejczak, *Textile Res. J.*, 59:665 (1989).

A-Acrilan® plus Custom Carpets and Rugs, Monsanto Textiles Co., Decatur, AL, 1975, pp. 27–34.

Drapery Fabrics of Acrilan Acrylic, New Product Bulletin NPR-11, Monsanto Textiles Co. (undated).

P. E. Morgan and F. Cognigni, *Textiles* Asia, (Dec. 1990), p. 61.

NFPA *701; Fire Tests for Flame-Resistant Textiles and Films*, National Fire Protection Association, Quincy, MA, 1977.

M. S. Reisch, *Chem. Eng. News, August* 3:16 (1992).

J. Studinka and P. E. Meier, U.K. Patent Application GB 2,095,298 (Sept. 29, 1982) to Redco.

T. Takahagi, I. Shimada, M. Fukehara, K. Morita, and A. Ishitani, *J. Polym. Sci. Polym. Chem. Ed.*, 24:3101 (1986).

J. Vogelsang and H. Boder, *Chemiefasern/Textilindustrie, 37/89*:T60 (1987).

L. M. Wallace, Monsanto, personal communication.

D. Ward, *Textile Month, Sept.*:11 (1983).

C. Weber, *Filtration and Separation, 25*:100 (1988).

M. B. Wheeler, N. Bush, and B. Krzesinski, *Textile* Asia, 22/8:126 (1991).

A. E. Wilkie, and A. Talley, U.S. Patent 4,839,211 (6/13/89) to Monsanto; A. Talley and A. E. Wilkie, U.S. Patent 4,882,222 (11/21/89) to Monsanto.

Index

Abrasion resistance, 317–318
 upholstery fabrics, 343–345
Acid or acid salt comonomers
 in carbon fiber precursor, 187–
  188
 in commercial processes, 78
 effect on color, 47
 effect on molecular weight determination, 249, 251–252
 itaconic acid, 46, 187, 361
 location on polymer chain, 55–
  56
 in modacrylic fibers, 169
 reactivity ratios, 42–43
 recovery and reuse, 59
 sodium methallyl sulfonate as a
  UV stabilizer, 277
 in solution polymerization, 61
 sulfonate salts, 44–45, 171, 174,
  190
 for water reversible crimp
  bicomponent, 171–174
Acid dyeable fiber, 46, 174–177,
  336
 in socks, 338
Acrilan:
 acid dyeable fiber, 174, 177
 Acrilan Plus producer dyed
  fiber, 19
 B-06 modacrylic for draperies,
  346
 B-37 pile fiber, 347

[Acrilan]
 bicomponent fibers, 173–174
 chemical exposure property
  retention, 367–368
 Duraspun sock fiber, 18, 338
 dye saturation constant, 289
 glass transition, 236
 history, 5–6
 HP low pill fiber, 193
 number of colored products, 179
 plant capacities, 29, 30, 32
 plant sales, 28, 29
 polymer process, 48, 63
 producer dyed fiber process,
  178–181
 SEF modacrylic, 20
  in battery plates, 357
  composition 169
 sunlight resistance in draperies,
  346–347
 Type 21 bicomponent 335–337
 Type 90 for carpets, 171
Acrylic fiber desirability index,
  317–318
Adsorption of dye, 287
Air gap spinning (*see* Dry jet wet
  spinning)
Air-jet spinning, 331
AKSA plant capacity, 29, 31
American Cyanamid (*see* Creslan)
American Viscose (*see* Acrilan,
  history)

Enichem (*see* Leacril)
Enthalpy of fusion, 224
  copolymer effect, 227
  vinyl acetate level effect, 232
Entropy of fusion, 224
Ethylene carbonate solvent:
  blanket stabilization process,
    348–349
  dope concentration, 78
  solution polymerization, 267
  use in commercial plants, 30–
    31, 74–77
Exlan:
  gel spun reinforcing fiber, 186
  history, 7
  plant capacity, 29, 31, 32
  polymer process, 48

Fake fur (*see* Pile, high)
Fiber desirability index, 317
Fibras Sineteticas:
  plant capacity, 29, 31
  plant equipment purchase, 32
Fibrillation, history, 5
Filament, continuous, history, 5
Filters:
  use in dry spinning, 112
  fabrics, 365–368
  history, 4
Finish:
  addition at dry spinning tower,
    120–121
  application to dry spun fiber
    after washing, 141
  carbon fiber, 364
  for reinforcing fibers, 184
  requirements, dry spun vs. wet
    spun, 163
Fisisa (*see* Fibras Sineteticas)
Flame resistance:
  antimony oxide synergists, 170

[Flame resistance]
  awning fabric, 360
  drapery composite structures,
    347
  nylon-acrylic blends, 351
  oxidized PAN, 363
  SEF modacrylic, 20
  Teklan modacrylic in uphol-
    stery, 345
  tests for, 168
Flat woven upholstery (*see* Woven)
Formosa Plastics, plant capacity,
    29, 31

Gel network:
  in DMF dope, 111–112
  dyeing in producer dye process,
    177–178
  melting behavior, 222–225
  in wet spinning, 83, 87–89
Gel spinning, 134–136
  of reinforcing fibers, 184
Glass transition, 233–242
  actinic degradation, 274
  of amorphous PAN, 209
  compared to atactic polypropy-
    lene, 200
  effect on drawing, 137–138
  in drying fiber, 83
  PAN, 285
  in stretching fiber, 97–98
  structure below, 216
Goodrich (*see* Darvan)
Graphite (*see* Carbon fiber)

Hand, pile, 348
Hand knitting yarn (*see* Craft yarn)
Hanil, plant capacity, 29, 31
Heat setting (see Relaxation)
High bulk fibers, 150–151
  for carpet fiber, 350–352

9 780367 401856